CAMBRIDGE LIBRARY COLLECTION

Books of enduring scholarly value

Technology

The focus of this series is engineering, broadly construed. It covers technological innovation from a range of periods and cultures, but centres on the
technological achievements of the industrial era in the West, particularly in
the nineteenth century, as understood by their contemporaries. Infra-structure
is one major focus, covering the building of railways and canals, bridges and
tunnels, land drainage, the laying of submarine cables, and the construction
of docks and lighthouses. Other key topics include developments in industrial
and manufacturing fields such as mining technology, the production of iron
and steel, the use of steam power, and chemical processes such as photography
and textile dyes.

Lives of the Engineers

A political and social reformer, Samuel Smiles (1812–1904) was also a noted
biographer in the Victorian period. Following the engineer's death in 1848,
Smiles published his highly successful *Life of George Stephenson* in 1857
(also reissued in this series). His interest in engineering evolved and he began
working on biographies of Britain's most notable engineers from the Roman
to the Victorian era. Originally published in three volumes between 1861 and
1862, this work contains detailed and lively accounts of the educations, careers
and pioneering work of seven of Britain's most accomplished engineers. These
volumes stand as a remarkable undertaking, advancing not only the genre, but
also the author's belief in what hard work could achieve. Volume 3 includes a
revised version of Smiles's biography of George Stephenson (1781–1848), as
well as a biography of his equally famous son, Robert (1803–59).

Cambridge University Press has long been a pioneer in the reissuing of out-of-print titles from its own backlist, producing digital reprints of books that are still sought after by scholars and students but could not be reprinted economically using traditional technology. The Cambridge Library Collection extends this activity to a wider range of books which are still of importance to researchers and professionals, either for the source material they contain, or as landmarks in the history of their academic discipline.

Drawing from the world-renowned collections in the Cambridge University Library and other partner libraries, and guided by the advice of experts in each subject area, Cambridge University Press is using state-of-the-art scanning machines in its own Printing House to capture the content of each book selected for inclusion. The files are processed to give a consistently clear, crisp image, and the books finished to the high quality standard for which the Press is recognised around the world. The latest print-on-demand technology ensures that the books will remain available indefinitely, and that orders for single or multiple copies can quickly be supplied.

The Cambridge Library Collection brings back to life books of enduring scholarly value (including out-of-copyright works originally issued by other publishers) across a wide range of disciplines in the humanities and social sciences and in science and technology.

Lives of the Engineers

With an Account of Their Principal Works

VOLUME 3

SAMUEL SMILES

CAMBRIDGE UNIVERSITY PRESS

Cambridge, New York, Melbourne, Madrid, Cape Town,
Singapore, São Paolo, Delhi, Mexico City

Published in the United States of America by Cambridge University Press, New York

www.cambridge.org
Information on this title: www.cambridge.org/9781108052948

© in this compilation Cambridge University Press 2012

This edition first published 1862
This digitally printed version 2012

ISBN 978-1-108-05294-8 Paperback

LIVES

OF

THE ENGINEERS.

Volume III.

a

George Stephenson.

Engraved by W. Holl, after the portrait by John Lucas.

London, John Murray, Albemarle Street, 1862.

LIVES

OF

THE ENGINEERS,

WITH

AN ACCOUNT OF THEIR PRINCIPAL WORKS;

COMPRISING ALSO

A HISTORY OF INLAND COMMUNICATION IN BRITAIN.

BY SAMUEL SMILES.

"Bid Harbours open, Public Ways extend;
Bid Temples, worthier of God, ascend;
Bid the broad Arch the dang'rous flood contain,
The Mole projected, break the roaring main;
Back to his bounds their subject sea command,
And roll obedient rivers through the land.
These honours, Peace to happy Britain brings;
These are imperial works, and worthy kings."
POPE.

WITH PORTRAITS AND NUMEROUS ILLUSTRATIONS.

VOL. III.

[GEORGE AND ROBERT STEPHENSON.]

LONDON:
JOHN MURRAY, ALBEMARLE STREET.
1862.

LONDON : PRINTED BY W. CLOWES AND SONS, STAMFORD STREET,
AND CHARING CROSS.

PREFACE.

THE following volume contains a revised edition of the Life of George Stephenson, with which is incorporated a Life of his son Robert, late President of the Institute of Civil Engineers. While complete in itself, this book also forms the continuation of the biographical history of British engineering—the earlier portions of which are comprised in the two volumes of 'Lives of the Engineers' already published,—and brings the subject down to the establishment of the railway system, in the course of which British engineers have displayed their highest skill and achieved their greatest triumphs.

Since the original appearance of the work some six years ago under the title of 'The Life of George Stephenson,' much additional information relative to the early history of railways and of the men principally concerned in establishing them, has been communicated to the author by the friends and pupils of the two Stephensons, as well as by the late Robert Stephenson himself, of which the author has availed himself in the present edition.

Although it is unusual to embody two biographies in one narrative, it will probably be admitted that in the case of the Stephensons such a combination is peculiarly appropriate,—the life and achievements of the son having been in a great measure the complement of the life and achievements of the father. The care with which the elder Stephenson, while occupying the position of an obscure workman, devoted himself to his son's education, and the zeal with which the latter repaid the affectionate self-denial of his father, are among the most effective illustrations of the personal character of both. As regards their professional history also, it will be found that the relations which existed between them, more particularly with reference to the improvement of the locomotive and the construction of the first passenger railways, were of so intimate a kind, that it is impossible to dissociate the history of the one engineer from that of the other.

These views were early formed by the author as to the proper treatment of the subject of George Stephenson's Life, and were carried out in the preparation of the original work, with the concurrence of Robert Stephenson, who supplied the requisite particulars relating to himself. Such portions of these were accordingly embodied in the narrative as could with propriety be published during the lifetime of the latter, and the remaining portions are now added, with the object of rendering the record of the son's life, as well as the early history of the railway system, more complete.

It may not be out of place to explain briefly the circumstances in which the book originated and was written, and the sources from which the facts it contains were derived, as a guarantee to the reader that every possible pains have been taken to secure due authenticity and accuracy of information.

The subject of a biography of George Stephenson was brought under the author's notice shortly after the death

of the engineer in 1848, by the present Mayor of Leeds,
James Kitson, Esq., a large locomotive-manufacturer in
that town, and an intimate friend of both the Stephensons. Mr. Kitson thought that the author's business
connection with railways, and his personal knowledge of
the elder Stephenson, with other qualifications, fitted him
for the task of writing his biography. The suggestion
was very tempting; but the preparation of such a work
involved too much labour to be lightly undertaken, and
beyond putting together a few memoranda, which were
published as an article in a London journal, nothing
further was then done in the matter.

In the mean time a very suggestive and able article
made its appearance in the *Athenæum* of December 8,
1849, urging the claims of the subject of railway enterprise and its early history upon the attention of literary
men. The reviewer pointed out that although there
then existed abundance of railway statistics, these would
be found of very little use to the historian who, a century
hence, looking to the extraordinary effects of the railway
system on the means and manners of Great Britain,
should try to relate how it arose, with what efforts and
influences, and by what manner of men it was brought
to pass within a few years—to discover, in short, something like what we now vainly seek and regret to find
untold of the great mechanical novelties of the last
century. "It is this," he observed, "which we now
desire to have collected, while the memory of the chief
facts is yet fresh, while many of the first authors are
still living, and while of those deceased—including a
principal author of the system, George Stephenson—
there are survivors able to supply authentic and lively

memorials. . . . . It is surely worth writing; and if the task be not soon accomplished, the materials requisite for its complete execution will have disappeared beyond recall. The real value of such records—the place due to their objects in the national annals—has hitherto been little regarded. Professed historians of the old school overlook them with dignified contempt; mere philosophical moderns at best admit them here and there to a summary notice made up of dry statistical matter, that reads but tamely among reports of party struggles and foreign disputes—of the vanities of courts and the achievements of armies. Our purpose here is to vindicate the claims of the subject and to show what part of it may well be preserved for the instruction of future times."

The only attempt made to work out the literary design so ably sketched in the *Athenæum*, was by Mr. Francis, in his ' History of the English Railway,' which, though an exceedingly graphic resumè of the early history of railways, failed in the main point of biographic interest in connection with the subject. A series of summary articles on the life and works of George Stephenson was also published by Mr. Hyde Clarke, C.E., in the ' Civil Engineer and Architect's Journal;' but, though valuable as a collection of facts and dates, it was not a biography, and the Life of George Stephenson, therefore, remained to be written.

To ascertain Robert Stephenson's views as to a Life of his father, the author called upon him at his office in Great George Street in March, 1851; Mr. Kitson having previously written him on the subject. Mr. Stephenson then said that a Memoir of his father had been frequently spoken of, but he had almost given up the hope of seeing it undertaken. He did not think the theme was one likely to attract the attention of literary men of eminence, nor did he seem to be at all sanguine as to its popular interest, though his views on this point afterwards underwent a change; but he promised that, in event of the author deciding to prosecute the proposed biography, he should give his best assistance in supplying the necessary facts.

Furnished by him with letters of introduction to several of his more intimate friends in Newcastle— among others to Mr. Budden, his business manager at the Forth Street Works—the author shortly after made a visit to that place, with the object of ascertaining what materials could be obtained for the purposes of the proposed memoir. After three or four days' diligent search it was found that the results, when reduced to shape, were of a very meagre kind. Books and newspapers were of no avail. The information wanted existed but in the memories of individuals, from whom it could only be gathered by direct personal intercourse and by slow degrees. Many of them were unlettered men, who, though they could communicate in conversation what they remembered, could not place it on written record. Others, possessing information and able to communicate it in writing, were too much engrossed by business affairs to give the requisite time for the purpose. Thus the author shortly became persuaded that to prepare a satisfactory Life of George Stephenson from authentic sources, required an actual residence of some period in the district where he had lived; and as the pursuits in which he was engaged at the time rendered this out of the question, he communicated to Robert Stephenson his regret at not being able, under these circumstances, to prosecute the proposed biography.

Thus three more years passed, during which nothing further was done.   No biographer of George Stephenson appeared; and the persons capable of furnishing information respecting him were being rapidly thinned off by death.   The author had himself almost dismissed the subject from his mind, when circumstances occurred in connection with his railway occupation which rendered it necessary for him to reside at Newcastle-upon-Tyne during the summer of 1854.   He was thus unexpectedly placed in a position to prosecute at his leisure the necessary inquiries relative to the Stephenson biography. Much of the desired information came directly in his way, and the rest he went in search of.   It became his recreation in the summer evenings to visit the places where George Stephenson had lived,—Wylam, where he was born,—Dewley, Callerton, Newburn, and Willington Quay, where he had worked as gin-driver, fireman, brakesman, and engineman by turns,—and Killingworth, where he had invented the safety-lamp and worked out the problem of the locomotive.   All these places were within easy reach of Newcastle by railway; and thus, helped by the recollections of the engineer's former associates, his life was traced from boyhood to manhood, from the cradle almost to the grave.

All who had known George Stephenson in his early years were proud to speak of him, and to communicate what they remembered of his history.   Though he had risen so much above them, there did not seem to mingle an atom of jealousy or envy in their recollections

of him.   They begrudged him neither his prosperity
nor his fame.   They spoke of "George" as if he had
been of their own kin, a member of their own family ;
and were as proud of his career as if it had been their
own.   There was much that was very graphic in their
relation of the incidents in "George's" early life, the
vividness of which, the author fears, may have escaped
in the process of reporting.   But so far as any merit
belongs to the earlier part of the narrative, he readily
acknowledges that it is in a great measure due to the
working men from whose lips he gathered it—colliers,
brakesmen, and enginemen, mostly old men, some of
them disabled by accidents and hard work—whom he
visited in succession at Wylam, Callerton, Newburn,
Willington, and Killingworth.

While residing at Newcastle, the author was also
enabled readily to visit Darlington, and to gather from
the lips of the venerable Edward Pease, to whom he
had been introduced by a letter from Robert Stephenson,
the interesting history of the Stockton and Darlington
Railway, of which Mr. Pease was the projector,—the
account of his employment of George Stephenson as
the engineer of that line,—and of his subsequent connection with him as partner in the locomotive foundry
in Forth Street, Newcastle-upon-Tyne.   At Darlington
also he obtained from John Dixon, C.E., many interesting facts relative to the survey of the Stockton and
Darlington Railway, and the construction of the Liverpool and Manchester Railway across Chat Moss, of
which portion of the line Mr. Dixon had been the
resident engineer.

Having thus gathered together the materials of what,

it was believed, would form an interesting and continuous narrative of George Stephenson's early career,
the author proceeded to communicate the result to
Robert Stephenson, and to express the hope of now
being able to proceed with the proposed biography of
his father.   To this communication a reply was received, dated " Dover, 26th Sept., 1854," in which
Mr. Stephenson said—" I am glad to hear that you
have not given up the idea of writing a memoir of
my late father; and now that I have more leisure, it
will afford me pleasure to assist you in many points
which are known only to myself, especially in reference
to the phases which the Locomotive Engine put on at
different periods of my father's active and remarkable
life—a life which spreads over a period comprising
probably one of the most astonishing pages in the history
of civilization.   I am about to visit Newcastle, when
I shall make a point of giving you my views as to the
form which the memoir, in my opinion, ought to take ;
and respecting the mechanical portions, I shall feel it
my duty to give every assistance."

Mr. Stephenson paid his promised visit to Newcastle
in the beginning of October, 1854, when he communicated his views as to the treatment of the proposed
biography, and took the author over the scenes of his
own and his father's early life, relating by the way
many interesting incidents which the sight of them
recalled to his memory.   The ride to Killingworth
will be found described at pp. 64-6 of the following
work.   The author afterwards read over to Mr. Stephenson the narrative he had by this time prepared
of his father's early life, much of which was entirely

new to him, though he was ready to admit its accuracy, considering the authentic sources from which it had been obtained. At a subsequent period the author enjoyed the advantage of much intimate personal intercourse with Mr. Stephenson, and obtained from him, either orally or in writing, many of the important facts embodied in the following narrative. Besides what was supplied directly by himself, much additional information was obtained through his instrumentality from other gentlemen well qualified to supply it — from Mr. Charles Parker, relative to the early history of the London and Birmingham Railway; from Mr. T. Sopwith, C.E., as to George Stephenson's visits to Belgium; and from Sir Joshua Walmsley as to his journey into Spain. Mr. Stephenson continued to furnish the author with corrections and additions from time to time as they occurred to him; and one of the last communications received from him, shortly before his death, was a letter accompanying a large bundle of the correspondence and papers of Mr. Joseph Sandars (since deceased), the projector of the Liverpool and Manchester Railway, of which due use has been made in the present edition. It has also been thought desirable to append Robert Stephenson's own narrative of his father's inventions and improvements in the form in which it was communicated to the author, the record being valuable as an authentic memorial of the early history of the Locomotive Engine and Railways.

Since the publication of the earlier editions of the Life of Stephenson, the author has been enabled to avail himself of the personal recollections of Mr. T. L. Gooch, C.E.; Mr. Vaughan, of Snibston; Mr. F. Swanwick, C.E.; and Mr. Binns, of Claycross, all of whom officiated as private secretaries to George Stephenson at different periods of his professional life, and afterwards held responsible offices either under him or in conjunction with him. The materials for the narrative of Robert Stephenson's career in Colombia have been kindly supplied by his friend Mr. R. S. Illingworth. Much of the valuable information communicated by these gentlemen is published for the first time in the present edition.

The same pains have been taken with the illustration of the book as in the case of the two volumes of 'Lives of the Engineers' already published. The author has had the advantage of being ably supported by his artists, Messrs. Leitch and Skelton, whose illustrations speak for themselves, and will, he believes, be found worthy of the subject.

London, October, 1862.

NOTE.—*End of the " Rocket."*—The important influence which this famous engine, which won the prize of 500*l.* at the Locomotive Competition at Rainhill in 1829, exercised on the general extension of the railway passenger system, led the author, in the early editions of the 'Life of George Stephenson,' to express the regret (repeated in the note to p. 274 of the following work) that pains had not been taken to ensure its preservation, in like manner as the French Government have preserved Cugnot's road locomotive of 1770 in the Conservatoire des Arts et Métiers at Paris. It is, therefore, with pleasure we have to state that, while these sheets are passing through the press, the " Rocket " is in course of removal to the Museum of Patents at Kensington, where it will find its appropriate place in that highly interesting national collection.

CONTENTS.

———◦◦◦———

CHAPTER V.

CHAPTER VI.

CHAPTER VII.

CHAPTER VIII.

CHAPTER IX.

CHAPTER X.

CHAPTER XI.

CHAPTER XII.

CHAPTER XIII.

CHAPTER XIV.

CHAPTER XV.

CHAPTER XVI.

CHAPTER XVII.

CHAPTER XVIII.

CHAPTER XIX.

CHAPTER XX.

CHAPTER XXI.

APPENDIX.

LIST OF ILLUSTRATIONS.

LIVES OF THE ENGINEERS.

LIVES

OF

GEORGE AND ROBERT STEPHENSON.

NEWCASTLE-UPON-TYNE AND THE HIGH LEVEL BRIDGE.

[By R. P. Leitch, after his original drawing.]

LIFE

OF

GEORGE STEPHENSON, &c.

CHAPTER I.

NEWCASTLE AND THE GREAT NORTHERN COAL-FIELD.

In no quarter of England have greater changes been wrought by the successive advances made in the practical science of engineering than in the extensive colliery districts of the North, of which Newcastle-upon-Tyne is the centre and the capital.

In ancient times the Romans planted a colony at Newcastle, bridging the Tyne by the Pons Ælii near the site of the present low-level bridge shown in the prefixed engraving, and erecting a strong fortification above it on the high ground now occupied by the Central Railway Station. North and north-west lay a wild and barren country, abounding in moors, mountains, and morasses, but occupied to a certain extent by fierce and barbarous tribes of Picts and Caledonians. To defend the young colony against the ravages of these dangerous neighbours, a strong wall was built by the Romans, extending from Wallsend on the north bank of the Tyne, a few miles below Newcastle, across the country to Burgh-upon-Sands on the shores of the Solway Frith. The remains of the wall are still to be traced in the less populous hill-districts of Northumberland. In the neighbourhood of Newcastle they have been gradually effaced by the works of succeeding

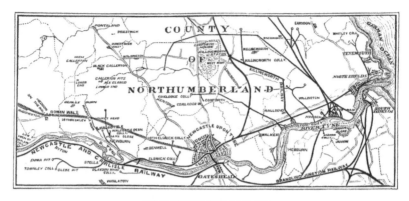

MAP OF NEWCASTLE DISTRICT.

generations, though the " Wallsend" coal consumed in
our household fires still serves to remind us of the great
Roman work.

A long period of obscurity followed the withdrawal
of these colonists, during which Northumbria became
planted by an entirely new race, principally Saxons
from North Germany and Norsemen from Scandinavia,
whose Eorls or Earls made Newcastle their principal
seat.   Then came the Normans, from whose *New* Castle,
built some eight hundred years since, the town derived
its present name. The keep of this venerable structure,
black with age and smoke, still stands entire at the
northern end of the noble high-level bridge — the
utilitarian work of modern times thus confronting the
warlike relic of the older civilisation.

The nearness of Newcastle to the Scotch Border was
a great hindrance to its security and progress in the
middle ages of English history. Indeed, the district
between it and Berwick continued to be ravaged by
mosstroopers long after the union of the Crowns.   The
gentry lived in their strong Peel castles; even the
larger farm-houses were fortified; and bloodhounds
were trained for the purpose of tracking the cattle-
reivers to their retreats in the hills. The Judges of
Assize rode from Carlisle to Newcastle guarded by an

escort armed to the teeth. A tribute called "dagger and protection money" was annually paid by the Sheriff of Newcastle for the purpose of providing daggers and other weapons for the escort; and, though the need of such protection has long since ceased, the tribute continues to be paid in broad gold pieces of the time of Charles the First.

Until about the middle of last century the roads across Northumberland were little better than horse-tracks, and not many years since the primitive agricultural cart with solid wooden wheels was almost as common in the western parts of the county as it is in Spain now. The track of the old Roman road continued to be the most practicable route between Newcastle and Carlisle, the traffic between the two towns having been carried along it upon pack-horses until a comparatively recent period. When Marshal Wade attempted to march westward in 1745, to intercept the Highland rebels on their way south, he was completely baffled by the state of the roads, which were impracticable for wheeled vehicles.[1] After the rebellion had been put down, the Marshal proceeded to construct a military road to connect Newcastle with Carlisle. He closely followed the line of the Roman wall for thirty miles west of Newcastle, and overthrew what remained of that work for the purpose of obtaining materials for his new "agger."

Since that time great changes have taken place on the Tyne. When wood for firing became scarce and dear, and the forests of the South of England were found inadequate to supply the increasing demand for fuel, attention was turned to the rich stores of coal lying underground in the neighbourhood of Newcastle and Durham. It then became an article of increasing export, and "seacoal" fires gradually supplanted those

---

[1] See 'Lives of the Engineers,' vol. i., Memoir of John Metcalf.

of wood. Hence an old writer described Newcastle as "the Eye of the North, and the Hearth that warmeth the South parts of this kingdom with Fire." Fuel has become the staple trade of the district, increasing from year to year, until at length the coal raised from these northern mines amounts to the extraordinary quantity of upwards of sixteen millions of tons a year, of which not less than nine millions of tons are annually conveyed away by sea.

Newcastle has in the mean time spread in all directions far beyond its ancient boundaries. From a walled mediæval town of monks and merchants, it has been converted into a busy centre of commerce and manufactures inhabited by nearly a hundred thousand people. It is no longer a Border fortress—a "shield and defence against the invasions and frequent insults of the Scots," as described in ancient charters—but a busy centre of peaceful industry, and the outlet for a vast amount of steam-power, which is exported in the form of coal to all parts of the world.

Newcastle is in many respects a town of singular and curious interest, especially in its older parts, which are full of crooked lanes and narrow streets, wynds, and chares,[1] formed by tall, antique houses, rising tier above tier along the steep northern bank of the Tyne, as the similarly precipitous streets of Gateshead crowd the opposite shore. A dense cloud of smoke constantly hangs over the place, almost obscuring the sun's light. North and south the atmosphere is similarly murky, and all over the coal region, which extends from the Coquet to the Tees, about fifty miles from north to south, the surface of the soil exhibits the signs of exten-

---

[1] In the Newcastle dialect, a chare is a narrow street or lane. At the local assizes some years since, one of the witnesses in a criminal trial swore that "*he saw three men come out of the foot of a chare.*" The judge cautioned the jury not to pay any regard to the man's evidence, as he must be insane. A little explanation by the foreman, however, satisfied his lordship that the original statement was correct.

sive underground workings. In every direction are to be
seen swollen heaps of ashes and refuse, coals and slag, the
rubbish of old abandoned pits, and the pumping-engines
and machinery of new. As you pass through the country
at night, the earth looks as if it were bursting with fire
at many points ; the blaze of coke-ovens, iron-furnaces,
and coal-heaps reddening the sky to such a distance that
the horizon seems to be a glowing belt of fire.

From the necessity which early existed for facilitating
the transport of coals from the pits to the shipping
places, it is easy to understand how the railway and
the locomotive should have first found their home in
the north. At an early period the coal was carried to
the boats in panniers, or in sacks upon horses' backs.
Then carts were used, to facilitate the progress of which
tramways of flag-stone were laid down. This led to the
enlargement of the vehicle, which became known as a
waggon, and was mounted on four wheels instead of
two. A local writer about the middle of the seventeenth
century says, "Many thousand people are engaged in
this trade of coals ; many live by working of them in
the pits ; and many live by conveying them in waggons
and wains to the river Tyne."[1]

Still further to facilitate the haulage of the waggons,
pieces of planking were laid parallel upon wooden
sleepers, or imbedded in the ordinary track, by which
friction was still further diminished. It is said that
these wooden rails were first employed by one Mr.
Beaumont,[2] about the year 1630 ; and on a road thus

---

[1] 'Chorographia ; or, a Survey
of Newcastle-upon-Tyne.' Newcastle,
1649.

[2] "Some South gentlemen have,
upon great hopes of benefit, come into
this country to hazard their monies in
coal-pits. Mr. Beaumont, a gentleman of great Ingenuity and rare Parts,
adventured into our mines, with his
thirty thousand Pounds, who brought
with him many rare Engines, not

known then in these Parts ; as, the
Art to bore with iron Rods, to try the
Deepness and Thickness of the Coal ;
rare Engines to draw the water out of
the Pits ; waggons, with one horse,
to carry the coals from the Pits to the
Stathes on the River, &c. Within a
few Years, he consumed all his Money,
and *rode Home upon his light Horse*."
—Harleian MS. vol. iii. 269.

laid, a single horse was capable of drawing a large
loaded waggon from the coal-pit to the shipping staith.
Roger North, in 1676, found the practice had become
extensively adopted, and he speaks of the large sum
then paid for way-leaves, that is, the permission granted
by the owners of lands lying between the coal-pit and
the river-side to lay down a tramway for the purpose
of connecting the one with the other.   A century later,
Arthur Young observed that not only had these roads
become greatly multiplied, but formidable works had
been constructed to carry them along upon the same
level.   " The coal-waggon roads from the pits to the
water," he says, " are great works, carried over all sorts
of inequalities of ground, so far as the distance of nine
or ten miles.   The tracks of the wheels are marked with
pieces of wood let into the road for the wheels of the
waggons to run on, by which one horse is enabled to
draw, and that with ease, fifty or sixty bushels of coals." [1]
Saint-Fond, the French traveller, who visited Newcastle
in 1791, spoke of the colliery waggon-ways in the
neighbourhood as superior to anything of the kind he
had seen.   He described the wooden rails as formed
with a rounded upper surface, like a projecting mould-
ing, and the waggon wheels as being " made of cast-iron,
and hollowed in the manner of a metal pulley," that they
might fit the rounded surface of the rails.   The economy
with which the coal was thus hauled to the shipping
places was urged by him as an inducement to his own
countrymen to adopt a similar method of transit. [2]

Similar waggon-roads were early laid down in the
coal districts of Wales, Cumberland, and Scotland.   At
the time of the Scotch rebellion, in 1745, a tramroad
existed between the Tranent coal-pits and the small
harbour of Cockenzie in East Lothian ; and a portion

----

[1] 'Six Months' Tour,' vol. iii. 9.
[2] 'Travels in England, Scotland, and the Hebrides,' vol. i. 142.

of the line was selected by General Cope as a position for his cannon at the battle of Prestonpans.

In these rude wooden tracks we find the germ of the modern railroad. Improvements were gradually made in them. Thus, at some collieries, thin plates of iron were nailed upon their upper surface, for the purpose of protecting the parts most exposed to friction. Cast-iron rails were also tried, the wooden rails having been found liable to rot. The first iron rails are supposed to have been laid down at Whitehaven as early as 1738. This cast-iron road was denominated a " plate-way," from the plate-like form in which the rails were cast. In 1767, as appears from the books of the Coalbrookdale Iron Works, in Shropshire, five or six tons of rails were cast, as an experiment, on the suggestion of Mr. Reynolds, one of the partners; and they were shortly after laid down to form a road.

In 1776, a cast-iron tramway, nailed to wooden sleepers, was laid down at the Duke of Norfolk's colliery near Sheffield. The person who designed and constructed this coal line was Mr. John Curr, whose son has erroneously claimed for him the invention of the cast-iron railway. He certainly adopted it early, and thereby met the fate of men before their age; for his plan was opposed by the labouring people of the colliery, who got up a riot in which they tore up the road and burnt the coal-staith, whilst Mr. Curr fled into a neighbouring wood for concealment, and lay there *perdu* for three days and nights, to escape the fury of the populace.[1] The plates of these early tramways had a ledge cast on their edge to guide the wheel along the road, after the manner shown in the annexed cut.

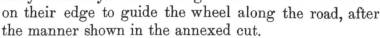

[1] 'Railway Locomotion and Steam Navigation, their Principles and Practice.' By John Curr. London, 1847.

In 1789, Mr. William Jessop constructed a railway at Loughborough, in Leicestershire, and there introduced the cast-iron edge-rail, with flanches cast upon the tire of the waggon-wheels to keep them on the track, instead of having the margin or flanch cast upon the rail itself; and this plan was shortly after adopted in other places. In 1800, Mr. Benjamin Outram, of Little Eaton, in Derbyshire (father of the distinguished General Outram), used stone props instead of timber for supporting the ends or joinings of the rails. Thus the use of railroads, in various forms, gradually extended, until they became generally adopted in the mining districts.

Such was the growth of the railway, which, it will be observed, originated in necessity, and was modified according to experience; progress in this, as in all departments of mechanics, having been effected by the exertions of many men, one generation entering upon the labours of that which preceded it, and carrying them onward to farther stages of improvement. We shall afterwards find that the invention of the locomotive was made by like successive steps. It was not the invention of one man, but of a succession of men, each working at the proper hour, and according to the needs of that hour; one inventor interpreting only the first word of the problem which his successors were to solve after long and laborious efforts and experiments. "The locomotive is not the invention of one man," said Robert Stephenson at Newcastle, "but of a nation of mechanical engineers." The same circumstances which led to the rapid extension of railways in the coal districts of the north, tended to direct the attention of the mining engineers to the early development of the powers of the steam-engine as an effective instrument of motive power. The necessity which existed for a more effective method of hauling the coals from the pits to the shipping places, was constantly present to many minds; and the daily pursuits of a large class of mechanics occupied in the management of steam power, by which the coal was

raised from the pits, and the mines were pumped clear
of water, had the effect of directing their attention to
the same agency as the most effective means of accomplishing that object.

Among the upper-ground workmen employed at
the coal-pits, the principal are the firemen, enginemen,
and brakesmen, who fire and work the engines, and
superintend the machinery by means of which the
collieries are worked. Previous to the introduction of
the steam-engine the usual machine employed for the
purpose was what is called a "gin." The gin consists
of a large drum placed horizontally, round which ropes
attached to buckets and corves are wound, which are
thus drawn up or sent down the shafts by a horse
travelling in a circular track or "gin race." This
method was employed for drawing up both coals and
water, and it is still used for the same purpose in small
collieries; but where the quantity of water to be raised
is great, pumps worked by steam power are called
into requisition.

Newcomen's atmospheric engine was first made use of
to work the pumps; and it continued to be so employed
long after the more powerful and economical condensing engine of Watt had been invented. In the
Newcomen or "fire engine," as it was called, the power
is produced by the pressure of the atmosphere forcing
down the piston in the cylinder, on a vacuum being
produced within it by condensation of the contained
steam by means of cold-water injection. The piston-rod
is attached to one end of a lever, whilst the pump-rod
works in connexion with the other,—the hydraulic
action employed to raise the water being exactly similar
to that of a common sucking-pump.

The working of a Newcomen engine is a clumsy and
apparently a very painful process, accompanied by an
extraordinary amount of wheezing, sighing, creaking,
and bumping. When the pump descends, there is heard

a plunge, a heavy sigh, and a loud bump : then, as it rises, and the sucker begins to act, there is heard a creak, a wheeze, another bump, and then a strong rush of water as it is lifted and poured out. Where engines of a more powerful and improved description are used, the quantity of water raised is enormous—as much as a million and a half gallons in the twenty-four hours.

The pitmen, who work out the coal below ground, or " the lads belaw," as they call themselves, are a peculiar class, quite distinct from the workmen employed on the surface. They are a people with peculiar habits, manners, and character, as much so as fishermen and sailors, to whom, indeed, they are supposed, perhaps from the dangerous nature of their calling, to bear a considerable resemblance. Some forty or fifty years since they were a very much rougher and worse-educated class than they are now ; hard workers, but very wild and uncouth ; much given to " steeks," or strikes ; and distinguished, in their hours of leisure and on pay-nights, for their love of cock-fighting, dog-fighting, hard drinking, and cuddy races. The pay-night was a fortnightly saturnalia, in which the pitman's character was fully brought out, especially when the " yel " was good. Though earning much higher wages than the ordinary labouring population of the upper soil, the latter did not mix nor intermarry with them ; so that they were left to form their own communities, and hence their marked peculiarities as a class. Indeed, a sort of traditional disrepute seems long to have clung to the pitmen, arising perhaps from the nature of their employment, and from the circumstance that the colliers were amongst the last classes enfranchised in England, as they were certainly the last in Scotland, where they continued bondmen down to the end of last century. The last thirty years, however, have worked a great improvement in the moral condition of the pitmen ; the abolition of the twelve months' bond to the mine, and the substitution of a month's notice previous to leaving, having given them greater freedom and opportunity for obtaining employment; and day-schools and Sunday-schools, together with the important influences of railways, have brought them fully up to a level with the other classes of the labouring population.

The coals, when raised from the pits, are emptied into the waggons placed alongside, from whence they are sent along the rails to the staiths erected by the river side, the waggons sometimes descending by their own gravity along inclined planes, the waggoner standing behind to check the speed by means of a convoy or wooden brake bearing upon the rims of the wheels. Arrived at the staiths, the waggons are emptied at once into the ships waiting alongside for cargo. Any one who has sailed down the Tyne from Newcastle Bridge cannot but have been struck with the appearance of the immense staiths, constructed of timber, which are erected at short distances from each other on both sides of the river.

COAL STAITH ON THE TYNE. [By R. P. Leitch.]

But a great deal of the coal shipped from the Tyne comes from above-bridge, where sea-going craft cannot reach, and is floated down the river in "keels," in which the coals are sometimes piled up according to convenience when large, or, when the coal is small or tender, it is conveyed in tubs to prevent breakage. These keels are of a very ancient model,—perhaps the oldest extant in England : they are even said to be of the same build as those in which the Norsemen navigated the Tyne centuries ago. The keel is a tubby, grimy-looking craft, rounded fore and aft, with a single large square sail, which the keel-bullies, as the Tyne watermen are called, manage with great dexterity ; the vessel being guided by the aid of the " swape," or great oar, which is used as a kind of rudder at the stern of the vessel. These keelmen are an exceedingly hardy class of workmen, not by any means so quarrelsome as their designation of " bully " would imply—the word being merely derived from the obsolete term " boolie," or beloved, an appellation still in familiar use amongst brother workers in the coal districts. One of the most curious sights upon the Tyne is the fleet of hundreds of these black-sailed, black-hulled keels, bringing down at each tide their black cargoes for the ships at anchor in the deep water at Shields and other parts of the river below Newcastle.

These preliminary observations will perhaps be sufficient to explain the meaning of many of the occupations alluded to, and the phrases employed, in the course of the following narrative, some of which might otherwise have been comparatively unintelligible to the general reader.

WYLAM. [By R. P. Leitch.]

CHAPTER II.

WYLAM AND DEWLEY BURN — GEORGE STEPHENSON'S EARLY YEARS.

THE colliery village of Wylam is situated on the north bank of the Tyne, about eight miles west of Newcastle. The Newcastle and Carlisle railway runs along the opposite bank; and the traveller by that line sees the usual signs of a colliery in the unsightly pumping-engines surrounded by heaps of ashes, coal-dust, and slag; whilst a neighbouring iron-furnace in full blast throws out dense smoke and loud jets of steam by day and lurid flames at night. These works form the nucleus of the village, which is almost entirely occupied by coal-miners and iron-furnacemen. The place is more remarkable for the amount of its population than for its cleanness or neatness as a village—the houses, as in most

colliery villages, being the property of the owners or
lessees, who employ them for the temporary purpose of
accommodating the workpeople, against whose earnings
there is a weekly set-off of so much for house and coals.
About the end of last century the estate of which
Wylam forms part, belonged to Mr. Blackett, a gentleman of considerable celebrity in coal-mining, then
more generally known as the proprietor of the 'Globe'
newspaper.

There is nothing to interest one in the village itself.
But a few hundred yards from its eastern extremity
stands a humble detached dwelling, which will be interesting to many as the birthplace of one of the most
remarkable men of our times—George Stephenson, the
Railway Engineer. It is a common two-storied, red-tiled,
rubble house, portioned off into four labourers' apart-
ments. It is known by the name of High Street House,
and was originally so called because it stands by the
side of what used to be the old riding post road or
street between Newcastle and Hexham, along which
the post was carried on horseback within the memory
of persons living.

The lower room in the west end of this house was
the home of the Stephenson family; and there George
Stephenson was born on the 9th of June, 1781. The
apartment is now, what it was then, an ordinary
labourer's dwelling,—its walls are unplastered, its
floor is of clay, and the bare rafters are exposed overhead.

Robert Stephenson, or "Old Bob," as the neighbours
familiarly called him, and his wife Mabel, were a respectable couple, careful and hard-working. They belonged
to the ancient and honourable family of Workers—that
extensive family which constitutes the backbone of our
country's greatness, the common working people of
England. A tradition is, indeed, preserved in the
family, that old Robert Stephenson's father and mother

HIGH-STREET HOUSE, WYLAM. [By R. P. Leitch.]

came across the Border from Scotland, on the loss of
considerable property there. Miss Stephenson, daughter
of Robert Stephenson's third son John, states that a suit
was even commenced for the recovery of the property,
but was dropped for want of means to prosecute it. Certain it is, however, that Robert Stephenson's position
throughout life was that of a humble workman. After
marrying at Walbottle, a village situated between
Wylam and Newcastle, he removed with his wife
Mabel to Wylam, where he found employment as fireman of the old pumping-engine at that colliery.

Mabel Stephenson was the only daughter of Robert
Carr, a dyer at Ovingham. The Carrs were for several generations the owners of a house in that village
adjoining the churchyard; and the family tombstone
may still be seen standing against the east end of the
chancel of the parish church, underneath the centre
lancet window; as the tombstone of Thomas Bewick, the

wood-engraver, occupies the western gable. The author, when engaged in tracing the early history of George Stephenson, casually entered into conversation one day with an old man near Dewley, a hamlet not far from Walbottle. Mabel Stephenson, he said, had been his mother's cousin; and all their "forbears" belonged to that neighbourhood. It appears that she was a woman of somewhat delicate constitution, and troubled occasionally, as her neighbours said, with "the vapours." But those who remembered her concurred in describing her as "a real canny body." And a woman of whom this is said by general consent in the Newcastle district may be pronounced a worthy person indeed; for it is about the highest praise of a woman which Northumbrians can express.

George Stephenson was the second of a family of six children. · The family Bible of Robert and Mabel Stephenson, which seems to have come into their possession in November, 1790, contains the following record of the births of these children, evidently written by one hand and at one time :—

"A Rechester of the children belonging Robert and Mabel Stepheson—

"James Stepheson Was Born March the 4 day 1779
"George Stepheson Was Born June 9 day 1781
"Elender Stepheson Was Born April the 16 day 1784
"Robert Stepheson Was Born March the 10 day 1788
"John Stepheson Was Born November the 4 day 1789
"Ann Stepheson Was Born July the 19 day 1792." [1]

It does not appear that the birth of any of these children was registered in the parish books, the author having made an unsuccessful search in the registers of Ovingham and Heddon-on-the-Wall to ascertain the

---

[1] Of the two daughters, Eleanor married Stephen Liddell, afterwards employed in the Locomotive Factory in Newcastle. Ann married John Nixon, with whom she emigrated to the United States; she died at Pittsburg, in 1860. John Stephenson was accidentally killed at the Locomotive Factory in January, 1831.

fact. Though the village of Wylam is within the parish of Ovingham, High Street House stands exactly beyond its boundary and within that of Heddon; the churches of both parishes being several miles distant.

Robert Stephenson, the father of this family, was a tall, gaunt man. A Wylam collier, who remembered him well, gave the following odd description of his personal appearance:—" Geordie's fayther war like a peer o' deals nailed thegither, an' a bit o' flesh i' th' inside; he war as queer as Dick's hatband—went thrice aboot, an' wudn't tie. His wife Mabel war a delicat' boddie, an' varry flighty. They war an honest family, but sair hadden doon i' th' world." Indeed the earnings of old Robert did not amount to more than twelve shillings a week; and, as there were six children to maintain, the family, during their stay at Wylam, were in very straitened circumstances. The father's wages being barely sufficient, even with the most rigid economy, for the sustenance of the household, there was little to spare for clothing, and nothing for education, so none of the children were sent to school.

Old Robert was a general favourite in the village, especially amongst the children, whom he was accustomed to draw about him whilst tending the engine-fire, and feast their young imaginations with tales of Sinbad the Sailor and Robinson Crusoe, besides others of his own invention; so that " Bob's engine-fire " came to be the most popular resort in the village. Another feature in his character, by which he was long remembered, was his affection for birds and animals; and he had many tame favourites of both sorts, which were as fond of resorting to his engine-fire as the boys and girls themselves. In the winter time he had usually a flock of tame robins about him; and they would come hopping familiarly to his feet to pick up the crumbs which he had saved for them out of his humble dinner. At his cottage he was rarely without one or more tame black-

birds, which flew about the house, or in and out at the
door.   In summer time he would go a-birdnesting with
his children ; and one day he took his little son George
to see a blackbird's nest for the first time.   Holding him
up in his arms, he let the wondering boy peep down,
through the branches held aside for the purpose, into a
nest full of young birds—a sight which the boy never
forgot, but used to speak of with delight to his intimate
friends when he himself had grown an old man.

The boy George led the ordinary life of working-
people's children.   He played about the doors; went
birdnesting when he could ; and ran errands to the village.   He was also an eager listener, with the other
children, to his father's curious tales ; and he early im-
bibed from him that affection for birds and animals
which continued throughout his life.   In course of time
he was promoted to the office of carrying his father's
dinner to him while at work, and it was on such occasions his great delight to see the robins fed.   At home
he helped to nurse, and that with a careful hand, his
younger brothers and sisters.   One of his duties was to
see that the other children were kept out of the way
of the chaldron waggons, which were then dragged by
horses along the wooden tramroad immediately in front
of the cottage-door.   This waggon-way was the first in
the northern district on which the experiment of a locomotive engine was tried.   But at the time of which
we speak, the locomotive had scarcely been dreamt
of in England as a practicable working power ; horses
only were used to haul the coal ; and one of the first
sights with which the boy was familiar was the coal-
waggons dragged by them along the wooden railway
at Wylam.

Thus eight years passed ; after which, the coal having
been worked out on the north side, the old engine,
which had grown " dismal to look at," as one of the
workmen described it, was pulled down ; and then

Robert, having obtained employment as a fireman at
the Dewley Burn Colliery, removed with his family to
that place. Dewley Burn, at this day, consists of a few
old-fashioned low-roofed cottages standing on either side
of a babbling little stream. They are connected by a
rustic wooden bridge, which spans the rift in front of
the doors. In the central one-roomed cottage of this
group, on the right bank, Robert Stephenson lived
for a time with his family; the pit at which he worked
standing in the rear of the cottages.

Young though he was, George was now of an age to
be able to contribute something towards the family
maintenance; for in a poor man's house, every child is
a burden until his little hands can be turned to profitable
account. That the boy was shrewd and active, and
possessed of a ready mother wit, will be evident enough
from the following incident. One day his sister Nell
went in to Newcastle to buy a bonnet; and Geordie
went with her "for company." At a draper's shop in
the Bigg Market, Nell found a "chip" quite to her
mind, but on pricing it, alas! it was found to be fifteen
pence beyond her means. Girl-like, she had set her
mind upon that bonnet, and no other would please
her. She accordingly left the shop disappointed and
very much dejected. But Geordie bravely said, "Never
heed, Nell; come wi' me, and I'll see if I canna win
siller enough to buy the bonnet; stand ye there, till
I come back." Away ran the boy and disappeared
amidst the throng of the market, leaving the girl to
wait his return. Long and long she waited, until it
grew dusk, and the market people had nearly all left.
She had begun to despair, and fears crossed her mind
that Geordie must have been run over and killed;
when at last up he came running, almost breathless.
"I've gotten the siller for the bonnet, Nell!" cried he.
"Eh, Geordie!" she said, "but hoo hae ye gotten it?"
"Hauddin the gentlemen's horses!" was the exultant

reply. The bonnet was forthwith bought, and the two returned to Dewley in triumph.

George's first regular employment was of a very humble sort.   A widow, named Grace Ainslie, then occupied the neighbouring farmhouse of Dewley.   She kept a number of cows, and had the privilege of grazing them along the waggon-road.   She needed a boy to herd the cows, to keep them out of the way of the waggons, and prevent their straying or trespassing on the neighbours' "liberties;" the boy's duty was also to bar the gates at night after all the waggons had passed. George petitioned for this post, and, to his great joy, he was appointed, at the wage of twopence a day.

It was light employment, and he had plenty of spare time on his hands, which he spent in birdnesting, making whistles out of reeds and scrannel straws, and erecting Lilliputian mills in the little water-streams that ran into the Dewley bog.   But his favourite amusement at this early age was erecting clay engines in conjunction with his chosen playmate, Bill Thirlwall.   The place is still pointed out, "just aboon the cut-end," as the people of the hamlet describe it, where the future engineers made their first essays in modelling.   The boys found the clay for their engines in the adjoining bog, and the hemlocks which grew about supplied them with imaginary steam-pipes.   They even proceeded to make a miniature winding machine in connexion with their engine, and the apparatus was erected upon a bench in front of the Thirlwalls' cottage.   Their corves were made out of hollowed corks; their ropes were supplied by twine; and a few bits of wood gleaned from the refuse of the carpenters' shop completed their materials.   With this apparatus the boys made a show of sending the corves down the pit and drawing them up again, much to the marvel of the pitmen.   But some mischievous person about the place seized the opportunity early one morning of smashing the fragile machinery, greatly to the sorrow of the young engineers. We may mention, in passing, that George's companion afterwards became a workman of repute, and creditably held the office of engineer at Shilbottle, near Alnwick, for a period of nearly thirty years.

As Stephenson grew older and abler to work, he was set to lead the horses when ploughing, though scarce big enough to stride across the furrows; and he used afterwards to say that he rode to his work in the mornings at an hour when most other children of his age were asleep in their beds. He was also employed to hoe turnips, and do similar farm-work, for which he was paid the advanced wage of fourpence a-day. But his highest ambition was to be taken on at the colliery where his father worked; and he shortly joined his elder brother James there as a " corf-bitter," or " picker," to clear the coal of stones, bats, and dross. His wages were then advanced to sixpence a-day, and afterwards to eightpence when he was set to drive the gin-horse.

Shortly after, he went to Black Callerton Colliery to drive the gin there; and as that colliery lies about two miles across the fields from Dewley Burn, the boy walked that distance early in the morning to his work, returning home late in the evening. One of the old residents at Black Callerton, who remembered him at that time, described him to the author as " a grit growing lad, with bare legs an' feet;" adding that he was " very quick-witted and full of fun and tricks : indeed, there was nothing under the sun but he tried to imitate." He was usually foremost also in the sports and pastimes of youth.

Among his first strongly developed tastes was the love of birds and animals, which he inherited from his father. Blackbirds were his special favourites. The hedges between Dewley and Black Callerton were capital birdnesting places; and there was not a nest there that he did not know of. When the young birds were old

enough, he would bring them home with him, feed
them, and teach them to fly about the cottage unconfined
by cages.   One of his blackbirds became so tame, that,
after flying about the doors all day, and in and out of
the cottage, it would take up its roost upon the bed-head
at night.   And most singular of all, the bird would dis-
appear in the spring and summer months, when it was
supposed to go into the woods to pair and rear its young,
after which it would reappear at the cottage, and resume
its social habits during the winter.   This went on for
several years.   George had also a stock of tame rabbits,
for which he built a little house behind the cottage, and
for many years he continued to pride himself upon the
superiority of his breed.

After he had driven the gin for some time at Dewley
and Black Callerton, he was taken on as an assistant to
his father in firing the engine at Dewley.   This was a
step of promotion which he had anxiously desired ; his
only fear being lest he should be found too young for
the work.   Indeed, he used afterwards to relate how he
was wont to hide himself when the owner of the colliery
went round, lest he should be thought too little a boy to
earn the wages paid him.   Since he had modelled his
clay engines in the bog, his young ambition was to be
an engineman ; and to be an assistant fireman was the
first step towards this position.   Great, therefore, was
his joy when, at about fourteen years of age, he was
appointed assistant fireman, at the wage of a shilling
a-day.

But the coal at Dewley Burn being at length worked
out, the pit was ordered to be " laid in," and old Robert
and his family were again under the necessity of shifting
their home ; for, to use the common phrase, they must
" follow the wark."

NEWBURN ON THE TYNE. [By R. P. Leitch.]

CHAPTER III.

Newburn and Callerton — Learns to be an Engineman.

On quitting their humble home at Dewley Burn, the
Stephenson family removed to a place called Jolly's
Close, a few miles to the south, close behind the village
of Newburn, where another coal-mine belonging to the
Duke of Northumberland, called "the Duke's Winnin,"
had recently been opened out.

One of the old persons in the neighbourhood, who
knew the family well, describes the dwelling in which
they lived as a poor cottage of only one room, in which
the father, mother, four sons, and two daughters lived
and slept. It was crowded with three low-poled beds.

The one apartment served for parlour, kitchen, sleeping-room, and all.

The children of the Stephenson family were now growing apace, and several of them were old enough to be able to earn money at various kinds of colliery work. James and George, the two eldest sons, worked as assistant-firemen; and the younger boys worked as wheelers or pickers on the bank-tops. The two girls helped their mother with the household work.

So far as weekly earnings went, the family were at this time pretty comfortable. Their united earnings amounted to from 35s. to 40s. a week, and they were enabled to command a fair share of the necessaries of life. But it will be remembered that in those days, from 1797 to 1802, it was much more difficult for the working classes to live than it is now. Money did not go nearly so far. The price of bread was excessive. Wheat, which for three years preceding 1795 had averaged only 54s., advanced to 76s. a quarter; and it continued to rise until, in December, 1800, it increased to 130s., and barley and oats in proportion. There was a great dearth of provisions, corn riots were of frequent occurrence, and the taxes on all articles of consumption were very heavy. The war with Napoleon was then raging; derangements of trade were frequent, causing occasional suspensions of employment in all departments of industry, from the pressure of which working people are always the first to suffer.

During this severe period George Stephenson continued to live with his parents at Jolly's Close. Other workings of the coal were opened out in the neighbourhood; and to one of these he was removed as fireman on his own account. This was called the "Mid Mill Winnin," where he had for his mate a young man named Bill Coe; and to these two was intrusted the working of the little engine put up at Mid Mill. They worked together there for about two years, by twelve-hour shifts, George firing the engine at the wage of a shilling a-day.

He was now fifteen years old. His ambition was as yet limited to attaining the standing of a full workman, at a man's wages; and with that view he endeavoured to attain such a knowledge of his engine as would eventually lead to his employment as an engineman, with its accompanying advantage of higher pay. He was a steady, sober, hardworking young man, but nothing more, in the estimation of his fellow-workmen.

One of his favourite pastimes in by-hours was trying feats of strength with his companions. Although in frame he was not particularly robust, yet he was big and bony, and considered very strong for his age. His principal competitor was Robert Hawthorn, with whom he had frequent trials of muscular strength and dexterity, such as lifting heavy weights, throwing the hammer, and putting the stone. At throwing the hammer George had no compeer; but there was a knack in putting the stone which he could never acquire, and there Hawthorn beat him. At lifting heavy weights off the ground from between his feet, by means of a bar of iron passed through them—placing the bar against his knees as a fulcrum, and then straightening his spine and lifting them sheer up—Stephenson was very successful. On one occasion, they relate, he lifted as much as sixty stones weight in this way—a striking indication of his strength of bone and muscle.

When the pit at Mid Mill was closed, George and his companion Coe were sent to work another pumping-engine erected near Throckley Bridge, where they continued for some months. It was while working at this place that his wages were raised to 12s. a week—an event to him of great importance. On coming out of the foreman's office that Saturday evening on which he received the advance, he announced the fact to his fellow-workmen, adding triumphantly, " I am now a made man for life ! "

The pit opened at Newburn, at which old Robert Stephenson worked, proving a failure, it was closed; and a new pit was sunk at Water-row, on a strip of land lying between the Wylam waggon-way and the river Tyne, about half-a-mile west of Newburn Church. A pumping-engine was erected there by Robert Hawthorn, the Duke's engineer at Walbottle; and old Stephenson went to work it as fireman, his son George acting as the engineman or plugman. At that time he was about seventeen years old—a very youthful age at which to fill so responsible a post. He had thus already got ahead of his father in his station as a workman; for the plugman holds a higher grade than the fireman, requiring more practical knowledge and skill, and usually receiving higher wages.

George's duty as plugman was to watch the engine, to see that it kept well in work, and that the pumps were efficient in drawing the water. When the water-level in the pit was lowered, and the suction became incomplete through the exposure of the suction-holes, it was then his duty to proceed to the bottom of the shaft and plug the tube so that the pump should draw : hence the designation of " plugman." If a stoppage in the engine took place through any defect in it which he was incapable of remedying, then it was for him to call in the aid of the chief engineer of the colliery to set the engine to rights.

But from the time when George Stephenson was appointed fireman, and more particularly afterwards as engineman, he applied himself so assiduously and so successfully to the study of the engine and its gearing —taking the machine to pieces in his leisure hours for the purpose of cleaning and understanding its various parts —that he soon acquired a thorough practical knowledge of its construction and mode of working, and very rarely needed to call to his aid the engineer of the colliery. His engine became a sort of pet with him, and he was never wearied of watching and inspecting it with admiration.

There is indeed a peculiar fascination about an engine to the person whose duty it is to watch and feed it. It is almost sublime in its untiring industry and quiet power : capable of performing the most gigantic work, yet so docile that a child's hand may guide it. No wonder, therefore, that the workman who is the daily companion of this life-like machine, and is constantly watching it with anxious care, at length comes to regard it with a degree of personal interest and regard. This daily contemplation of the steam-engine, and the sight of its steady action, is an education of itself to an ingenious and thoughtful man. And it is a remarkable fact, that nearly all that has been done for the improvement of this machine has been accomplished, not by philosophers and scientific men, but by labourers, mechanics, and enginemen. It would appear as if this were one of the departments of practical science in which the higher powers of the human mind must bend to mechanical instinct. The steam-engine was but a mere toy until it was taken in hand by workmen. Savery was originally a working miner, Newcomen a blacksmith, and his partner Cawley a glazier. In the hands of Watt, the instrument-maker, who devoted almost a life to the study of the subject, the condensing-engine acquired gigantic strength ; and George Stephenson, the colliery engineman, was certainly not the least of those who have assisted to bring the high-pressure engine to its present power.

Although the progress made by our young mechanic was unusually rapid—helped as he was by native shrewdness, quick perception, and assiduous application —he had not yet even begun his literary culture. He was eighteen years old before he learnt to read; and, having the charge of an engine which occupied his time to the extent of twelve hours every day, he had thus very few leisure moments that he could call his own. But the busiest man will find them if he watch for them; and if he be careful in turning these moments to useful account, he will prove them to be the very " gold-dust of time," as Young has so beautifully described them.

Not many of his fellow-workmen had learnt to read; but those who could do so were placed under frequent contribution by George and the other labourers at the pit. It was one of their greatest treats to induce some one to read to them by the engine-fire, out of any book or stray newspaper which found its way into the village of Newburn. Buonaparte was then overrunning Italy, and astounding Europe by his brilliant succession of victories; and there was no more eager auditor of these exploits, when read from the newspaper accounts, than the young engine-man at the Water-row Pit.

There were also numerous stray bits of information and intelligence contained in these papers, which excited Stephenson's interest. One of these related to the Egyptian art of hatching birds' eggs by means of artificial heat. Curious about everything relating to birds, he determined to test the art by experiment. It was spring time, and he forthwith went a birdnesting in the adjoining woods and hedges, where there were few birds' nests of which he did not know. He brought a collection of eggs of all kinds into the engine-house, set them in flour in a warm place, covering the whole over with wool, and then waited the issue of his experiment. But though the heat was kept as steady as possible, and the eggs were carefully turned every twelve hours, they never hatched. The eggs chipped, and some of them exhibited well-grown chicks; but none of the birds came forth alive, and thus the experiment failed. The incident, however, serves to show that the inquiring mind of the youth was fairly at work.

Another of his favourite occupations continued to be the modelling of clay engines. He not only made models of engines which he had seen, but he also tried to make models of others which were described to him. These attempts no doubt showed considerable improvement upon his first trials in the clay of Dewley Burn bog, when occupied there as a herd-boy. He was told, however, that all the wonderful engines of Watt and Boulton, about which he was so anxious to know, were to be found described in books, and that he must satisfy his curiosity by searching them for a complete description of the machines which he desired to model. But, alas! Stephenson could not read; he had not yet learnt even his letters.

Thus he shortly found, when gazing wistfully in the direction of knowledge, that to advance further as a skilled workman, he must master this wonderful art of reading—the key to so many other arts. Only thus could he gain an access to books, the depositories of the wisdom and experience of the past. Although a grown man and doing the work of a man, he was not ashamed to confess his ignorance, and go to school, big as he was, to learn his letters. Perhaps, too, he foresaw that, in laying out a little of his spare earnings for this purpose, he was investing money judiciously, and that, in every hour he spent at school, he was really working for better wages. He determined, therefore, to learn this useful art of reading, and to make a beginning—a small beginning, it is true, but still a right one, and a pledge and assurance that he was in earnest in the work of self-culture. He desired thus to open for himself a road into knowledge; and no man can sincerely desire this but he will eventually succeed in finding it.

His first schoolmaster was Robin Cowens, a poor

teacher in the village of Walbottle. He kept a night-
school, which was attended by a few of the colliers and
labourers' sons in the neighbourhood. George took
lessons in spelling and reading three nights in the week.
Tommy Musgrove, the lad who "sled out" the engine
at the Water-row Pit, usually went with him to the
evening lesson. Robin Cowen's teaching cost three-
pence a week; and though it was not very good, yet
George, being hungry for knowledge and eager to
acquire it, soon learnt to read. He also practised "pothooks," and at the age of nineteen he was proud to be
able to write his own name.

A Scotch dominie, named Andrew Robertson, set up
a night-school in the village of Newburn, in the winter
of 1799. It was more convenient for George to attend
this school, as it was nearer to his work, and not more
than a few minutes' walk from Jolly's Close. Besides,
Andrew had the reputation of being a skilled arithmetician; and this was a branch of knowledge that Stephenson was desirous of acquiring. He accordingly began
taking lessons from him, paying fourpence a week.
Robert Gray, the junior fireman at the Water-row Pit,
began arithmetic at the same time; and Gray afterwards
told the author that George learnt "figuring" so much
faster than he did, that he could not make out how it
was—"he took to figures so wonderful." Although
the two started together from the same point, at the
end of the winter George had mastered "reduction,"
while Robert Gray was still struggling with the difficulties of simple division. But George's secret was his
perseverance. He worked out the sums in his bye-
hours, improving every minute of his spare time by
the engine-fire, and there solving the arithmetical
problems set for him upon his slate by the master. In
the evenings he took to Robertson the sums which he
had thus "worked," and new ones were "set" for him
to study out the following day. Thus his progress was

rapid, and, with a willing heart and mind, he soon became well advanced in arithmetic. Indeed, Andrew Robertson became somewhat proud of his scholar; and shortly afterwards, when the Water-row Pit was closed, and George removed to Black Callerton to work there, the poor schoolmaster, not having a very extensive connexion in Newburn, went with his pupils, and set up his night-school at Black Callerton, where he continued his instruction to them.

George still found time to attend to his favourite animals while working at the Water-row Pit. Like his father, he used to tempt the robin-redbreasts to hop and fly about him at the engine-fire, by the bait of bread-crumbs saved from his dinner. But his favourite animal was his dog—so sagacious that he performed the office of a servant, in almost daily carrying his dinner to him at the pit. The tin containing the meal was suspended from the dog's neck, and, thus laden, he proudly walked the road from Jolly's Close to Water-row Pit, quite through the village of Newburn. He turned neither to left nor right, nor minded for the time the barking of curs at his heels. But his course was not unattended with perils. One day the big strange dog of a passing butcher espied the engineman's messenger, ran after him, and fell upon him with the tin can about his neck. There was a terrible tussle and worrying between the dogs, which lasted for a brief while, and, shortly after, the dog's master, anxious for his dinner, saw his faithful servant approaching, bleeding but triumphant. The tin can was still round his neck, but the dinner had escaped in the struggle. Though George went without his dinner that day, when the circumstances of the combat were related to him by the villagers who had seen it, he was prouder of his dog than ever.

It was while working at the Water-row Pit that Stephenson first learnt the art of brakeing an engine. This being one of the higher departments of colliery labour,

and amongst the best paid, George was very anxious to
learn it.   A small winding-engine having been put up
for the purpose of drawing the coals from the pit, Bill
Coe, his friend and fellow-workman, was appointed the
brakesman.   He frequently allowed George to try his
hand at the brake, and instructed him how to proceed.
Coe was, however, opposed in this by several of the
other workmen—one of whom, a brakesman named
William Locke,[1] went so far as to stop the working of
the pit because Stephenson had been called in to the
brake.   But one day as Mr. Charles Nixon, the manager
of the pit, was observed approaching, Coe adopted an
expedient which had the effect of putting a stop to
the opposition.   He called upon George Stephenson
to "come into the brake-house, and take hold of the
machine."   No sooner had he done this, than Locke, as
usual, sat down, and the working of the pit was stopped.
Locke, when requested by the manager to give an
explanation, said that " young Stephenson couldn't
brake, and, what was more, never would learn to brake,
he was so clumsy."   Mr. Nixon, however, ordered
Locke to go on with the work, which he did ; and
Stephenson, after some further practice, acquired the
art of brakeing.

After working at the Water-row Pit and at other
engines in the neighbourhood of Newburn, for about
three years, George, with his companion Coe, went to
work at Black Callerton early in 1801.   Though only
twenty years of age, his employers thought so well of
him that they appointed him to the responsible office of
brakesman at the Dolly Pit.   For convenience' sake,
he took lodgings at a small farmer's in the village,
finding his own victuals, and paying so much a week
for lodging and attendance.   In the locality this was
called " picklin in his awn poke neuk."   It not unfre-

---

[1] Father of Mr. Locke, M.P., the engineer.   He afterwards removed to
Barnsley, in Yorkshire.

quently happens that the young workman about the collieries, when selecting a lodging, contrives to pitch his tent where the daughter of the house ultimately becomes his wife. This is often the real attraction that draws the youth from home, though a very different one may be pretended.

George Stephenson's duties as brakesman may be briefly described. The work was somewhat monotonous, and consisted in superintending the working of the engine and machinery by means of which the coals were drawn out of the pit. Brakesmen are almost invariably selected from those who have had considerable experience as engine-firemen, and borne a good character for steadiness, punctuality, watchfulness, and "mother wit." In George Stephenson's day the coals were drawn out of the pit in corves, or large baskets made of hazel rods. The corves were placed two together in a cage, between which and the pit-ropes there was usually from fifteen to twenty feet of chain. The approach of the corves towards the pit mouth was signalled by a bell, brought into action by a piece of mechanism worked from the shaft of the engine. When the bell sounded, the brakesman checked the speed, by taking hold of the hand-gear connected with the steam-valves, which were so arranged that by their means he could regulate the speed of the engine, and stop or set it in motion when required. Connected with the fly-wheel was a powerful wooden brake, acting by pressure against its rim, something like the brake of a railway-carriage against its wheels. On catching sight of the chain attached to the ascending corve-cage, the brakesman, by pressing his foot upon a foot-step near him, was enabled, with great precision, to stop the revolutions of the wheel, and arrest the ascent of the corves at the pit mouth, when they were forthwith landed on the "settle board." On the full corves being replaced by empty ones, it was then the duty of the brakesman to reverse the engine, and send the corves down the pit to be filled again.

The monotony of George Stephenson's occupation as a brakesman was somewhat varied by the change which he made, in his turn, from the day to the night shift. His duty, on the latter occasions, consisted chiefly in sending men and materials into the mine, and in drawing other men and materials out. Most of the workmen enter the pit during the night shift, and leave it in the latter part of the day, whilst coal-drawing is proceeding. The requirements of the work at night are such, that the brakesman has a good deal of spare time on his hands, which he is at liberty to employ in his own way. From an early period, George was accustomed to employ those vacant night hours in working the sums set for him by Andrew Robertson upon his slate, practising writing in his copy-book, and mending the shoes of his fellow-workmen. His wages while working at the Dolly Pit amounted to from 1*l*. 15*s*. to 2*l*. in the fortnight; but he gradually added to them as he became more expert at shoe-mending, and afterwards at shoe-making. Probably he was stimulated to take in hand this extra work by the attachment he had by this time formed for a young woman named Fanny Henderson, who officiated as servant in the small farmer's house in which he lodged. The personal attractions of Fanny, though these were considerable, were the least of her charms. Her temper was of the sweetest; and those who knew her were accustomed to speak of the charming modesty of her demeanour, her kindness of disposition, and withal her sound good sense.

Amongst his various mendings of old shoes at Callerton, George was on one occasion favoured with the shoes of his sweetheart to sole. One can imagine the pleasure with which he would linger over such a piece of work, and the pride with which he would execute it.

A friend of his, still living, relates that; after he had finished the shoes, he carried them about with him in his pocket on the Sunday afternoon, and that from time to time he would whip them out and hold them up, exclaiming, "what a capital job he had made of them!" Other lovers have carried about with them a lock of their fair one's hair, a glove, or a handkerchief; but none could have been prouder of their cherished love-token than was George Stephenson of his Fanny's shoes, which he had just soled, and of which he had made such a " capital job."

Out of his earnings by shoe-mending at Callerton, George contrived to save his first guinea. The first guinea saved by a working man is no trivial thing. If, as in Stephenson's case, it has been the result of prudent self-denial, of extra labour at bye-hours, and of the honest resolution to save and economise for worthy purposes, the first guinea saved is an earnest of better things. When Stephenson had saved this guinea he was not a little elated at the achievement, and expressed the opinion to a friend, who many years after reminded him of it, that he was " now a rich man."

Not long after he began to work at Black Callerton as brakesman, he had a quarrel with a pitman named Ned Nelson, a roistering bully, who was the terror of the village. Nelson was a great fighter; and it was therefore considered dangerous to quarrel with him. Stephenson was so unfortunate as not to be able to please this pitman by the way in which he drew him out of the pit; and Nelson swore at him grossly because of the alleged clumsiness of his brakeing. George defended himself, and appealed to the testimony of the other workmen. But Nelson had not been accustomed to George's style of self-assertion ; and, after a great deal of abuse, he threatened to kick the brakesman, who defied him to do so. Nelson ended by challenging Stephenson to a pitched battle ; and the latter accepted

the challenge, when a day was fixed on which the fight
was to come off.

Great was the excitement at Black Callerton when
it was known that George Stephenson had accepted
Nelson's challenge. Everybody said he would be killed.
The villagers, the young men, and especially the boys
of the place, with whom George was a great favourite,
all wished that he might beat Nelson, but they scarcely
dared to say so. They came about him while he was at
work in the engine-house to inquire if it was really true
that he was "goin to fight Nelson?" "Ay; never
fear for me; I'll fight him." And fight him he did.
For some days previous to the appointed day of battle,
Nelson went entirely off work for the purpose of keeping himself fresh and strong, whereas Stephenson went
on doing his daily work as usual, and appeared not in
the least disconcerted by the prospect of the affair. So,
on the evening appointed, after George had done his
day's labour, he went into the Dolly Pit Field, where
his already exulting rival was ready to meet him.
George stripped, and "went in" like a practised pugilist—though it was his first and last fight. After a few
rounds, George's wiry muscles and practised strength
enabled him severely to punish his adversary, and to
secure an easy victory.

This circumstance is related in illustration of Stephenson's personal pluck and courage; and it was thoroughly
characteristic of the man. He was no pugilist, and the
very reverse of quarrelsome. But he would not be put
down by the bully of the colliery, and he fought him.
There his pugilism ended; they afterwards shook hands,
and continued good friends. In after life, Stephenson's
mettle was often as hardly tried, though in a different
way; and he did not fail to exhibit the same resolute
courage in contending with the bullies of the railway
world, as he showed in his encounter with Ned Nelson,
the fighting pitman of Callerton.

STEPHENSON'S COTTAGE AT WILLINGTON QUAY. [By R. P. Leitch.]

CHAPTER IV.

ENGINEMAN AT WILLINGTON QUAY AND KILLINGWORTH.

GEORGE STEPHENSON had now acquired the character of an expert workman. He was diligent and observant while at work, and sober and studious when the day's work was over. His friend Coe described him to the author as " a standing example of manly character." On pay-Saturday afternoons, when the pitmen held their fortnightly holiday, occupying themselves chiefly in cock-fighting and dog-fighting in the adjoining fields, followed by adjournments to the " yell-house," George was accustomed to take his engine to pieces, for the purpose of obtaining " insight," and he cleaned all the parts and put the machine in thorough working order before leaving her. His amusements continued to be principally of the athletic kind ; and he found few that could beat him at lifting heavy weights, leaping, and throwing the hammer.

In the evenings he improved himself in the arts of reading and writing, and occasionally took a turn at modelling. It was at Callerton, his son Robert informed us, that he began to try his hand at original invention; and for some time he applied his attention to a machine of the nature of an engine-brake, which reversed itself by its own action. But nothing came of the contrivance, and it was eventually thrown aside as useless. Yet not altogether so; for even the highest skill must undergo the inevitable discipline of experiment, and submit to the wholesome correction of occasional failure.

After working at Callerton for about two years, he received an offer to take charge of the engine on Willington Ballast Hill at an advanced wage. He determined to accept it, and at the same time to marry Fanny Henderson, and begin housekeeping on his own account. Though he was only twenty-one years old, he had contrived, by thrift, steadiness, and industry, to save as much money as enabled him to take a cottage-dwelling at Willington Quay, and furnish it in a humble but comfortable style for the reception of his young bride.

Willington Quay lies on the north bank of the Tyne, about six miles below Newcastle. It consists of a line of houses straggling along the river side; and high behind it towers up the huge mound of ballast emptied out of the ships which resort to the quay for their cargoes of coal for the London market. The ballast is thrown out of the ships' holds into waggons laid alongside. When filled, a train of these is dragged to the summit of the Ballast Hill, where they are run out, and their contents emptied on to the monstrous accumulation of earth, chalk, and Thames mud already laid there, probably to form a puzzle for future antiquaries

and geologists, when the origin of these immense hills
along the Tyne has been forgotten.   On the summit of
the Willington Ballast Hill was a fixed engine, which
drew the trains of laden waggons up the incline; and
of this engine George Stephenson acted as brakesman.

The cottage in which he took up his abode was a
small two-storied dwelling, standing a little back from
the quay, with a bit of garden ground in front.[1]   The
Stephenson family occupied the upper room in the west
end of the cottage.   Close behind rose the Ballast Hill.

When the cottage-dwelling had been made snug, and
was ready for the young wife's reception, the marriage
took place.   It was celebrated in Newburn Church, on
the 28th of November, 1802.   George Stephenson's
signature, as it stands in the register, is that of a person
who seems to have just learnt to write.   Yet it is the
signature of a man, written slowly and deliberately,
in strong round hand.   With all his care, however, he
had not been able to avoid a blotch; the word "Stephenson" seems to have been brushed over before the
ink was dry.

George Stephenson

Frances Henderson

After the ceremony, George and his newly-wedded
wife proceeded to the house of old Robert Stephenson
and his wife Mabel at Jolly's Close.   The old man was
now becoming infirm, though he still worked as an

---

[1] The Stephenson Memorial Schools
have since been erected on the site
of the old cottage at Willington Quay
represented in the engraving at the
head of this chapter.   A vignette of
the Memorial Schools will be found at
the end of the volume.

engine-fireman, and contrived with difficulty " to keep
his head above water." When the visit had been paid,
the bridal party prepared to set out for their new home
at Willington Quay. They went in a style which was
quite common before travelling by railway had been
invented. Two farm horses were borrowed from Mr.
Burn, of the Red House Farm, Wolsingham, where
Anne Henderson, the bride's sister, lived as servant.
The two horses were each provided with a saddle and
a pillion, and George having mounted one, his wife
seated herself behind him, holding on by her arms round
his waist. Robert Gray and Anne Henderson in like
manner mounted the other horse ; and in this wise the
wedding party rode across the country, passing through
the old streets of Newcastle, and then by Wallsend to
Willington Quay—a long ride of about fifteen miles.

George Stephenson's daily life at Willington was
that of a steady workman. By the manner, however,
in which he continued to improve his spare hours in
the evening, he was silently and surely paving the
way for being something more than a manual labourer.
He diligently set himself to study the principles of
mechanics, and to master the laws by which his engine
worked. For a workman, he was even at that time
more than ordinarily speculative — often taking up
strange theories, and trying to sift out the truth that
was in them. While sitting by the side of his young
wife in his cottage-dwelling in the winter evenings, he
was usually occupied in studying mechanical subjects,
or in modelling experimental machines. Amongst his
various speculations while at Willington, he tried to
discover a means of Perpetual Motion. Although he
failed, as so many others had done before him, the very
efforts he made tended to whet his inventive faculties, and
to call forth his dormant powers. He actually went so far
as to construct the model of a machine for the purpose.
It consisted of a wooden wheel, the periphery of which

was furnished with glass tubes filled with quicksilver; as the wheel rotated, the quicksilver poured itself down into the lower tubes, and thus a sort of self-acting motion was kept up in the apparatus, which, however, did not prove to be perpetual. Where he had first obtained the idea of this machine—whether from conversation, or reading, or his own thoughts, is not known; but his son Robert was of opinion that he had heard of an apparatus of the kind described in the "History of Inventions." As he had then no access to books, and indeed could barely read with ease, it is probable that he had been told of the invention, and set about testing its value according to his own methods.

Much of his spare time continued to be occupied by labour more immediately profitable, regarded in a pecuniary point of view. In the evenings, after his day's labour at his engine, he would occasionally employ himself for a few hours in casting ballast out of the collier ships, by which means he was enabled to earn a few shillings extra weekly. Mr. William Fairbairn of Manchester has informed the author that while Stephenson was employed at the Willington Ballast Hill he himself was working in the neighbourhood as an engine apprentice at the Percy Main Colliery. He was very fond of George, who was a fine, hearty fellow, besides being a capital workman. In the summer evenings young Fairbairn was accustomed to go down to Willington to see his friend, and on such occasions he would frequently take charge of George's engine for a few hours to enable him to take a two or three hours' turn at heaving ballast out of the collier vessels. It is pleasant to think of the future President of the British Association thus helping the future Railway Engineer to earn a few extra shillings by overwork in the evenings, at a time when both occupied the rank but of humble working men in an obscure northern village.

Mr. Fairbairn was also a frequent visitor at George's

cottage on the Quay, where, though there was no luxury, there was comfort, cleanliness, and a pervading spirit of industry. Even at home George was never for a moment idle. When there was no ballast to heave out, he took in shoes to mend; and from mending he proceeded to making them, as well as shoe-lasts, in which he was admitted to be very expert. William Coe, who continued to live at Willington in 1851, informed the author that he bought a pair of shoes from George Stephenson for 7s. 6d., and he remembered that they were a capital fit, and wore very well.

But an accident occurred in Stephenson's household about this time, which had the effect of directing his industry into a new and still more profitable channel. The cottage chimney took fire one day in his absence, when the alarmed neighbours, rushing in, threw quantities of water upon the flames; and some, in their zeal, even mounted the ridge of the house, and poured buckets of water down the chimney. The fire was soon put out, but the house was thoroughly soaked. When George came home he found the water running out of the door, everything in disorder, and his new furniture covered with soot. The eight-day clock, which hung against the wall—one of the most highly-prized articles in the house—was grievously injured by the steam with which the room had been filled. Its wheels were so clogged by the dust and soot, that it was brought to a complete stand-still. George was always ready to turn his hand to anything, and his ingenuity, never at fault, immediately set to work for the repair of the unfortunate clock. He was advised to send it to the clockmaker, but that would have cost money; and he declared that he would repair it himself—at least he would try. The clock was accordingly taken to pieces and cleaned; the tools which he had been accumulating for the purpose of constructing his Perpetual Motion machine, readily enabled him to do this; and he succeeded so well that, shortly after, the neighbours sent him their clocks to clean, and he soon became one of the most famous clock-doctors in the neighbourhood.

It was while living at Willington Quay that George Stephenson's only son was born, on the 16th of October, 1803.[1] The child was from the first, as may well be imagined, a great favourite with his father, whose evening hours were made happier by his presence. George Stephenson's strong "philoprogenitiveness," as phrenologists call it, had in his boyhood expended itself on birds, and dogs, and rabbits, and even on the poor old gin-horses which he had driven at the Callerton Pit; and now he found in his child a more genial object on which to expend the warmth of his affection.

The christening of the boy took place in the school-house at Wallsend, the old parish church being at the time in so dilapidated a condition from the "creeping" or subsidence of the ground, consequent upon the excavation of the coal, that it was considered dangerous to enter it. On this occasion, Robert Gray and Anne Henderson, who had officiated as bridesman and bridesmaid at the wedding, came over again to Willington, and stood as godfather and godmother to little Robert, as the child was named, after his grandfather.

After working for about three years as a brakesman at the Willington machine, George Stephenson was induced to leave his situation there for a similar one at the West Moor Colliery, Killingworth. It was not without considerable persuasion that he was induced to leave the Quay, as he knew that he should thereby give

---

[1] No register was made of Robert Stephenson's birth, and he himself was in doubt whether he was born in October, November, or December. For instance, a dinner was given to him by the contractors of the London and Birmingham Railway on the 16th November, 1839, that day being then supposed by his father to have been his birthday. When preparing the 'Life of George Stephenson,' Robert stated to the author that the 16th of December was the correct day. But after the book had passed through four editions he desired the date to be corrected to the 16th of October, which on the whole he thought the right date, and it was so altered accordingly.

WEST MOOR COLLIERY. [By R. P. Leitch]

up the chance of earning extra money by casting ballast from the keels.   At last, however, he consented, in the hope of making up the loss in some other way.   The village of Killingworth lies about seven miles north of Newcastle, and is one of the best-known collieries in that neighbourhood.   The workings of the coal are of vast extent, and give employment to a large number of workpeople.   The place stands high, and commands an extensive view of the adjacent country; it overlooks the valley of the Tyne on the south, and the pinnacles of the Newcastle spires may be discerned in the distance, when not obscured by the clouds of smoke which rise up from that hive of manufacturing industry.

To this place George Stephenson first came as a brakesman in the year 1804.   He had scarcely settled down in his new home, ere he sustained a heavy loss in the death of his wife, for whom he cherished the sincerest affection.   Their married life had been happy, sweetened as it was by daily successful toil.   The husband was sober and hard-working, and his young wife made his hearth so bright and his home so snug, that no attraction could draw him from her side in the evening hours. But this domestic happiness was all to pass away; and the twinkling feet, for which the lover had made those tiny shoes at Callerton, were now to be hidden for evermore from his eyes. It was a terrible blow to him, and he long lamented the bereavement.

Shortly after this event, while his grief was still fresh, he received an invitation from some gentlemen concerned in large spinning works near Montrose in Scotland, to proceed thither and superintend the working of one of Boulton and Watt's engines. He accepted the offer, and made arrangements to leave Killingworth for a time.

Having left his boy in charge of his father and mother, still living at Jolly's Close, near Newburn, he set out upon his long journey to Scotland on foot, with his kit upon his back. It was while working at Montrose that he first gave proofs of that practical ability in contrivance for which he was afterwards so distinguished. It appears that the water required for the purposes of his engine, as well as for the use of the works, was pumped from a considerable depth, being supplied from the adjacent extensive sand strata. The pumps frequently got choked by the sand drawn in at the bottom of the well through the snore-holes, or apertures through which the water to be raised is admitted. The barrels soon became worn, and the bucket and clack leathers destroyed, so that it became necessary to devise a remedy; and with this object the engine-man proceeded to adopt the following simple but original expedient. He had a wooden box or boot made, twelve feet high, which he placed in the sump or well, and into this he inserted the lower end of the pump. The result was, that the water flowed clear from the outer part of the well over into the boot, and was drawn up without any admixture of sand, and the difficulty was thus conquered.[1]

During his short stay, being paid good wages, Stephenson contrived to save a sum of 28*l*., which he took back with him to Killingworth, after an absence of about a year. Longing to get back to his own kindred, his heart yearning for his son whom he had left behind, our engine-man took leave of his Montrose employers, and trudged back to Killingworth on foot as he had gone. He related to his friend Coe, on his return, that when on the borders of Northumberland, late one evening, footsore and wearied with his long day's journey, he knocked at a small farmer's cottage door, and requested shelter for the night. It was refused, and then he entreated that, being sore tired and unable to proceed any further, they would permit him to lie down in the outhouse, for that a little clean straw would serve him. The farmer's wife appeared at the door, looked at the traveller, then retiring with her husband, the two confabulated a little apart, and finally they invited Stephenson into the cottage. Always full of conversation and anecdote, he soon made himself at home in the farmer's family, and spent with them a few pleasant hours. He was hospitably entertained for the night, and when he left the cottage in the morning, he pressed them to make some charge for his lodging, but they " would not hear of such a thing." They asked him to remember them kindly, and if he ever came that way, to be sure and call again. Many years after, when

---

[1] This incident was related by Robert Stephenson during a voyage to the north of Scotland in 1857, when off Montrose, on board his yacht *Titania;* and the reminiscence was immediately communicated to the author by the late Mr. William Kell of Gateshead, who was present, at Mr. Stephenson's request, as being worthy of insertion in his father's biography. Mr. George Elliott, one of the most skilled coal-viewers in the North, was of the party, and expressed his admiration at the ready skill with which the difficulty had been overcome, the expedient of the boot being then unknown in the Northumberland and Durham mines. He acknowledged it to be " a wrinkle," adding that its application would, in several instances within his own knowledge, have been of great practical value.

Stephenson had become a thriving man, he did not forget the humble pair who had thus succoured and entertained him on his way; he sought their cottage again, when age had silvered their hair; and when he left the aged couple, on that occasion, they may have been reminded of the old saying that we may sometimes " entertain angels unawares."

Reaching home, Stephenson found that his father had met with a serious accident at the Blucher Pit, which had reduced him to great distress and poverty. While engaged in the inside of an engine, making some repairs, a fellow-workman accidentally let in the steam upon him. The blast struck him full in the face; he was terribly scorched, and his eyesight was irretrievably lost. The helpless and infirm man had struggled for a time with poverty; his sons who were at home, poor as himself, were little able to help him, while George was at a distance in Scotland. On his return, however, with his savings in his pocket, his first step was to pay off his father's debts, amounting to about 15*l.*; and shortly after he removed the aged pair from Jolly's Close to a comfortable cottage adjoining the tramroad near the West Moor at Killingworth, where the old man lived for many years, supported entirely by his son.

Stephenson was again taken on as a brakesman at the West Moor Pit. He does not seem to have been very hopeful as to his prospects in life about the time (1807-8). Indeed the condition of the working class generally was then very discouraging. England was engaged in a great war, which pressed upon the industry, and severely tried the resources, of the country. Heavy taxes were imposed upon all the articles of consumption that would bear them. There was a constant demand for men to fill the army, navy, and militia. Never before had England witnessed such drumming and fifing for recruits. In 1805, the gross forces of the United Kingdom amounted to nearly

700,000 men, and early in 1808 Lord Castlereagh
carried a measure for the establishment of a local
militia of 200,000 men. These measures produced great
and general distress amongst the labouring classes.
There were riots in Manchester, Newcastle, and elsewhere, through scarcity of work and lowness of wages.
The working people were also liable to be pressed for
the navy, or drawn for the militia; and though men
could not fail to be discontented under such circumstances, they scarcely dared, in those perilous times,
even to mutter their discontent to their neighbours.

George Stephenson was one of those drawn for the
militia. He must therefore either quit his work and go
a-soldiering, or find a substitute. He adopted the latter
course, and borrowed 6*l.*, which, with the remainder of
his savings, enabled him to provide a militia-man to
serve in his stead. Thus the whole of his hard-won
earnings were swept away at a stroke. He was almost
in despair, and contemplated the idea of leaving the
country, and emigrating to the United States. Although
a voyage there was then a much more formidable thing
for a working man to accomplish than a voyage to Aus-
tralia is now, he seriously entertained the project, and
had all but made up his mind to go. His sister Ann, with
her husband, emigrated about that time, but George
could not raise the requisite money, and they departed
without him. After all, it went sore against his heart
to leave his home and his kindred, the scenes of his
youth and the friends of his boyhood; and he struggled
long with the idea, brooding over it in sorrow. Speaking afterwards to a friend of his thoughts at the time,
he said: "You know the road from my house at the
West Moor to Killingworth. I remember once when I
went along that road I wept bitterly, for I knew not
where my lot in life would be cast." But Providence
had better things in store for George Stephenson than
the lot of a settler in the wilds of America. It was

well that his poverty prevented him from prosecuting
further the idea of emigration, and rooted him to the
place where he afterwards worked out his great career
so manfully and victoriously.

In 1808, Stephenson, with two other brakesmen,
named Robert Wedderburn and George Dodds, took a
small contract under the colliery lessees for brakeing
the engines at the West Moor Pit. The brakesmen
found the oil and tallow; they divided the work amongst
them, and were paid so much per score for their labour.
There being two engines working night and day, two
of the three men were always at work; the average
earnings of each amounting to from 18s. to 20s. a week.
It was the interest of the brakesmen to economise the
working as much as possible, and George no sooner
entered upon the contract than he proceeded to devise
ways and means of making the contract "pay." He
observed that the ropes with which the coal was drawn
out of the pit by the winding-engine were badly arranged; they "glued," and wore each other to tatters
by the perpetual friction. There was thus great wear
and tear, and a serious increase in the expenses of the
pit. George found that the ropes which, at other pits
in the neighbourhood, lasted about three months, at the
West Moor Pit became worn out in about a month.
He accordingly set himself to ascertain the cause of
the defect; after which he proceeded, with the sanction
of the head engine-wright and of the colliery owners,
to shift the pulley-wheels so that they worked immediately over the centre of the pit, and by an entire
rearrangement of the gearing of the machine, he shortly
succeeded in greatly lessening the wear and tear of the
ropes, to the advantage of the owners as well as of
the workmen, who were thus enabled to labour more
continuously and profitably.

About the same time he attempted an improvement
in the winding-engine which he worked, by placing a

valve between the air-pump and condenser. This expedient, although it led to no practical result, showed that his mind was actively at work in mechanical adaptations. It continued to be his regular habit, on Saturdays, to take his engine to pieces, for the purpose, at the same time, of familiarising himself with its action, and of placing it in a state of thorough working order. And by diligently mastering the details of the engine, he was enabled, as opportunity occurred, to turn to practical account the knowledge thus acquired.

Such an opportunity was not long in presenting itself. In the year 1810, a pit was sunk by the "Grand Allies" (the lessees of the mines) at the village of Killingworth, now known as the Killingworth High Pit. An atmospheric or Newcomen engine, originally made by Smeaton, was fixed there for the purpose of pumping out the water from the shaft; but, somehow or other, the engine failed to clear the pit. As one of the workmen has since described the circumstance— "She couldn't keep her jack-head in water: all the enginemen in the neighbourhood were tried, as well as Crowther of the Ouseburn, but they were clean bet." The engine had been fruitlessly pumping for nearly twelve months, and began to be spoken of as a total failure. Stephenson had gone to look at it when in course of erection, and then observed to the over-man that he thought it was defective; he also gave it as his opinion that, if there were much water in the mine, the engine would never keep it under. Of course, as he was only a brakesman, his opinion was considered to be worth very little on such a point, and no more was thought about it. He continued, however, to make frequent visits to the engine, to see "how she was getting on." From the bank-head where he worked his brake he could see the chimney smoking at the High Pit; and as the workmen were passing to and from their work, he would call out and inquire "if they had gotten to the bottom yet?" And the reply was always to the same effect—the pumping made no progress, and the workmen were still "drowned out."

One Saturday afternoon he went over to the High Pit to examine the engine more carefully than he had yet done. He had been turning the subject over in his mind; and after a long examination, he seemed to satisfy himself as to the cause of the failure. Kit Heppel, who was a sinker at the pit, said to him: "Weel, George, what do you mak' o' her? Do you think you could do anything to improve her?" "Man," said George in reply, "I could alter her and make her draw: in a week's time from this I could send you to the bottom."

Forthwith Heppel reported this conversation to Ralph Dodds, the head viewer; and Dodds, being now quite in despair, and hopeless of succeeding with the engine, determined to give George's skill a trial. George had already acquired the character of a very clever and ingenious workman; and at the worst he could only fail, as the rest had done. In the evening, Mr. Dodds went towards Stephenson's cottage in search of him. He met him on the road, dressed in his Sunday's suit, about to proceed to "the preaching" in the Methodist Chapel, which he at that time attended. "Well, George," said Mr. Dodds, accosting him, "they tell me you think you can put the engine at the High Pit to rights." "Yes, sir," said George, "I think I could." "If that's the case, I'll give you a fair trial, and you must set to work immediately. We are clean drowned out, and cannot get a step further. The engineers hereabouts are all bet; and if you really succeed in accomplishing what they cannot do, you may depend upon it I will make you a man for life."

Stephenson began his operations early next morning. The only condition that he made, before setting to work, was that he should select his own workmen. There was, as he knew, a good deal of jealousy amongst the "regular" men that a colliery brakesman should pretend to know more about their engine than they themselves did, and attempt to remedy defects which the most skilled men of their craft, including the engineer of the colliery, had failed to do. But George made the condition a *sine quâ non*. "The workmen," said he, "must either be all Whigs or all Tories." There was no help for it, so Dodds ordered the old hands to stand aside. The men grumbled, but gave way; and then George and his party went in.

The engine was taken entirely to pieces. The cistern containing the injection water was raised ten feet; the injection cock, being too small, was enlarged to nearly double its former size, and it was so arranged that it should be shut off quickly at the beginning of the stroke. These and other alterations were necessarily performed in a rough way, but, as the result proved, on true principles. Stephenson also, finding that the boiler would bear a greater pressure than five pounds to the inch, determined to work it at a pressure of ten pounds, though this was contrary to the directions of both Newcomen and Smeaton. The necessary alterations were made in about three days, and many persons came to see the engine start, including the men who had put her up. The pit being nearly full of water, she had little to do on starting, and, to use George's words, "came bounce into the house." Dodds exclaimed, "Why, she was better as she was; now, she will knock the house down." After a short time, however, the engine got fairly to work, and by ten o'clock that night the water was lower in the pit than it had ever been before. The engine was kept pumping all Thursday, and by the Friday afternoon the pit was cleared of water, and the workmen were "sent to the bottom," as

Stephenson had promised. Thus the alterations effected in the pumping apparatus proved completely successful.[1]

Mr. Dodds was particularly gratified with the manner in which the job had been done, and he made Stephenson a present of ten pounds, which, though very inadequate when compared with the value of the work performed, was accepted by him with gratitude. He was proud of the gift as the first marked recognition of his skill as a workman; and he used afterwards to say that it was the biggest sum of money he had up to that time earned in one lump. Ralph Dodds, however, did more than this. He released the brakesman from the handles of the engine at West Moor, and appointed him engineman at the High Pit, at good wages, during the time the pit was sinking,—the job lasting for about a year; and he also kept him in mind for further advancement.

Stephenson's skill as an engine-doctor soon became noised abroad, and he was called upon to prescribe remedies for all the old, wheezy, and ineffective pumping machines in the neighbourhood. In this capacity he soon left the " regular " men far behind, though they in their turn were very much disposed to treat the Killingworth brakesman as no better than a quack. Nevertheless, his practice was really founded upon a close study of the principles of mechanics, and on an intimate practical acquaintance with the details of the pumping-engine.

Another of his smaller achievements in the same line is still told by the people of the district. At the corner of the road leading to Long Benton, there was a quarry from which a peculiar and scarce kind of ochre was

---

[1] As different versions have been given of this affair, it may be mentioned that the above statement is made on the authority of the late Robert Stephenson, and of George Stephenson himself, as communicated by the latter to his friend Thomas L. Gooch, C.E., who has kindly supplied the author with his memoranda on the subject.

taken. In the course of working it out, the water had collected in considerable quantities ; and there being no means of draining it off, it accumulated to such an extent that the further working of the ochre was almost entirely stopt. Ordinary pumps were tried, and failed ; and then a windmill was tried, and failed too. On this, George was asked what ought to be done to clear the quarry of the water. He said "he would set up for them an engine little bigger than a kail-pot, that would clear them out in a week." And he did so. A little engine was speedily erected, by means of which the quarry was pumped dry in the course of a few days. Thus his skill as a pump-doctor soon became the marvel of the district.

In elastic muscular vigour, Stephenson was now in his prime, and he still continued to be zealous in measuring his strength and agility with his fellow workmen. The competitive element in his nature was always strong ; and his success in these feats of rivalry was certainly remarkable. Few, if any, could lift such weights, throw the hammer and putt the stone so far, or cover so great a space at a standing or running leap. One day between the engine hour and the rope-rolling hour, Kit Heppel challenged him to leap from one high wall to another, with a deep gap between them. To Heppel's surprise and dismay, George took the standing leap, and cleared the eleven feet at a bound. Had his eye been less accurate, or his limbs less agile and sure, the feat must have cost him his life.

But so full of redundant muscular vigour was he, that leaping, putting, or throwing the hammer were not enough for him. He was also ambitious of riding on horseback, and, as he had not yet been promoted to an office enabling him to keep a horse of his own, he sometimes borrowed one of the gin-horses for a ride. On one of these occasions, he brought the animal back reeking ; when Tommy Mitcheson, the bank horse-keeper,

a rough-spoken fellow, exclaimed to him : " Set such
fellows as you on horseback, and you'll soon ride to the
De'il."   But Tommy Mitcheson lived to tell the joke,
and to confess that, after all, there had been a better
issue to George's horsemanship than that which he
predicted.

Old Cree, the engine-wright at Killingworth High
Pit, having been killed by an accident, George Stephenson was, in 1812, appointed engine-wright of the colliery
at the salary of 100*l.* a year.   He was also allowed the
use of a galloway to ride upon in his visits of inspection
to the collieries leased by the " Grand Allies " in that
neighbourhood.   The " Grand Allies " were a company
of gentlemen, consisting of Sir Thomas Liddell (afterwards Lord Ravensworth), the Earl of Strathmore, and
Mr. Stuart Wortley (afterwards Lord Wharncliffe), the
lessees of the Killingworth collieries.   Having been
informed of the merits of Stephenson, of his indefatigable
industry, and the skill which he had displayed in the
repairs of the pumping-engines, they readily acceded to
Mr. Dodds' recommendation that he should be appointed
the colliery engine-wright ; and, as we shall afterwards
find, they continued to honour him by distinguished
marks of their approval.

KILLINGWORTH HIGH PIT.

GLEBE FARM HOUSE, BENTON   [By R. P. Leitch.]

CHAPTER V.

The Stephensons at Killingworth — Education and Self-Education of Father and Son.

GEORGE STEPHENSON had now been diligently employed for many years in the work of self-improvement, and he experienced the usual results in increasing mental strength, capability, and skill. Perhaps the secret of every man's best success in life is to be found in the alacrity and industry with which he takes advantage of the opportunities which present themselves for well-doing. Our engineman was an eminent illustration of the importance of cultivating this habit of life. Every spare moment was laid under contribution by him, either for the purpose of adding to his earnings, or to his knowledge. He missed no opportunity of extending his observations, especially in his own department of work, aiming at improvement, and trying to turn all that he did know to useful practical account.

He continued his attempts to solve the mystery of Perpetual Motion, and contrived several model machines with the object of embodying his ideas in a practical working shape. He afterwards used to lament the time he had lost in these futile efforts, and said that if he had enjoyed the opportunity which most young men now have, of learning from books what previous experimenters had accomplished, he would have been spared much labour and mortification. Not being acquainted with what other mechanics had done, he groped his way in pursuit of some idea originated by his own independent thinking and observation ; and, when he had brought it into some definite form, lo ! he found that his supposed invention had long been known and recorded in scientific books. Often he thought he had hit upon discoveries, which he subsequently found were but old and exploded fallacies. Yet his very struggle to overcome the difficulties which lay in his way, was of itself an education of the best sort. By wrestling with them, he strengthened his judgment and sharpened his skill, stimulating and cultivating his inventiveness and mechanical ingenuity. Being very much in earnest, he was compelled to consider the subject of his special inquiry in all its relations ; and the necessity for thoroughness would not suffer him to be superficial. Thus he gradually acquired practical ability even through his very efforts after the impracticable.

Many of his evenings were spent in the society of John Wigham, whose father occupied the Glebe farm at Benton, close at hand. John was a fair penman and a sound arithmetician, and Stephenson frequented his society chiefly for the purpose of improving himself in writing and "figures." Under Andrew Robertson, he had never quite mastered the Rule of Three, and it was only when Wigham took him in hand that he made much progress in the higher branches of arithmetic. He generally took his slate with him to the Wighams' cottage, when he had his sums set, that he might work them out while tending the engine on the following day. When too busy with other work to be able to call upon Wigham in person, he sent the slate by a fellow-workman to have the former sums corrected and new ones set. Sometimes also, at leisure moments, he was enabled to do a little "figuring" with chalk upon the sides of the coal-waggons. So much patient perseverance could not but eventually succeed; and by dint of practice and study, Stephenson was enabled successively to master the various rules of arithmetic.

John Wigham was of great use to his pupil in many ways. He was a good talker, fond of argument, an extensive reader as country reading went in those days, and a very suggestive thinker. Though his store of information might be comparatively small when measured with that of more highly-cultivated minds, much of it was entirely new to Stephenson, who regarded him as a very clever and extraordinary person. Young as John Wigham was, he could give much useful assistance to Stephenson at the time, and his neighbourly services were worth untold gold to the eager pupil. Wigham taught him to draw plans and sections; though in this branch Stephenson proved so apt that he soon surpassed his master. Wigham also possessed some knowledge of Natural Philosophy, and a volume of 'Ferguson's Lectures on Mechanics' which he possessed was a great treasure to both the students. One who remembers their evening occupations says he used to wonder what they meant by weighing the air and water in so odd a way. They were trying the specific gravities of objects; and the devices which they employed, the mechanical shifts to which they were put, were often of the rudest kind. In these evening entertainments, the mechanical contrivances were supplied by Stephenson, whilst Wigham found the scientific rationale. The opportunity thus afforded to the former of cultivating his mind by contact with one wiser than himself proved of great value, and in after-life Stephenson gratefully

remembered the assistance which, when a humble workman, he had derived from John Wigham, the farmer's
son.

His leisure moments thus carefully improved, it will
be inferred that Stephenson continued a sober man.
Though his notions were never extreme on this point,
he was systematically temperate. It appears that on
the invitation of his master, Ralph Dodds—and an invitation from a master to a workman is not easy to resist
—he had, on one or two occasions, been induced to join
him in a forenoon glass of ale in the public-house of the
village. But one day, about noon, when Mr. Dodds had
got him as far as the public-house door, on his invitation
to "come in and take a glass o' yel," Stephenson made a
dead stop, and said, firmly, "No, sir, you must excuse
me; I have made a resolution to drink no more at this
time of day." And he went back. He desired to retain
the character of a steady workman; and the instances
of men about him who had made shipwreck of their
character through intemperance, were then, as now,
unhappily but too frequent.

But another consideration besides his own self-improvement had already begun to exercise an important
influence upon his life. This was the training and
education of his son Robert, now growing up a healthy,
intelligent boy, as full of fun and tricks as his father had
been, but like him also possessing an abundant capacity
for knowledge. When a little fellow, not big enough
to reach so high as to put a clock-head on when placed
upon the table, his father would make him mount a chair
for the purpose; and to "help father" was the proudest
work which the boy then, and ever after, could take part
in. When the little engine was set up at the Ochre
Quarry to pump it dry, Robert was scarcely absent for
an hour. He watched the machine very eagerly when
it was set to work; and he was very much annoyed at
the fire burning away the grates. The man who fired

the engine was a sort of wag, and thinking to get a laugh at the boy, he said, " Those bars are getting varra bad, Robert ; I think we maun cut up some of that hard wood, and put it in instead."   " What would be the use of that, you fool ? " said the boy quickly.   " You would no sooner have put them in than they would be burnt out again ! "

So soon as Robert was of a proper age, his father sent him over to the road-side school at Long Benton, kept by Rutter, the parish clerk.   But the education which Rutter could give was of a very limited kind, scarcely extending beyond the primer and pothooks.   While working as a brakesman on the pit-head at Killingworth, the father had often bethought him of the obstructions he had himself encountered in life through his own want of schooling ; and he formed the noble determination that no labour, nor pains, nor self-denial on his part should be spared to furnish his son with the best education that it was in his power to bestow.

RUTTER'S SCHOOL HOUSE, LONG BENTON.   [By R. P Leitch]

It is true his earnings were comparatively small at that time. He was still maintaining his infirm parents; and the cost of living continued excessive. But he fell back, as before, upon his old expedient of working up his spare time in the evenings at home, or during the night shifts when it was his turn to tend the engine, in mending and making shoes, cleaning clocks and watches, making shoe-lasts for the shoemakers of the neighbourhood, and cutting out the pitmen's clothes for their wives; and we have been told that to this day there are clothes worn at Killingworth made after "Geordy Steevie's cut." To give his own words :—"In the earlier period of my career," said he, "when Robert was a little boy, I saw how deficient I was in education, and I made up my mind that he should not labour under the same defect, but that I would put him to a good school, and give him a liberal training. I was, however, a poor man; and how do you think I managed? I betook myself to mending my neighbours' clocks and watches at nights, after my daily labour was done, and thus I procured the means of educating my son." [1]

By dint of such extra labour in his bye-hours, with this object, Stephenson contrived to save a sum of 100*l*., which he accumulated in *guineas*, each of which he afterwards sold to Jews who went about buying up gold coins (then dearer than silver), at twenty-six shillings apiece; and he lent out the proceeds at interest. He was now, therefore, a comparatively thriving man. The first guinea which he had saved with so much difficulty at Black Callerton had proved the nest-egg of future guineas; and the habits of economy and sobriety which he had so early cultivated, now enabled him to secure a firmer foothold in the world, and to command the increased esteem and respect of his fellow-workmen and employers.

---

[1] Speech at Newcastle, on the 18th of June, 1844, at the meeting held in celebration of the opening of the Newcastle and Darlington Railway.

Carrying out the resolution as to his boy's education,
Robert was sent to Mr. Bruce's school in Percy Street,
Newcastle, at Midsummer, 1815, when he was about
twelve years old. His father bought for him a donkey,
on which he rode into Newcastle and back daily; and
there are many still living who remember the little boy,
dressed in his suit of homely grey stuff, cut out by his
father, cantering along to school upon the "cuddy," with
his wallet of provisions for the day and his bag of books
slung over his shoulder.

BRUCE'S SCHOOL, NEWCASTLE. [By R. P. Leitch.]

When Robert went to Mr. Bruce's school, he was a
shy, unpolished country lad, speaking the broad dialect
of the pitmen; and the other boys would occasionally
tease him, for the purpose of provoking an outburst
of his Killingworth Doric. As the shyness got rubbed
off by familiarity, his love of fun began to show itself,
and he was found able enough to hold his own amongst
the other boys. As a scholar he was steady and diligent,

and his master was accustomed to hold him up to the laggards of the school as an example of good conduct and industry.   But his progress, though satisfactory, was by no means extraordinary.   He used in after-life to pride himself on his achievements in mensuration, though another boy, John Taylor, beat him at arithmetic.   He also made considerable progress in mathematics; and in a letter written to the son of his teacher, many years after, he said, "It was to Mr. Bruce's tuition and methods of modelling the mind that I attribute much of my success as an engineer; for it was from him that I derived my taste for mathematical pursuits and the facility I possess of applying this kind of knowledge to practical purposes and modifying it according to circumstances."

During the time Robert attended school at Newcastle, his father made the boy's education instrumental to his own.   Robert was accustomed to spend some of his spare time at the rooms of the Literary and Philosophical Institute; and when he went home in the evening, he would recount to his father the results of his reading. Sometimes he was allowed to take with him to Killingworth a volume of the 'Repertory of Arts and Sciences,' which father and son studied together.   But many of the most valuable works belonging to the Newcastle Library were not permitted to be removed from the rooms; these Robert was instructed to read and study, and bring away with him descriptions and sketches for his father's information.   His father also practised him in the reading of plans and drawings without at all referring to the written descriptions.   He used to observe to his son, "A good drawing or plan should always explain itself;" and, placing a drawing of an engine or machine before the youth, he would say, "There, now, describe that to me—the arrangement and the action."   Thus he taught him to read a drawing

as easily as he would read a page of a book. Both
father and son profited by this excellent practice, which
shortly enabled them to apprehend with the greatest
facility the details of even the most difficult and complicated mechanical drawing.

While Robert went on with his lessons in the evenings,
his father was usually occupied with his watch and clock
cleaning; or in contriving models of pumping engines;
or endeavouring to embody in a tangible shape the
mechanical inventions which he found described in
the odd volumes on Mechanics which fell in his way.
This daily and unceasing example of industry and application, working on before the boy's eyes in the person
of a loving and beloved father, imprinted itself deeply
upon his mind in characters never to be effaced. A spirit
of self-improvement was thus early and carefully planted
and fostered in Robert's mind, which continued to influence him through life; and to the close of his career, he
was proud to confess that if his professional success had
been great, it was mainly to the example and training of
his father that he owed it.

Robert was not, however, exclusively devoted to
study, but, like most boys full of animal spirits, he was
very fond of fun and play, and sometimes of mischief.
Dr. Bruce relates that an old Killingworth labourer, when
asked by Robert, on one of his last visits to Newcastle,
if he remembered him, replied with emotion, "Ay,
indeed! Haven't I paid your head many a time when
you came with your father's bait, for you were always
a sad hempy?"

The author had the pleasure, in the year 1854, of
accompanying Robert Stephenson on a visit to his old
home and haunts at Killingworth. He had so often
travelled the road upon his donkey to and from school,
that every foot of it was familiar to him; and each
turn in it served to recall to mind some incident of

his boyish days.[1]   His eyes glistened when he came
in sight of Killingworth pit head.   Pointing to a humble
red-tiled house by the road-side at Benton, he said,
" You see that house—that was Rutter's, where I learnt
my A B C, and made a beginning of my school learning.   And there," pointing to a colliery chimney on the
left, " there is Long Benton, where my father put up
his first pumping-engine; and a great success it was.
And this humble clay-floored cottage you see here, is
where my grandfather lived till the close of his life.
Many a time have I ridden straight into the house,
mounted on my cuddy, and called upon grandfather to
admire his points.   I remember the old man feeling the
animal all over—he was then quite blind—after which
he would dilate upon the shape of his ears, fetlocks, and
quarters, and usually end by pronouncing him to be a
' real blood.'   I was a great favourite with the old man,
who continued very fond of animals, and cheerful to the
last; and I believe nothing gave him greater pleasure
than a visit from me and my cuddy."

On the way from Benton to High Killingworth,
Mr. Stephenson pointed to a corner of the road where
he had once played a boyish trick upon a Killingworth
collier.   " Straker," said he, " was a great bully, a
coarse, swearing fellow, and a perfect tyrant amongst
the women and children.   He would go tearing into old
Nanny the huxter's shop in the village, and demand in
a savage voice, ' What's ye'r best ham the pund?'
' What's floor the hunder?'   ' What d'ye ax for prime
bacon?'—his questions often ending with the miserable
order, accompanied with a tremendous oath, of ' Gie's a
penny rrow (roll) an'. a baubee herrin!'   The poor
woman was usually set ' all of a shake' by a visit from

---

[1] At one part of the road he was
once pulled off his donkey by some
mischievous boys, and released by a
young man named James Burnet.
Many years after, Burnet was taken on
as a workman at the Newcastle factory,
probably owing his selection in some
measure to the above circumstance.

this fellow. He was also a great boaster, and used to crow over the robbers whom he had put to flight; mere men in buckram, as everybody knew. We boys," he continued, " believed him to be a great coward, and determined to play him a trick. Two other boys joined me in waylaying Straker one night at that corner," pointing to it. " We sprang out and called upon him, in as gruff voices as we could assume, to 'stand and deliver!' He dropped down upon his knees in the dirt, declaring he was a poor man, with a sma' family, asking for 'mercy,' and imploring us, as ' gentlemen, for God's sake, t' let him a-be!' We couldn't stand this any longer, and set up a shout of laughter. Recognizing our boys' voices, he sprang to his feet again and rattled out a volley of oaths; on which we cut through the hedge, .and heard him shortly after swearing his way along the road to the yill-house."

On another occasion, he played a series of tricks of a somewhat different character. Like his father, he was very fond of reducing his scientific reading to practice; and after studying Franklin's description of the lightning experiment, he proceeded to expend his store of Saturday pennies in purchasing about half-a-mile of copper wire at a brazier's shop in Newcastle. Having prepared his kite, he sent it up in the field opposite his father's door, and bringing the wire, insulated by means of a few feet of silk cord, over the backs of some of Farmer Wigham's cows, he soon had them skipping about the field in all directions with their tails up. One day he had his kite flying at the cottage-door as his father's galloway was hanging by the bridle to the paling, waiting for the master to mount. Bringing the end of the wire just over the pony's crupper, so smart an electric shock was given it, that the brute was almost knocked down. At this juncture the father issued from the door, riding-whip in hand, and was witness to the scientific trick just played off upon his galloway.

STEPHENSON'S COTTAGE, WEST MOOR. [By R. P. Leitch.]

" Ah! you mischievous scoondrel!" cried he to the boy,
who ran off.　He inwardly chuckled with pride, nevertheless, at Robert's successful experiment.

At this time, and for many years after, Stephenson
dwelt in a cottage standing by the side of the road
leading from the West Moor colliery to Killingworth.
The railway from the West Moor Pit crosses this road
close by the easternmost end of the cottage.　The
dwelling originally consisted of but one apartment on
the ground-floor, with a garret over-head, to which
access was obtained by means of a step-ladder.　But
with his own hands Stephenson built an oven, and in
the course of time he added rooms to the cottage, until
it became a comfortable four-roomed dwelling, in which
he remained as long as he lived at Killingworth.

He continued as fond of birds and animals as ever,
and seemed to have the power of attaching them to him
in a remarkable degree.　He had a blackbird at Killingworth so fond of him, that it would fly about the

F 2

cottage, and on holding out his finger, the bird would come and perch upon it directly. A cage was built for " blackie " in the partition between the passage and the room, a square of glass forming its outer wall; and Robert used afterwards to take pleasure in describing the oddity of the bird, imitating the manner in which it would cock its head on his father's entering the house, and follow him with its eye into the inner apartment.

Neighbours were accustomed to call at the cottage and have their clocks and watches set to rights when they went wrong. One day, after looking at the works of a watch left by a pitman's wife, George handed it to his son; " Put her in the oven, Robert," said he, " for a quarter of an hour or so." It seemed an odd way of repairing a watch; nevertheless, the watch was put into the oven, and at the end of the appointed time it was taken out, going all right. The wheels had merely got clogged by the oil congealed by the cold; which at once explains the rationale of the remedy adopted.

There was a little garden attached to the cottage, in which, while a workman, Stephenson took a pride in growing gigantic leeks and astounding cabbages. There was great competition amongst the villagers in the growth of vegetables, all of whom he excelled, excepting one of his neighbours, whose cabbages some- times outshone his. In the protection of his garden- crops from the ravages of the birds, he invented a strange sort of " fley-craw," which moved its arms with the wind; and he fastened his garden-door by means of a piece of ingenious mechanism, so that no one but him- self could enter it. His cottage was quite a curiosity- shop of models of engines, self-acting planes, and perpetual-motion machines, — which last contrivance, however, as effectually baffled him as it had done hundreds of preceding inventors. His odd and eccentric contrivances often excited great wonder amongst the Killingworth villagers. He won the women's admiration by connecting their cradles with the smoke-jack, and making them self-acting. Then he astonished the pitmen by attaching an alarum to the clock of the watchman whose duty it was to call them betimes in the morning. He also contrived a wonderful lamp which burned under water, with which he was after-wards wont to amuse the Brandling family at Gosforth, —going into the fish-pond at night, lamp in hand, attracting and catching the fish, which rushed wildly towards the flame.

Dr. Bruce tells of a competition which Stephenson had with the joiner at Killingworth, as to which of them could make the best shoe-last; and when the former had done his work, either for the humour of the thing, or to secure fair play from the appointed judge, he took it to the Morrisons in Newcastle, and got them to put their stamp upon it. So that it is possible the Killingworth brakesman, afterwards the inventor of the safety-lamp and the originator of the railway system, and John Morrison, the last-maker, afterwards the translator of the Scriptures into the Chinese language, may have confronted each other in solemn contem-plation over the successful last, which won the verdict coveted by its maker.

Sometimes he would endeavour to impart to his fellow-workmen the results of his scientific reading. Everything that he learnt from books was so new and so wonderful to him, that he regarded the facts he drew from them in the light of discoveries, as if they had been made but yesterday. Once he tried to explain to some of the pitmen how the earth was round, and kept turning round. But his auditors flatly declared the thing to be impossible, as it was clear that "at the bottom side they must fall off!" "Ah!" said George, "you don't quite understand it yet." His son Robert also early endeavoured to communicate to others the information which he had gathered at school; and Dr.

Bruce has related that, when visiting Killingworth on
one occasion, he found him engaged in teaching algebra
to such of the pitmen's boys as would become his pupils.

While Robert was still at school, his father proposed
to him during the holidays that he should construct a
sun-dial, to be placed over their cottage-door at West
Moor. " I expostulated with him at first," said Robert,
" that I had not learnt sufficient astronomy and mathematics to enable me to make the necessary calculations.
But he would have no denial. ' The thing is to be
done,' said he; ' so just set about it at once.' Well;
we got a ' Ferguson's Astronomy,' and studied the
subject together. Many a sore head I had while making
the necessary calculations to adapt the dial to the latitude of Killingworth. But at length it was fairly
drawn out on paper, and then my father got a stone,
and we hewed, and carved, and polished it, until we
made a very respectable dial of it; and there it is, you
see," pointing to it over the cottage-door, " still quietly

numbering the hours
when the sun is shining.
I assure you, not a little
was thought of that piece
of work by the pitmen
when it was put up, and
began to tell its tale of
time." The date carved
upon the dial is "August
11th, MDCCCXVI." Both
father and son were in
after-life very proud of
their joint production.
Many years after, George
took a party of savans, when attending the meeting of
the British Association at Newcastle, over to Killingworth to see the pits, and he did not fail to direct their
attention to the sun-dial; and Robert, on the last visit

which he made to the place, a short time before his death, took a friend into the cottage, and pointed out to him the very desk, still there, at which he had sat while making his calculations of the latitude of Killingworth.

From the time of his appointment as engineer at the Killingworth Pit, George Stephenson was in a measure relieved from the daily routine of manual labour, having, as we have seen, advanced himself to the grade of a higher class workman.   He had not ceased to be a worker, though he employed his industry in a different way.   It might, indeed, be inferred that he had now the command of greater leisure; but his spare hours were as much as ever given to work, either necessary or self-imposed.   So far as regarded his social position, he had already reached the summit of his ambition; and when he had got his hundred a year, and his dun galloway to ride on, he said he never wanted to be any higher.   When Robert Wetherly offered to give him an old gig, his travelling having so much increased of late, he accepted it with great reluctance, observing, that he should be ashamed to get into it, "people would think him so proud."

When the High Pit had been sunk, and the coal was ready for working, Stephenson erected his first winding-engine to draw the coals out of the pit, and also a pumping-engine for Long Benton colliery, both of which proved quite successful.   Amongst other works of this time, he projected and laid down a self-acting incline along the declivity which fell towards the coal-loading place near Willington, where he had formerly officiated as brakesman; and he so arranged it, that the full waggons descending drew the empty waggons up the incline.   This was one of the first self-acting inclines laid down in the district.

Stephenson had now many more opportunities for improving himself in mechanics than he had hitherto possessed.   His familiar acquaintance with the steam-

engine proved of great value to him. His shrewd
insight, together with his intimate practical acquaintance with its mechanism, enabled him to apprehend, as
if by intuition, its most abstruse and difficult combina-
tions. The practical study which he had given to it
when a workman, and the patient manner in which he
had groped his way through all the details of the
machine, gave him the power of a master in dealing
with it as applied to colliery purposes.

Sir Thomas Liddell was frequently about the works,
and took pleasure in giving every encouragement to
the engine-wright in his efforts after improvement.
The subject of the locomotive engine was already
closely occupying Stephenson's attention; although it
was still regarded in the light of a curious and costly
toy, of comparatively little real use. But he had at an
early period detected the practical value of the machine,
and formed an adequate conception of the might which
as yet slumbered within it; and he was not slow in
bending the whole faculties of his mind to the development of its extraordinary powers.

COLLIERS' COTTAGES AT LONG BENTON.

CHAPTER VI.

EARLY HISTORY OF THE LOCOMOTIVE — GEORGE STEPHENSON
BEGINS ITS IMPROVEMENT.

THE rapid increase in the coal trade of the Tyne about the
beginning of the present century had the effect of stimulating the ingenuity of mechanics, and encouraging them
to devise improved methods of transporting the coal from
the pits to the shipping places.   From our introductory
chapter, it will have been observed that the improvements
which had thus far been effected were confined almost
entirely to the road.   The railway waggons still continued to be drawn by horses.   By improving and flat-
tening the tramway, considerable economy in horse-power
had indeed been secured ; but unless some more effective
method of mechanical traction could be devised, it was
clear that railway improvement had almost reached its
limits.

Many expedients had been tried with this object.
One of the earliest was that of hoisting sails upon the
waggons, and driving them along the waggon-way, as
a ship is driven through the water by the wind.   This
method seems to have been employed by Sir Humphrey
Mackworth, an ingenious coal-miner at Neath in
Glamorganshire, about the end of the seventeenth century.   In Waller's ' Essay on Mines,' published in 1698,
the writer highly eulogises Sir Humphrey's " new
sailing-waggons, for the cheap carriage of his coal to
the waterside, whereby one horse does the work of ten
at all times ; but when any wind is stirring (which is

seldom wanting near the sea) one man and a small sail do the work of twenty." [1]

This method of impelling coal-waggons, however, could not have come into general use, as it was lost sight of for more than a century, when it was again proposed as a new method of transit by Richard Lovell Edgworth, with the addition of a portable railway,[2] since revived in Boydell's patent. But although Mr. Edgworth devoted himself to the subject for many years, he failed in securing the adoption of his sailing carriage.[3] He made numerous experiments with his machines on Hare Hatch Common, but they were abandoned in consequence of the dangerous results which threatened to attend them. It is indeed quite clear that a power so uncertain as wind could never be relied on for ordinary traffic, and Mr. Edgworth's project was consequently left to repose in the limbo of the Patent Office, with thousands of other equally useless though ingenious contrivances.

A much more favourite scheme was the application of steam power for the purpose of carriage traction. Savery, the inventor of the working steam-engine, was the first to propose its employment to propel vehicles along common roads; and in 1759 Dr. Robison, then a young man studying at Glasgow College, threw out the same idea to his friend James Watt; but the scheme

---

[1] The writer adds, "I believe he (Sir Humphrey Mackworth) is the first gentleman in this part of the world that hath set up sailing-engines on land, driven by the wind; not for any curiosity, or vain applause, but for real profit, whereby he could not fail of Bishop Melkin's blessing on his undertakings, in case he were in a capacity to bestow it."—'An Essay on the Value of the Mines late of Sir Carberry Price.' By William Waller, Gent., Steward of the said Mines. London, 1698.

[2] Specification of patent, No. 953.

[3] Mr. Edgworth says in his 'Memoirs,' that he devoted himself to improving his scheme for a period of not less than forty years, during which he made above a hundred working models in a great variety of forms; and he adds, that he gained far more in amusement than he lost by unsuccessful labour. "Indeed," he says, "the only mortification that affected me was my discovery, many years after I had taken out my patent, that the rudiments of my whole scheme were mentioned in an obscure memoir of the French Academy."

was not matured. Watt afterwards, in the specification of his patent of 1769, described an engine of the kind suggested by his friend Robison, in which the expansive force of steam was proposed as the motive power; but no steps were taken to reduce the invention to practice.

The first locomotive steam-carriage was built at Paris by a French engineer named Cugnot, a native of Lorraine. It is said to have been invented for the purpose of dragging cannon into the field independent of the help of horses.[1] The first model of this machine was made in 1763. Marshal Saxe was so much pleased with it that on his recommendation a full-sized engine was constructed at the Arsenal at the cost of the French monarch, and in 1769 it was tried in the presence of the Duc de Choiseul, Minister of War, General Gribeauval, and other officers. At one of the experiments it ran onward with such force that it knocked down a wall which stood in its way. It was found, however, that the new vehicle, loaded with four persons, could not travel faster than two miles and a half in the hour. The size of the boiler not being sufficient to keep up the steam, it could only work for about fifteen minutes at a time; after which it was necessary to wait until

CUGNOT'S ENGINE.

the steam had again risen to a sufficient pressure. To remedy this defect, Cugnot constructed a new machine in 1770, which gave somewhat more satisfactory results.

---

[1] 'Le Vieux-Neuf: Histoire Ancienne des Inventions et Découvertes | Modernes.' Par Edouard Fournier. Paris, 1859.

It was composed of two parts—the fore part consisting
of a small steam-engine, formed of a round copper boiler,
with a furnace inside, provided with two small chim-
neys and two single-acting brass steam cylinders, whose
pistons acted alternately upon the single driving wheel.
The hinder part consisted merely of a rude carriage on
two wheels to carry the load, furnished with a seat in
front for the conductor. This engine was tried in the
streets of Paris; but when passing near where the
Madeleine now stands, it overbalanced itself on turning a corner, and fell over with a crash; after which,
its employment being thought dangerous, it was locked
up in the Arsenal to prevent further mischief. The
machine is, however, still to be seen in the collection
of the Conservatoire des Arts et Métiers at Paris. It
has very much the look of a long brewer's cart,
with the addition of the circular boiler hung on at one
end. Nevertheless it was a highly creditable piece of
work, considering the period at which it was executed;
and as the first actual machine constructed for the purpose of travelling on ordinary roads by the power of
steam, it is certainly a most curious and interesting
mechanical relic, well worthy of preservation.

But though Cugnot's road locomotive remained locked
up from public sight, the subject was not dead; for we
find inventors from time to time employing themselves
in attempting to solve the problem of steam locomotion
in places far remote from Paris. The idea had taken
root, and was striving to grow into a reality. Thus
Oliver Evans, the American, invented a steam-carriage
in 1772 to travel on common roads; in 1787 he obtained from the State of Maryland an exclusive right
to make and use steam-carriages; but his invention
never came into use. Then, in 1784, William Symington, one of the early inventors of the steamboat, was
similarly occupied in Scotland in endeavouring to develope the latent powers of the steam-carriage. He had

a working model of one constructed, which he exhibited in 1786 to the professors of Edinburgh College; but the state of the Scotch roads was then so bad that he found it impracticable to proceed further with his scheme, which he shortly after abandoned in favour of steam navigation.[1]

The very same year in which Symington was occupied upon his steam-carriage, William Murdock, the friend and assistant of Watt, constructed his model of a locomotive at the opposite end of the island—at Redruth in Cornwall. His model was of small dimensions, standing little more than a foot high; and it was until recently in the possession of the son of the inventor, at whose house we saw it a few years ago. The annexed section will give an idea of the arrangements of this machine.

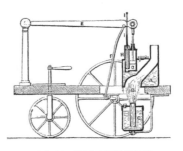

SECTION OF MURDOCK'S MODEL.

It acted on the high-pressure principle, and, like Cugnot's engine, ran upon three wheels, the boiler being heated by a spirit-lamp. Small though the machine was, it went so fast on one occasion that it fairly outran the speed of its inventor. It seems that one night, after returning from his duties at the Redruth mine, Murdock determined to try the working of his model locomotive. For this purpose he had recourse to the walk leading to the church, about a mile from the town. The walk was rather narrow, and was bounded on either side by high hedges. It was a dark night, and Murdock set out alone to try his experiment. Having

---

[1] See a pamphlet entitled 'A brief Narrative, proving the right of the late William Symington, Civil Engineer, to be considered the Inventor of Steam Land-Carriage Locomotion; and also the Inventor and Introducer of Steam Navigation.' By Robert Bowie. London, 1833.

lit his lamp, the water shortly began to boil, and off started the engine with the inventor after it. He soon heard distant shouts of despair. It was too dark to perceive objects; but he shortly found, on following up the machine, that the cries for assistance proceeded from the worthy pastor of the parish, who, going towards the town on business, was met on this lonely road by the hissing and fiery little monster, which he subsequently declared he had taken to be the Evil One *in propriâ personâ*. No further steps, however, were taken by Murdock to embody his idea of a locomotive carriage in a more practical form.

We next find the discussion of steam-power as a means of haulage of heavy articles taken up in the colliery districts of the North, where the want of some more effective means of transport than horse-power was most generally felt. One Thomas Allen took out a patent in 1789 for conveying goods from one place to another by the power of steam only. From his plan, which is in the possession of the Society of Antiquaries at Newcastle-upon-Tyne, it appears that he intended the wheels of his machine to be cogged; yet he anticipated a speed on a common road of about ten miles an hour. It will be observed that no one had yet proposed to apply steam-carriages to railways, but only to common roads, though it is easy to see how the steam-engine and the iron-road should have come together in the ordinary course of things. The use of tramroads and railways had now become quite general in the mining districts, and their extension throughout the country for the conveyance of general merchandise began to be seriously discussed. Thus, in 1800, we find Mr. Thomas, of Denton, Northumberland, reading a paper before the Literary and Philosophical Society of Newcastle, in which he urged "the propriety of introducing roads on the principle of the coal-waggon ways, for the general carriage of goods and mer-

chandise throughout the country." In the course of
the following year the same idea was taken up and
strongly advocated by Dr. James Anderson, of Edinburgh, in his 'Recreations of Agriculture.' He held
that if railways were laid along the existing turnpike roads, and worked by horse-power, the cost of
most articles of consumption would be diminished,
whilst all departments of human industry would be
greatly benefited. Mr. Edgworth, also, continued his
enthusiastic advocacy of railways, and urged their
general employment for the conveyance of passengers
as well as goods. " Stage-coaches," he said, "might be
made to go at six miles an hour, and post-chaises and
gentlemen's travelling carriages at eight—both with
one horse ; and small stationary steam-engines, placed
from distance to distance, might be made, by means of
circulating chains, to draw the carriages, with a great
diminution of horse-labour and expense."

While this discussion was going forward, Richard
Trevithick, a captain in a Cornish tin-mine, and a pupil
of William Murdock's—influenced, no doubt, by the
successful action of the model engine which the latter
had constructed—determined to build a steam-carriage
adapted for use on common roads as well as on railways. He took out a patent to secure the right of his
invention in the year 1802. Andrew Vivian, his
cousin, joined with him in the patent[1]—Vivian finding
the money, and Trevithick the brains. The steam-
carriage built on this patent presented the appearance
of an ordinary stage-coach on four wheels. The engine
had one horizontal cylinder, which, together with the
boiler and the furnace-box, was placed in the rear of

---

[1] The patent was dated the 24th
March, 1802, and described as "A
grant unto Richard Trevithick and
Andrew Vivian, of the parish of Cran-
bourne, in the county of Cornwall,
engineers and miners, for their invented methods of improving the construction of steam-engines, and the
application thereof for driving carriages, and for other purposes." No.
of the patent 2599.

the hind axle. The motion of the piston was transmitted to a separate crank-axle, from which, through
the medium of spur-gear, the axle of the driving-wheel
(which was mounted with a fly-wheel) derived its motion. The steam-cocks and the force-pump, as also the
bellows used for the purpose of quickening combustion
in the furnace, were worked off the same crank-axle.

John Petherick, of Camborne, has related that he
remembers this first English steam-coach passing along
the principal street of his native town. Considerable
difficulty was experienced in keeping up the pressure of
steam; but when there was pressure enough, Trevithick
would call upon the people to "jump up," so as to
create a load upon the engine. It was soon covered
with men attracted by the novelty, nor did their number seem to make any difference in the speed of the
engine so long as the steam could be kept up; but it
was constantly running short, and the horizontal bellows failed to keep it up.

This road-locomotive of Trevithick's was one of the
first high-pressure working engines constructed on the
principle of moving a piston by the elasticity of steam
against the pressure only of the atmosphere. Such an
engine had been described by Leopold, though in his
apparatus it was proposed that the pressure should act
only on one side of the piston. In Trevithick's engine
the piston was not only raised, but was also depressed
by the action of the steam, being in this respect an
entirely original invention, and of great merit. The
steam was admitted from the boiler under the piston
moving in a cylinder, impelling it upward. When the
motion had reached its limit, the communication between
the piston and the under side was shut off, and the
steam allowed to escape into the atmosphere. A passage being then opened between the boiler and the
upper side of the piston, which was pressed downwards, the steam was again allowed to escape as before.

Thus the power of the engine was equal to the difference between the pressure of the atmosphere and the elasticity of the steam in the boiler.

This steam-carriage excited considerable interest in the remote district near the Land's End where it had been erected. Being so far removed from the great movements and enterprise of the commercial world, Trevithick and Vivian determined upon exhibiting their machine in the metropolis. They accordingly set out with it to Plymouth, whence it was conveyed by sea to London.

The carriage safely reached the metropolis, and excited much public interest. It also attracted the notice of scientific men, amongst others of Mr. Davies Gilbert and Sir Humphry Davy, both Cornishmen like Trevithick, who went to see the private performances of the engine, and were greatly pleased with it. Writing to a Cornish friend shortly after its arrival in town, Sir Humphry said : " I shall soon hope to hear that the roads of England are the haunts of Captain Trevithick's dragons—a characteristic name." The machine was afterwards publicly exhibited in an enclosed piece of ground near Euston Square, where the London and North-Western Station now stands, and it dragged behind it a wheel-carriage full of passengers. On the second day of the performance, crowds flocked to see it ; but Trevithick, in one of his odd freaks, shut up the place, and shortly after removed the engine. It is, however, probable that the inventor came to the conclusion that the state of the roads at that time was such as to preclude its coming into general use for purposes of ordinary traffic.

While the steam-carriage was being exhibited, a gentleman was laying heavy wagers as to the weight which could be hauled by a single horse on the Wandsworth and Croydon iron tramway ; and the number and weight of waggons drawn by the horse were some-

thing surprising. Trevithick very probably put the
two things together—the steam-horse and the iron-way
—and kept the performance in mind when he proceeded to construct his second or railway locomotive.
The idea was not, however, entirely new to him; for,
although his first engine had been constructed with a
view to its employment upon common roads, the specification of his patent distinctly alludes to the application
of his engine to travelling on railroads.   Having been
employed at the iron-works of Pen-y-darran, in South
Wales, to erect a forge engine for the Company, a convenient opportunity presented itself, on the completion
of this work, for carrying out his design of a locomotive to haul the minerals along the Pen-y-darran
tramway.   Such an engine was erected by him in
1803, in the blacksmiths' shop at the Company's works,
and it was finished and ready for trial before the end
of the year.

The boiler of this second engine was cylindrical in
form, flat at the ends, and made of wrought iron.[1]   The
furnace and flue were inside the boiler, within which
the single cylinder, eight inches in diameter and four
feet six inches stroke, was placed horizontally.   As in
the first engine, the motion of the wheels was produced
by spur gear, to which was also added a fly-wheel on
one side to secure a rotatory motion in the crank at the
end of each stroke of the piston in the single cylinder.
The waste steam was thrown into the chimney through
a tube inserted into it at right angles; but it will be
obvious that this arrangement was not calculated to
produce any result in the way of a steam-blast in the

---

[1] It is not, however, quite clear
whether the boiler of this engine was
of wrought or cast iron. The statement that it was of wrought iron is
made on the authority of Rees Jones,
who worked at the fitting of the engine under Trevithick, and was alive
in 1858.   But other accounts state
that the boiler was of cast iron, as
that of the next engine built after
Trevithick's patent certainly was. We
allude to the engine erected by Whinfield of Gateshead, for Mr. Blackett of
Wylam, in 1804, after Trevithick's
own plans.

chimney. In fact, the waste steam seems to have been turned into the chimney in order to get rid of the nuisance caused by throwing the jet directly into the air. Trevithick was here hovering on the verge of a great discovery; but that he was not aware of the action of the blast in contributing to increase the draught and thus quicken combustion, is clear from the fact that he employed bellows for this special purpose; and at a much later date (1815) he took out a patent which included a method of urging the fire by means of fanners.

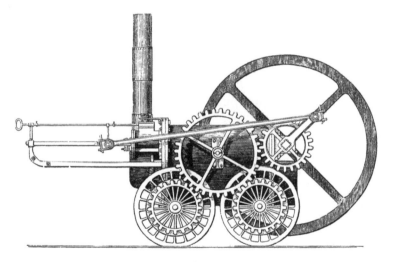

TREVITHICK'S HIGH PRESSURE TRAM-ENGINE

At the first trial of this engine it succeeded in dragging after it several waggons, containing ten tons of bar-iron, at the rate of about five miles an hour. Rees Jones, who worked at the fitting of the engine and remembers its performances, says, " She was used for bringing down metal from the furnaces to the Old Forge. She worked very well; but frequently, from her weight, broke the tram-plates and the hooks between the trams. After working for some time in this way, she took a load of iron from Pen-y-darran down the

G 2

Basin-road, upon which road she was intended to work. On the journey she broke a great many of the tram-plates; and before reaching the basin she ran off the road, and was brought back to Pen-y-darran by horses. The engine was never after used as a locomotive." [1]

It seems to have been felt that unless the road were entirely reconstructed so as to bear the heavy weight of the locomotive—so much greater than that of the tram-waggons, to carry which the original rails had been laid down — the regular employment of Trevithick's high-pressure tram-engine was altogether impracticable; and as the owners of the works were not prepared to incur the heavy cost of such reconstruction, it was determined to take the locomotive off the road, and use the engine for other purposes. It was accordingly dismounted from its wheels, and fixed and used for some time after as a pumping-engine, for which purpose it was found well adapted. Trevithick himself seems from this time to have given up the locomotive as an impracticable engine, and took no further steps to bring it into use. We find him, shortly after, engaged upon schemes of a more promising character, leaving the locomotive to take care of itself, and no further progress was made with it for several years. An imaginary difficulty seems to have tended, amongst other obstacles, to prevent its adoption and improvement. This was the idea that, if any heavy weight were placed behind the engine, the "grip" or "bite" of the smooth wheels of the locomotive upon the equally smooth iron rail must necessarily be so slight that the wheels would slip round upon the rail, and, consequently, that the machine would not make any progress.[2] Hence Trevithick, in

---

[1] Statement of Rees Jones to Mr. Menelaus, Dowlais Iron-works, made 9th September, 1858, and published in the 'Mining Journal.'

[2] The same fallacy seems long to have held its ground in France; for M. Granier tells us that some time after the first of George Stephenson's locomotives had been placed on the Liverpool and Manchester line, a model of one was exhibited before the Academy. After it had been examined,

his patent, provided that the periphery of the driving-wheels should be made rough by the projection of bolts or cross-grooves, so that the adhesion of the wheels to the road might be secured.[1]

Following up the presumed necessity for a more effectual adhesion between the wheels and the rails, Mr. Blenkinsop of Leeds, in 1811, took out a patent for a racked or tooth-rail laid along one side of the road, into which the toothed-wheel of his locomotive worked as pinions work into a rack. The boiler of his engine was supported by a carriage with four wheels without teeth, and rested immediately upon the axles. These wheels were entirely independent of the working parts of the engine, and therefore merely supported its weight upon the rails, the progress being effected by means of the cogged-wheel working into the cogged-rail. The engine had two cylinders instead of one, as in Trevithick's engine. The invention of the double cylinder was due to Matthew Murray, of Leeds, one of the best mechanical engineers of his time ; Mr. Blenkinsop, who was not himself a mechanic, having consulted him as to all the practical arrangements of his locomotive. The connecting-rods gave the motion to two pinions by cranks at right angles to each other; these pinions communicating the motion to the wheel which worked into the cogged-rail.

Mr. Blenkinsop's engines began running on the railway extending from the Middleton collieries to the town of Leeds, a distance of about three miles and a half, on the 12th of August, 1812. They continued for

---

a member of that learned body said, smiling, " Yes, this is all very ingenious, no doubt, but unfortunately the machine will never move; it is too heavy. The wheels will turn round and round in the same place."

[1] The following is the description given in the specification of the patent :—" It is to be noticed that we do occasionally, or in certain cases, make the external periphery of the wheels uneven, by projecting heads of nails, or bolts, or cross groves, or fittings to railroads, when required ; and that in case of hard pull we cause a lever, bolt, or claw to project through the rim of one or both of the said wheels, so as to take hold of the ground."

many years to be one of the principal curiosities of the place, and were visited by strangers from all parts. In the year 1816, the Grand Duke Nicholas (afterwards Emperor) of Russia observed the working of Blenkinsop's locomotive with curious interest and expressions of no slight admiration. An engine dragged behind it as many as thirty coal-waggons at a speed of about three miles and a quarter per hour. These engines continued for many years to be thus employed in the haulage of coal, and furnished the first instance of the regular employment of locomotive power for commercial purposes.

The Messrs. Chapman, of Newcastle, in 1812, endeavoured to overcome the same fictitious difficulty of the want of adhesion between the wheel and the rail, by patenting a locomotive to work along the road by means of a chain stretched from one end of it to the other. This chain was passed once round a grooved barrel-wheel under the centre of the engine : so that, when the wheel turned, the locomotive, as it were, dragged itself along the railway. An engine, constructed after this plan, was tried on the Heaton Railway, near Newcastle ; but it was so clumsy in its action, there was so great a loss of power by friction, and it was found to be so expensive and difficult to keep in repair, that it was very soon abandoned. Another remarkable expedient was adopted by Mr. Brunton, of the Butterley Works, Derbyshire, who, in 1813, patented his Mechanical Traveller, to go *upon legs*, working alternately like those of a horse.[1] But this engine never got beyond the experimental state, for, at its very first trial, the driver, to make sure of a good start, overloaded the safety-valve, when the boiler burst, and

---

[1] The specification of patent (No. 3700) is dated the 22nd May, 1813. Other machines, with legs, were patented in the following year by Lewis Gompertz and by Thomas Tindall (Nos. 3804 and 3817). In Tindall's specification it is provided that the power of the engine is to be assisted by a *horizontal windmill;* and the four pushers, or legs, are to be caused to come successively in contact with the ground, and impel the carriage!

killed a number of the bystanders, wounding many more. These, and other contrivances with the same object, projected about the same time, show that invention was actively at work, and that many minds were now anxiously labouring to solve the important problem of locomotive traction upon railways.

But the difficulties contended with by these early inventors, and the step-by-step progress which they made, will probably be best illustrated by the experiments conducted by Mr. Blackett, of Wylam, which are all the more worthy of notice, as the persevering efforts of this gentleman in a great measure paved the way for the labours of George Stephenson, who, shortly after, took up the question of steam locomotion, and brought it to a successful issue.

The Wylam waggon-way is one of the oldest in the north of England. Down to the year 1807 it was formed of wooden spars or rails, laid down between the colliery at Wylam—where old Robert Stephenson had worked—and the village of Lemington, some four miles down the Tyne, where the coals were loaded into keels or barges, and floated down the river past Newcastle, to be shipped for the London market. Each chaldron-waggon was originally drawn by one horse, with a man to each horse and waggon. The rate at which the waggons were hauled was so slow that only two journeys were performed by each man and horse in one day, and three on the day following, the driver being allowed sevenpence for each journey. This primitive waggon-way passed, as before stated, close in front of the cottage in which George Stephenson was born ; and one of the earliest sights which met his infant eyes was this wooden tramroad worked by horses.

Mr. Blackett was the first colliery owner in the North who took an active interest in the locomotive. Having formed the acquaintance of Trevithick in London, and inspected the performances of his engine, he determined to repeat the Pen-y-darran experiment upon the Wylam waggon-way. He accordingly obtained from Trevithick, in October, 1804, a plan of his engine, provided with "friction-wheels," and employed Mr. John Whinfield, of Pipewellgate, Gateshead, to construct it at his foundry there. The engine was constructed under the superintendence of one John Steele,[1] an ingenious mechanic, who had been in Wales, and worked under Trevithick in fitting the engine at Pen-y-darran. When the Gateshead locomotive was finished, a temporary way was laid down in the works, on which it was run backwards and forwards many times. For some reason or other, however—it is said because the engine was deemed too light for drawing the coal-trains—it never left the works, but was dismounted from the wheels, and set to blow the cupola of the foundry, in which service it was employed for many years.

Several years elapsed before Mr. Blackett took any further steps to carry out his idea. The final abandonment of Trevithick's locomotive at Pen-y-darran perhaps contributed to deter him from proceeding further; but he had the wooden tramway taken up in 1808, and a plate-way of cast-iron laid down instead—a single line

---

[1] John Steele was one of the many "born mechanics" of the Northumberland district. When a boy at Colliery Dykes, his native place, he was noted for his "turn for machinery." He used to take his playfellows home to see and admire his imitations of pit-engines. While a mere youth he lost his leg by an accident; and those who remember him at Whinfield's speak of his hopping about the locomotive, of which he was very proud, upon his wooden leg. It was a great disappointment to him when Mr. Blackett refused to take the engine. One day he took a friend to look at it when reduced to its degraded office of blowing the cupola bellows; and, referring to the cause of its rejection, he observed that he was certain it would succeed, if made sufficiently heavy. "Our master," he continued, "will not be at the expense of following it up; but depend upon it the day will come when such an engine will be fairly tried, and then it will be found to answer." Steele was afterwards extensively employed by the British Government in raising sunken ships; and later in life he established engine-works at Rouen, where he made marine-engines for the French Government. He was unfortunately killed by the explosion of an engine-boiler (with the safety-valve of which something had gone wrong), when upon an experimental trip with one of the steamers fitted up by himself, and while on his way to England to visit his family near Newcastle.

furnished with sidings to enable the laden waggons to pass the empty ones. The new iron road proved so much smoother than the old wooden one, that a single horse, instead of drawing one laden waggon, was now enabled to draw two, or even three waggons.

Encouraged by the success of Mr. Blenkinsop's experiment, Mr. Blackett eventually determined to follow his example; and in 1812 he ordered a second engine, to work with a toothed driving wheel upon a rack-rail as at Leeds. This locomotive was constructed by Thomas Waters, of Gateshead, under the superintendence of Jonathan Foster, Mr. Blackett's principal engine-wright. It was a combination of Trevithick's and Blenkinsop's engines; but it was of a more awkward construction than either. The boiler was of cast-iron. The engine was provided with a single cylinder six inches in diameter, with a flywheel working at one side to carry the crank over the dead points. Jonathan Foster described it to the author in 1854, as " a strange machine, with lots of pumps, cog-wheels, and plugs, requiring constant attention while at work." The weight of the whole was about six tons.

When finished, it was conveyed to Wylam on a waggon, and there mounted upon a wooden frame supported by four pairs of wheels, which had been constructed for its reception. A barrel of water, placed on another frame upon wheels, was attached to it as a tender. After a great deal of labour, the cumbrous machine was got upon the road. At first it would not move an inch. Its maker, Tommy Waters, became impatient, and at length enraged, and taking hold of the lever of the safety valve, declared in his desperation, that " either *she* or *he* should go." At length the machinery was set in motion, on which, as Jonathan Foster described to the author, "she flew all to pieces, and it was the biggest wonder i' the world that we were not all blewn up." The incompetent and useless engine was declared to be a failure; it was shortly after dismounted and sold; and Mr. Blackett's praiseworthy efforts thus far proved in vain.

He was still, however, desirous of testing the practicability of employing locomotive power in working the coal down to Lemington, and he determined on making yet another trial. He accordingly directed his engine-wright, Jonathan Foster, to proceed with the building of a third engine in the Wylam workshops. This new locomotive had a single eight-inch cylinder, was provided with a flywheel like its predecessor, and the driving wheel was cogged on one side to enable it to travel in the rack-rail laid along the road. This engine proved more successful than the former one; and it was found capable of dragging eight or nine loaded waggons, though at the rate of little more than a mile an hour, from the colliery to the shipping-place. It sometimes took six hours, as Jonathan Foster informed us, to perform the journey of five miles. Its weight was found too great for the road, and the cast-iron plates were constantly breaking. It was also very apt to get off the rack-rail, and then it stood still. The driver was one day asked how he got on? "Get on?" said he, "we don't get on; we only get off!" On such occasions, horses had to be sent out to drag along the waggons as before, and others to haul the engine back to the workshops. It was constantly getting out of order; its plugs, pumps, or cranks, got wrong; it was under repair as often as at work; at length it became so cranky that the horses were usually sent out after it to drag it along when it gave up; and the workmen generally declared it to be a "perfect plague." Mr. Blackett did not obtain credit amongst his neighbours for these experiments. Many laughed at his machines, regarding them only in the light of crotchets,—frequently quoting the proverb of "a fool and his money." Others regarded them as absurd innovations on the established method of hauling coal; and pronounced that they would " never answer."

Notwithstanding, however, the comparative failure of this second locomotive, Mr. Blackett persevered with his experiments. He was zealously assisted by Jonathan Foster, the engine-wright, and William Hedley, the viewer of Wylam Colliery. The latter was a highly ingenious person, and proved of great use in carrying out the experiments to a successful issue. One of the chief causes of failure being the rack-rail, the idea occurred to him that it might be possible to secure sufficient adhesion between the wheel and the rail by the mere weight of the engine, and he proceeded to make a series of experiments for the purpose of determining this problem. He had a frame placed on four wheels, and fitted up with windlasses attached by gearing to the several wheels. The frame having been properly weighted, six men were set to work the windlasses; when it was found that the adhesion of the smooth wheels on the smooth rails was quite sufficient to enable them to propel the machine without slipping. Having thus found the proportion which the power bore to the weight, he demonstrated by successive experiments that the weight of the engine would of itself produce sufficient adhesion to enable it to draw upon a smooth railroad the requisite number of waggons in all kinds of weather. And thus was the fallacy which had heretofore prevailed on this subject completely exploded, and it was satisfactorily proved that rack-rails, toothed wheels, endless chains, and legs, were alike unnecessary for the efficient traction of loaded waggons upon a moderately level road.[1]

From this time forward considerably less difficulty was experienced in working the coal trains upon the Wylam tramroad. At length the rack-rail was dis-

---

[1] Mr. Hedley took out a patent to secure his invention, dated the 13th March, 1813. Specification No. 3666.

pensed with. The road was laid with heavier rails; the working of the old engine was improved; and a new engine was shortly after built and placed upon the road, still on eight wheels, driven by seven rack-wheels working inside them—with a wrought-iron boiler through which the flue was returned so as largely to increase the heating surface, and thus give increased power to the engine.[1] The following is a representation of this improved Wylam engine.

WYLAM ENGINE.

As may readily be imagined, the jets of steam from the piston, blowing off into the air at high pressure while the engine was in motion, caused considerable annoyance to horses passing along the Wylam road, at

---

[1] By the year 1825, the progress made on the Wylam railroad was thus described by Mr. Mackenzie in his 'History of Northumberland:' "A stranger," said he, "is struck with surprise and astonishment on seeing a locomotive engine moving majestically along the road at the rate of four or five miles an hour, drawing along from ten to fourteen loaded waggons, weighing about 21½ tons; and his surprise is increased on witnessing the extraordinary facility with which the engine is managed. This invention is a noble triumph of science."

that time a public highway. The nuisance was felt to
be almost intolerable, and a neighbouring gentleman
threatened to have it put down. To diminish the noise
as much as possible, Mr. Blackett gave orders that so
soon as any horse, or vehicle drawn by horses, came in
sight, the locomotive was to be stopped, and the frightful
blast of the engine thus suspended until the passing
animals had got out of sight. Much interruption was
caused to the working of the railway by this measure;
and it excited considerable dissatisfaction amongst the
workmen. The following plan was adopted to abate
the nuisance: a reservoir was provided immediately
behind the chimney (as shown in the preceding cut)
into which the waste steam was thrown after it had
performed its office in the cylinder; and from this
reservoir, the steam gradually escaped into the atmosphere without noise. This arrangement was devised
with the express object of preventing a blast in the
chimney, the value of which, as we shall subsequently
find, was not detected until George Stephenson, adopting it with a preconceived design and purpose, demonstrated its importance and value,—as being, in fact, the
very life-breath of the locomotive engine.

While Mr. Blackett was thus experimenting and
building locomotives at Wylam, George Stephenson
was anxiously studying the same subject at Killingworth.
He was no sooner appointed engine-wright of the
collieries than his attention was directed to the means
of more economically hauling the coal from the pits to
the river side. We have seen that one of the first
important improvements which he made, after being
placed in charge of the colliery machinery, was to apply
the surplus power of a pumping steam-engine, fixed
underground, for the purpose of drawing the coals out
of the deeper workings of the Killingworth mines, — by
which he succeeded in effecting a large reduction in
the expenditure on manual and horse labour.

The coals, when brought above ground, had next to be laboriously dragged by means of horses to the shipping staiths on the Tyne, several miles distant. The adoption of a tramroad, it is true, had tended to facilitate their transit : nevertheless the haulage was both tedious and expensive. With the view of economising labour, Stephenson laid down inclined planes where the nature of the ground would admit of this expedient being adopted. Thus, a train of full waggons let down the incline by means of a rope running over wheels laid along the tramroad, the other end of which was attached to a train of empty waggons placed at the bottom of the parallel road on the same incline, dragged them up by the simple power of gravity. But this applied only to a comparatively small part of the road. An economical method of working the coal trains, instead of by means of horses—the keep of which was at that time very costly in consequence of the high price of corn,—was still a great desideratum ; and the best practical minds in the collieries were actively engaged in the attempt to solve the problem.

In the first place Stephenson resolved to make himself thoroughly acquainted with what had already been done. Mr. Blackett's engines were working daily at Wylam, past the cottage where he had been born ; and thither he frequently went [1] to inspect the improvements made by Mr. Blackett from time to time both in the locomotive and in the plateway along which it worked. Jonathan

[1] At the Stephenson Memorial meeting at Newcastle-on-Tyne, 26th October, 1858, Mr. Hugh Taylor, Chairman of the Northern Coal-owners, gave the following· account of one of such visits made by Stephenson to Wylam, in the company of Mr. Nicholas Wood and himself :— "It was, I think, in 1812, that Mr. Stephenson and Mr. Wood came to my house, then at Newburn, and after we had dined, we went and examined the locomotive then on Mr. Blackett's waggon-way. At that early date it went by a sort of cog-wheel; there was also something of a chain to it. There was no idea that the machine would be sufficiently adhesive to the rails by the action of its own weight; but I remember a man going before— that was after the chain was abrogated—and scattering ashes on the rails, in order to give it adhesiveness, and two or three miles an hour was about the rate of progress."

Foster informed us that, after one of these visits, Stephenson declared to him his conviction that a much more effective engine might be made, that should work more steadily and draw the load more effectively.

He had also the advantage, about the same time, of seeing one of Blenkinsop's Leeds engines, which was placed on the tramway leading from the collieries of Kenton and Coxlodge, on the 2nd of September, 1813. This locomotive drew sixteen chaldron waggons containing an aggregate weight of seventy tons, at the rate of about three miles an hour. George Stephenson and several of the Killingworth men were amongst the crowd of spectators that day; and after examining the engine and observing its performances, he observed to his companions, that "he thought he could make a better engine than that, to go upon legs." Probably he had heard of the invention of Brunton, whose patent had by this time been published, and proved the subject of much curious speculation in the colliery districts. Certain it is, that, shortly after the inspection of the Coxlodge engine, he contemplated the construction of a new locomotive, which was to surpass all that had preceded it. He observed that those engines which had been constructed up to this time, however ingenious in their arrangements, had proved practical failures. Mr. Blackett's was as yet both clumsy and expensive. Chapman's had been removed from the Heaton tramway in 1812, and was regarded as a total failure. And the Blenkinsop engine at Coxlodge was found very unsteady and costly in its working; besides, it pulled the rails to pieces, the entire strain being upon the rack-rail on one side of the road. The boiler, however, having shortly blown up, there was an end of that engine; and the colliery owners did not feel encouraged to try any further experiment.

An efficient and economical working locomotive engine, therefore, still remained to be invented; and to accomplish this object Mr. Stephenson now applied himself. Profiting by what his predecessors had done, warned by their failures and encouraged by their partial successes, he commenced his labours. There was still wanting the man who should accomplish for the locomotive what James Watt had done for the steam-engine, and combine in a complete form the separate plans of others, embodying with them such original inventions and adaptations of his own as to entitle him to the merit of inventing the working locomotive, in the same manner as James Watt is to be regarded as the inventor of the working condensing engine. This was the great work upon which George Stephenson now entered, though probably without any adequate idea of the ultimate importance of his labours to society and civilization.

He proceeded to bring the subject of constructing a "Travelling Engine," as he then denominated the locomotive, under the notice of the lessees of the Killingworth Colliery, in the year 1813. Lord Ravensworth, the principal partner, had already formed a very favourable opinion of the new colliery engine-wright, from the improvements which he had effected in the colliery engines, both above and below ground ; and, after considering the matter, and hearing Stephenson's explanations, he authorised him to proceed with the construction of a locomotive,—though his lordship was, by some, called a fool for advancing money for such a purpose. "The first locomotive that I made," said Mr. Stephenson, many years after,[1] when speaking of his early career at a public meeting in Newcastle, " was at Killingworth Colliery, and with Lord Ravensworth's money. Yes ; Lord Ravensworth and partners were the first to entrust me, thirty-two years since, with money

---

[1] Speech at the opening of the Newcastle and Darlington Railway, June 18, 1844.

to make a locomotive engine. I said to my friends, there was no limit to the speed of such an engine, if the works could be made to stand."

Our engine-wright had, however, many obstacles to encounter before he could get fairly to work with the erection of his locomotive. His chief difficulty was in finding workmen sufficiently skilled in mechanics, and in the use of tools, to follow his instructions and embody his designs in a practical shape. The tools then in use about the collieries were rude and clumsy; and there were no such facilities as now exist for turning out machinery of an entirely new character. Stephenson was under the necessity of working with such men and tools as were at his command; and he had in a great measure to train and instruct the workmen himself. The engine was built in the workshops at the West Moor, the leading mechanic being John Thirlwall, the colliery blacksmith, an excellent workman in his way, though quite new to the work now entrusted to him.

In this first locomotive constructed at Killingworth, Stephenson to some extent followed the plan of Blenkinsop's engine. The wrought-iron boiler was cylindrical, eight feet in length and thirty-four inches in diameter, with an internal flue tube twenty inches wide passing through it. The engine had two vertical cylinders of eight inches diameter, and two feet stroke, let into the boiler, working the propelling gear with cross heads and connecting rods. The power of the two cylinders was combined by means of spurwheels, which communicated the motive power to the wheels supporting the engine on the rail, instead of, as in Blenkinsop's engine,

THE SPUR-GEAR.

to cogwheels which acted on the cogged rail independent of the four supporting wheels. The engine thus worked upon what is termed the second motion. The chimney was of wrought iron, round which was a chamber extending back to the feed-pumps, for the purpose of heating the water previous to its injection into the boiler. The engine had no springs whatever, and was mounted on a wooden frame supported on four wheels. In order, however, to neutralise as much as possible the jolts and shocks which such an engine would necessarily encounter from the obstacles and inequalities of the then very imperfect plateway, the water-barrel which served for a tender was fixed to the end of a lever and weighted, the other end of the lever being connected with the frame of the locomotive carriage. By this means the weight of the two was more equally distributed, though the contrivance did not by any means compensate for the total absence of springs.

The wheels of the locomotive were all smooth, Mr. Stephenson having satisfied himself by experiment that the adhesion between the wheels of a loaded engine and the rail would be sufficient for the purpose of traction. Robert Stephenson informed us that his father caused a number of workmen to mount upon the wheels of a waggon moderately loaded, and throw their entire weight upon the spokes on one side, when he found that the waggon could thus be easily propelled forward without the wheels slipping. This, together with other experiments, satisfied him of the expediency of adopting smooth wheels on his engine, and it was so finished accordingly.

The engine was, after much labour and anxiety, and frequent alterations of parts, at length brought to completion, having been about ten months in hand. It was placed upon the Killingworth Railway on the 25th of July, 1814; and its powers were tried on the same day. On an ascending gradient of 1 in 450, the engine succeeded in drawing after it eight loaded carriages of thirty tons' weight at about four miles an hour; and for some time after it continued regularly at work.

Although a considerable advance upon previous locomotives, "Blutcher" (as the engine was popularly
called) was nevertheless a somewhat cumbrous and
clumsy machine. The parts were huddled together.
The boiler constituted the principal feature; and being
the foundation of the other parts, it was made to do
duty not only as a generator of steam, but also as a basis
for the fixings of the machinery and for the bearings of
the wheels and axles. The want of springs was seriously
felt; and the progress of the engine was a succession of
jolts, causing considerable derangement to the machinery.
The mode of communicating the motive power to the
wheels by means of the spur gear also caused frequent
jerks, each cylinder alternately propelling or becoming
propelled by the other, as the pressure of the one upon
the wheels became greater or less than the pressure of
the other; and, when the teeth of the cogwheels became
at all worn, a rattling noise was produced during the
travelling of the engine.

As the principal test of the success of the locomotive
was its economy as compared with horse power, careful
calculations were made with the view of ascertaining this
important point. The result was, that it was found the
working of the engine was at first barely economical;
and at the end of the year the steam power and the
horse power were ascertained to be as nearly as possible
upon a par in point of cost. The fate of the locomotive
in a great measure depended on this very engine. Its
speed was not beyond that of a horse's walk, and the
heating surface presented to the fire being comparatively
small, sufficient steam could not be raised to enable it to
accomplish more on an average than about four miles
an hour. The result was anything but decisive; and
the locomotive might have been condemned as useless,
had not Mr. Stephenson at this juncture applied the
steam-blast, and by its means carried his experiment to
a triumphant issue.

The steam, after performing its duty in the cylinders, was at first allowed to escape into the open atmosphere with a hissing blast, to the terror of horses and cattle. It was complained of as a nuisance; and a neighbouring squire threatened to commence an action against the colliery lessees unless it was put a stop to. But Mr. Stephenson's attention had already been drawn to the circumstance of the much greater velocity with which the steam issued from the exit pipe compared with that at which the smoke escaped from the chimney of the engine. He then thought that, by conveying the eduction steam into the chimney by means of a small pipe after it had performed its office in the cylinders, and allowing it to escape in a vertical direction, its velocity would be imparted to the smoke from the fire, or to the ascending current of air in the chimney,[1] thereby increasing the draft, and consequently the intensity of combustion in the furnace.

The experiment was no sooner made than the power of the engine was at once more than doubled; combustion was stimulated by the blast; consequently the capability of the boiler to generate steam was greatly increased, and the effective power of the engine augmented in precisely the same proportion, without in any way adding to its weight. This simple but beautiful expedient was really fraught with the most important consequences to railway communication; and it is not too much to say that the success of the locomotive depended upon its adoption. Without the steam-blast, by which the intensity of combustion, and the consequent evolution of steam, were maintained at their highest point, high rates of speed could not have been maintained, the advantages of the multitubular boiler (afterwards invented) could never have been fairly

---

[1] The subject of the Steam Blast, and the various claims which have been made as to its invention, will be found discussed at some length in the Appendix to this work.

tested, and locomotives might still have been dragging themselves unwieldily along at little more than five or six miles an hour.

The steam-blast had scarcely been adopted, with so decided a success, when Mr. Stephenson, observing the numerous defects in his engine, and profiting by the experience which he had already acquired, determined to construct a second engine, in which to embody his improvements in their best form.   Careful and cautious observation of the working of his locomotive had convinced him that the complication arising out of the action of the two cylinders being combined by spur-wheels would prevent its coming into practical use. He accordingly directed his attention to an entire change in the construction and mechanical arrangements of the machine; and in the following year, conjointly with Mr. Dodds, who provided the necessary funds, he took out a patent, dated the 28th of February, 1815,[1] for an engine which combined in a remarkable degree the essential requisites of an economical locomotive; that is to say, few parts, simplicity in their action, and directness in the mode by which the power was communicated to the wheels supporting the engine.

This locomotive, like the first, had two vertical cylinders, which communicated *directly* with each pair of the four wheels that supported the engine, by means of a cross head and a pair of connecting rods.   But, in attempting to establish a direct communication between the cylinders and the wheels that rolled upon the rails, considerable difficulties presented themselves.   The ordinary joints could not be employed to unite the parts of the engine, which was a rigid mass, with the wheels rolling upon the irregular surface of the rails; for it was evident that the two rails of the line of way—more especially in those early days of imperfect construction

---

[1] Specification of patent, No. 3887.

of the permanent road—could not always be maintained
at the same level,—that the wheel at one end of the
axle might be depressed into one part of the line which
had subsided, whilst the other wheel would be comparatively elevated ; and, in such a position of the axle
and wheels, it was obvious that a rigid communication
between the cross head and the wheels was impracticable.
Hence it became necessary to form a joint at the top of
the piston-rod where it united with the cross head, so
as to permit the cross head to preserve complete parallelism with the axle of the wheels with which it was in
communication.

In order to obtain that degree of flexibility combined
with direct action, which was essential for ensuring
power and avoiding needless friction and jars from
irregularities in the road, Mr. Stephenson made use of
the " ball and socket " joint for effecting a union between
the ends of the cross heads where they united with the
connecting rods, and between the ends of the connecting
rods where they were united with the crank-pins attached
to each driving wheel.   By this arrangement the parallelism between the cross head and the axle was at all
times maintained and preserved, without producing any
serious jar or friction on any part of the machine.
Another important point was, to combine each pair of
wheels by means of some simple mechanism instead of
by the cogwheels which had formerly been used.   And,
with this object, Mr. Stephenson began by making in
each axle cranks at right angles to each other, with rods
communicating horizontally between them.

A locomotive was accordingly constructed upon this
plan in the year 1815, and it was found to answer
extremely well.   But at that period the mechanical
skill of the country was not equal to the task of forging
cranked axles of the soundness and strength necessary
to stand the jars incident to locomotive work.   Mr.
Stephenson was accordingly compelled to fall back upon

a substitute, which, although less simple and efficient, was within the mechanical capabilities of· the workmen of that day, in respect of construction as well as repair. He adopted a chain which rolled over indented wheels placed on the centre of each axle, and was so arranged that the two pairs of wheels were effectually coupled and made to keep pace with each other. The chain, however, after a few years' use, became stretched ; and then the engines were liable to irregularity in their working, especially in changing from working back to working forward again. Eventually the chain was laid aside, and the front and hind wheels were united by rods on the outside, instead of by rods and crank axles inside, as specified in the original patent. This expedient completely answered the purpose required, without involving any expensive or difficult workmanship.

Thus, in the year 1815, Mr. Stephenson, by dint of patient and persevering labour,—by careful observation of the works of others, and never neglecting to avail himself of their suggestions,—had succeeded in manufacturing an engine which included the following important improvements on all previous attempts in the same direction:—viz., simple and direct communication between the cylinders and the wheels rolling upon the rails ; joint adhesion of all the wheels, attained by the use of horizontal connecting rods ; and finally, a beautiful method of exciting the combustion of the fuel by employing the waste steam, which had formerly been allowed uselessly to escape into the air. Although many improvements in detail were afterwards introduced in the locomotive by Mr. Stephenson himself, as well as by his equally distinguished son, it is perhaps not too much to say that this engine, as a mechanical contrivance, contained the germ of all that has since been effected. It may in fact be regarded as the type of the present locomotive engine.

CHAPTER VII.

Invention of the " Geordy " Safety-Lamp.

Explosions of fire-damp were unusually frequent in the coal mines of Northumberland and Durham about the time when George Stephenson was engaged in the construction of his first locomotives. These explosions were often attended with fearful loss of life and dreadful suffering to the workpeople. Killingworth Colliery was not free from such deplorable calamities; and during the time that Stephenson was employed as a brakesman at the West Moor, several "blasts" took place in the pit, by which many workmen were scorched and killed, and the owners of the colliery sustained heavy losses. One of the most serious of these accidents occurred in 1806, not long after he had been appointed brakesman, by which ten persons were killed. Stephenson was working at the mouth of the pit at the time, and the circumstances connected with the accident made a deep impression on his mind.[1]

Another explosion took place in the same pit in 1809, by which twelve persons lost their lives. The blast did not reach the shaft as in the former case; the unfortunate persons in the pit having been suffocated by the after-damp. More calamitous still were the explosions which took place in neighbouring collieries; one of the worst being that of 1812, in the Felling Pit, near Gateshead, a mine belonging to Mr. Brandling, by which no fewer than ninety men and boys were suffocated or burnt to death. And a similar accident occurred in the same pit in the year following, by which twenty-two men and boys perished.

---

[1] See evidence given by him before the Select Committee on Accidents in Mines, 26th June, 1835.

It was natural that George Stephenson should devote his attention to the cause of these deplorable accidents, and to the means by which they might if possible be prevented. His daily occupation led him to think much and deeply on the subject. As engine-wright of a colliery so extensive as that of Killingworth, where there were nearly 160 miles of gallery excavation, in which he personally superintended the working of inclined planes for the conveyance of the coal to the pit entrance, he was necessarily very often underground, and brought face to face with the dangers of fire-damp. From fissures in the roofs of the galleries, carburetted hydrogen gas was constantly flowing; in some of the more dangerous places it might be heard escaping from the crevices of the coal with a hissing noise. Ventilation, firing, and all conceivable modes of drawing out the foul air had been adopted, and the more dangerous parts of the galleries were built up. Still the danger could not be wholly prevented. The miners must necessarily guide their steps through the extensive underground ways with lighted lamps or candles, the naked flame of which, coming in contact with the inflammable air, daily exposed them and their fellow-workers in the pit to the risk of death in one of its most dreadful forms.

One day, in the year 1814, a workman hurried into Stephenson's cottage with the startling information that the deepest main of the colliery was on fire! He immediately hastened to the pit-head, about a hundred yards off, whither the women and children of the colliery were running, with wildness and terror depicted in every face. In an energetic voice Stephenson ordered the engineman to lower him down the shaft in the corve. There was danger, it might be death, before him, but he must go. As those about the pit-mouth saw him descend rapidly out of sight, and heard from the depths of the shaft the mingled cries of despair and agony

THE PIT HEAD, WEST MOOR. [By R. P. Leitch.]

rising from the workpeople below, they gazed on the heroic man with breathless amazement.

He was soon at the bottom, and in the midst of the men, who were paralysed at the danger which threatened the lives of all in the pit.   Leaping from the corve on its touching the ground, he called out : " Are there six men among you who have courage to follow me ? If so, come, and we will put the fire out."   The Killingworth pitmen had the most perfect confidence in their engine-wright, and they readily volunteered to follow him. Silence succeeded the frantic tumult of the previous minute, and the men set to work with a will.   In every mine, bricks, mortar, and tools enough are at hand, and by Stephenson's direction the materials were forth-with carried to the required spot, where, in a very short time, a wall was raised at the entrance to the main, he himself taking the most active part in the work.   The atmospheric air was by this means excluded, the fire was extinguished, the people were saved from death, and the mine was preserved.

This anecdote of Stephenson was related to the writer, near the pit-mouth, by one of the men, Kit

Heppel, who had been present and helped to build up
the brick wall by which the fire was stayed, though
several workmen were suffocated. Heppel relates that,
when down the pit some days after, seeking out the
dead bodies, the cause of the accident was the subject
of some conversation between himself and Stephenson,
and Heppel then asked him, " Can nothing be done
to prevent such awful occurrences ? " Stephenson
replied that he thought something might be done.
" Then," said Heppel, " the sooner you start the better ;
for the price of coal-mining now is *pitmen's lives.*"

Fifty years since, many of the best pits were so full
of the inflammable gas given forth by the coal, that
they could not be worked without the greatest danger ;
and for this reason some were altogether abandoned.
The rudest possible methods were adopted of producing
light sufficient to enable the pitmen to work by. The
phosphorescence of decayed fish-skins was tried ; but this,
though safe, was very inefficient. The most common
method employed was what was called a steel mill, the
notched wheel of which, being made to revolve against
a flint, struck a succession of sparks, which scarcely
served to do more than make the darkness visible. A
boy carried the apparatus after the miner, working the
wheel, and by the imperfect light thus given forth he
plied his dangerous trade. Candles were only used in
those parts of the pit where gas was not abundant.
Under this rude system not more than one-third of the
coal could be worked ; and two-thirds were left.

What the workmen, not less than the coal-owners,
eagerly desired was, a lamp that should give forth sufficient light, without communicating flame to the inflammable gas which accumulated in certain parts of the
pit. Something had already been attempted towards
the invention of such a lamp by Dr. Clanny, of Sunderland, who, in 1813, contrived an apparatus to which he
gave air from the mine through water, by means of

bellows.   This lamp went out of itself in inflammable
gas.   It was found, however, too unwieldy to be used
by the miners for the purposes of their work, and did
not come into general use.   A committee of gentlemen
was formed at Sunderland to investigate the causes of
the explosions, and to devise, if possible, some means of
preventing them.   At the invitation of that Committee,
Sir Humphry Davy, then in the full zenith of his reputation, was requested to turn his attention to the subject.
He accordingly visited the collieries near Newcastle on
the 24th of August, 1815; and on the 9th of November
following, he read his celebrated paper " On the Fire-
Damp of Coal Mines, and on Methods of lighting the
Mine so as to prevent its Explosion," before the Royal
Society of London.

But a humbler though not less diligent and original
thinker had been at work before him, and had already
practically solved the problem of the Safety-Lamp.
Stephenson was of course well aware of the anxiety
which prevailed in the colliery districts as to the invention of a lamp which should give light enough for
the miners to work by without exploding the fire-damp.
The painful incidents above described only served to
quicken his eagerness to master the difficulty.

For several years he had been engaged, in his own
rude way, in making experiments with the fire-damp in
the Killingworth mine.   The pitmen used to expostulate
with him on these occasions, believing his experiments
to be fraught with danger.   One of the sinkers, called
M'Crie, observing him holding up lighted candles to the
windward of the " blower " or fissure from which the
inflammable gas escaped, entreated him to desist; but
Stephenson's answer was, that " he was busy with a
plan by which he hoped to make his experiments useful
for preserving men's lives."   On these occasions the
miners usually got out of the way before he lit the gas.

In 1815, although he was very much occupied with

the business of the collieries and the improvement of
his locomotive engine, he was also busily engaged in
making experiments upon inflammable gas in the Killingworth pit. As he himself afterwards related to the
Committee of the House of Commons which sat on the
subject of Accidents in Mines in 1835, he imagined that
if he could construct a lamp with a chimney at the
top, so as to cause a strong current, it would not fire
at the top of the chimney; as the burnt air would
ascend with such a velocity as to prevent the inflammable air of the pit from descending towards the flame;
and such a lamp, he thought, might be taken into an
explosive atmosphere without risk of exploding.

Such was Stephenson's theory, when he proceeded to
embody his idea of a miner's safety-lamp in a practical
form. In the month of August, 1815, he requested his
friend Nicholas Wood, the head viewer, to prepare a
drawing of a lamp according to the description which
he gave him. After several evenings' careful delibe-
rations, the drawing was prepared, and it was shown to
several of the head men about the works. "My first
lamp," said Mr. Stephenson, describing it to the Committee above referred to, "had a chimney at the top of
the lamp, and a tube at the bottom, to admit the atmospheric air, or fire-damp and air, to feed the burner or
combustion of the lamp. I was not aware of the precise
quantity required to feed the combustion; but to know
what quantity was necessary, I had a slide at the bottom
of the first tube in my lamp, to admit such a quantity
of air as might eventually be found necessary to keep
up the combustion."

Accompanied by his friend Wood, Stephenson went
into Newcastle, and ordered a lamp to be made according to his plan, by the Messrs. Hogg, tinmen, at the
head of the Side—a well-known street in Newcastle.
At the same time he ordered a glass to be made for the
lamp at the Northumberland Glass House, in the same

town.   This lamp was received from the makers on the
21st of October, and was taken to Killingworth for the
purpose of immediate experiment.

"I remember that evening as distinctly as if it had
been but yesterday," said Robert Stephenson, describing
the circumstances to the author in 1857.   "Moodie
came to our cottage about dusk, and asked, 'if father
had got back yet with the lamp?' 'No.' 'Then I'll
wait till he comes,' said Moodie, 'he can't be long now.'
In about half-an-hour, in came my father, his face all
radiant.   He had the lamp with him!   It was at once
uncovered and shown to Moodie.   Then it was filled
with oil, trimmed, and lighted.   All was ready, only
the head viewer hadn't arrived.   'Run over to Benton
for Nichol, Robert,' said my father to me, 'and ask him
to come directly; say we're going down the pit to try
the lamp.'   By this time it was quite dark; and off I
ran to bring Nicholas Wood. His house was at Benton,
about a mile off.   There was a short cut through Benton
Churchyard, but just as I was about to pass the wicket,
I saw what I thought was a white figure moving about
amongst the grave-stones.   I took it for a ghost!
My heart fluttered, and I was in a great fright, but to
Nichol's house I must get, so I made the circuit of the
Churchyard; and when I got round to the other side I
looked, and lo! the figure was still there.   But what do
you think it was? Only the grave-digger, plying his
work at that late hour by the light of his lanthorn set
upon one of the gravestones!   I found Wood at home,
and in a few minutes he was mounted and off to my
father's.   When I got back, I was told they had just
left — it was then about eleven — and gone down the
shaft to try the lamp in one of the most dangerous parts
of the mine."

Arrived at the bottom of the shaft with the lamp, the
party directed their steps towards one of the foulest
galleries in the pit, where the explosive gas was issuing

through a blower in the roof of the mine with a loud hissing noise. By erecting some deal boarding round that part of the gallery into which the gas was escaping, the air was thus made more foul for the purpose of the experiment. After waiting about an hour, Moodie, whose practical experience of fire-damp in pits was greater than that of either Stephenson or Wood, was requested to go into the place which had thus been made foul; and, having done so, he returned, and told them that the smell of the air was such, that if a lighted candle were now introduced, an explosion must inevitably take place. He cautioned Stephenson as to the danger both to themselves and to the pit, if the gas took fire. But Stephenson declared his confidence in the safety of his lamp, and, having lit the wick, he boldly proceeded with it towards the explosive air. The others, more timid and doubtful, hung back when they came within hearing of the blower; and apprehensive of the danger, they retired into a safe place, out of sight of the lamp, which gradually disappeared with its bearer in the recesses of the mine. It was a critical moment; and the danger was such as would have tried the stoutest heart. Stephenson advancing alone, with his yet untried lamp, in the depths of those underground workings, —calmly venturing his life in the determination to discover a mode by which the lives of many might be saved, and death disarmed in these fatal caverns,—presented an example of intrepid nerve and manly courage, more noble even than that which, in the excitement of battle and the collective impetuosity of a charge, carries a man up to the cannon's mouth.

Advancing to the place of danger, and entering within the fouled air, his lighted lamp in hand, Stephenson held it firmly out, in the full current of the blower, and within a few inches of its mouth! Thus exposed, the flame of the lamp at first increased, then flickered, and then went out; but there was no explo-

sion of the gas. Returning to his companions, who
were still at a distance, he told them what had occurred.
Having now acquired somewhat more confidence, they
advanced with him to a point from which they could
observe him repeat his experiment, but still at a safe
distance. They saw that when the lighted lamp was
held within the explosive mixture, there was a great
flame; the lamp was almost full of fire; and then it
smothered out. Again returning to his companions, he
relighted the lamp, and repeated the experiment. This
he did several times, with the same result. At length
Wood and Moodie ventured to advance close to the
fouled part of the pit; and, in making some of the later
trials, Mr. Wood himself held up the lighted lamp to
the blower. Such was the result of the first experiments with the *first practical Miner's Safety-Lamp*; and
such the daring resolution of its inventor in testing its
qualities.

Before leaving the pit, Stephenson expressed his
opinion that, by an alteration of the lamp, which he
then contemplated, he could make it burn better. This
was by a change in the slide through which the air was
admitted into the lower part of the lamp, under the
flame. After making some experiments on the air
collected at the blower, by means of bladders which
were mounted with tubes of various diameters, he satisfied himself that, when the tube was reduced to a
certain diameter, the explosion would not pass through;
and he fashioned his slide accordingly, reducing the
diameter of the tube until he conceived it was quite
safe. In the course of about a fortnight the experiments were repeated in the pit, in a place purposely
made foul as before. On this occasion a larger number
of persons ventured to witness the experiments, which
again proved successful. The lamp was not yet, however, so efficient as the inventor desired. It required,
he observed, to be kept very steady when burning in

the inflammable gas, otherwise it was liable to go out, in consequence, as he imagined, of the contact of the burnt air (as he then called it), or azotic gas, which lodged round the exterior of the flame. If the lamp was moved backwards and forwards, the azote came in contact with the flame and extinguished it. "It struck me," said he, "that if I put more tubes in, I should discharge the poisonous matter that hung round the flame, by admitting the air to its exterior part." Although he had then no access to scientific works, nor intercourse with scientific men, nor anything that could assist him in his inquiries on the subject, besides his own indefatigable spirit of inquiry, he contrived a rude apparatus, by means of which he proceeded to test the explosive properties of the gas and the velocity of current (for this was the direction of his inquiries) necessary to enable the explosion to pass through tubes of different diameters. In making these experiments in his humble cottage at the West Moor, Nicholas Wood and George's son Robert usually acted as his assistants, and sometimes the gentlemen of the neighbourhood—amongst others William Brandling and Matthew Bell, interested in coal-mining—attended as spectators. One who was present on such an occasion remembers that, when an experiment was about to be performed, and all was ready, George called to Mr. Wood, who worked the stop-cocks of the gasometer, "Wise on [turn on] the hydrogen, Nichol!"

These experiments were not performed without risk, for on one occasion the experimenting party had nearly blown off the roof of the cottage. One of these "blows up" was described by Stephenson himself before the Committee on Accidents in Coal Mines in 1835: "I made several experiments," said he, "as to the velocity required in tubes of different diameters, to prevent explosion from fire-damp. We made the mixtures in all proportions of light carburetted hydrogen with atmospheric air in the receiver, and we found by the experiments that when a current of the most explosive mixture that we could make was forced up a tube four-tenths of an inch in diameter, the necessary current was nine inches in a second to prevent its coming down that tube. These experiments were repeated several times. We had two or three blows up in making the experiments, by the flame getting down into the receiver, though we had a piece of very fine wire-gauze put at the bottom of the pipe, between the receiver and the pipe through which we were forcing the current. In one of these experiments I was watching the flame in the tube, my son was taking the vibrations of the pendulum of the clock, and Mr. Wood was attending to give me the column of water as I called for it, to keep the current up to a certain point. As I saw the flame descending in the tube I called for more water, and Wood unfortunately turned the cock the wrong way; the current ceased, the flame went down the tube, and all our implements were blown to pieces, which at the time we were not very well able to replace."

The explosion of this glass receiver, which had been borrowed from the stores of the Philosophical Society at Newcastle for the purpose of making the experiments, caused the greatest possible dismay among the party, and they dreaded to inform Mr. Turner, the Secretary,[1] of the calamity which had occurred. For-

---

[1] The early connexion of Robert with the Philosophical and Literary Society of Newcastle had brought him into communication with the Rev. William Turner, one of the secretaries of the institution. That gentleman was always ready to assist the inquirer after knowledge, and took an early interest in the studious youth from Killingworth, with whose father he also became acquainted. Mr. Turner cheerfully helped them in their joint inquiries, and excited while he endeavoured to satisfy their thirst for scientific information. Towards the close of his life, Mr. Stephenson often spoke of the gratitude and esteem he felt towards his revered instructor. "Mr. Turner," he said, "was always ready to assist me with books, with instruments, and with counsel, gratuitously and cheerfully. He gave me the most valuable assistance and instruction, and to my dying day I can never forget the obligations which I owe to my venerable friend."
Mr.

tunately none of the experimenters were injured by the explosion.

Stephenson followed up those experiments by others of a similar kind, with the view of ascertaining whether ordinary flame would pass through tubes of a small diameter, and with this object he filed off the barrels of several small keys. Placing these together, he held them perpendicularly over a strong flame, and ascertained that it did not pass upward. This was a further proof to him of the soundness of the principle he was pursuing.

In order to correct the defect of his first lamp, he accordingly resolved to alter it so as to admit the air to the flame by several tubes of reduced diameter, instead of by a single tube. He inferred that a sufficient quantity of air would thus be introduced into the lamp for the purposes of combustion, whilst the smallness of the apertures would still prevent the explosion passing downwards, at the same time that the "burnt air" (the cause, in his opinion, of the lamp going out) would be more effectually dislodged. He accordingly took the lamp to the shop of Mr. Matthews, a tinman in Newcastle, and had it altered so that the air was admitted by three small tubes inserted in the bottom of the lamp, the openings of which were placed on the outside of the burner, instead of having (as in

---

Mr. Turner's conduct towards George Stephenson was all the more worthy of admiration, because at that time the object of his friendly instruction and counsel occupied but the position of a comparatively obscure workman, of no means or influence, who had become known to him only through his anxious desire for information on scientific subjects. He could little have dreamt that the object of his almost fatherly attention would achieve a reputation so distinguished as that which he afterwards obtained, and that he would revolu-tionise by his inventions and improvements the internal communications of the civilised world. The circumstance is encouraging to those who, like Mr. Turner, are still daily devoting themselves with equal disinterestedness to the education of the working classes in our schools and mechanics' institutes. Though the opportunity of lending a helping hand to such men as George Stephenson may but rarely occur, the labours of such teachers are never without the most valuable results.

the original lamp) the one tube opening directly under the flame.

This second or altered lamp was tried in the Killingworth pit on the 4th of November, and was found to burn better than the first lamp, and to be perfectly safe. But as it did not yet come up entirely to the inventor's expectations, he proceeded to contrive a third lamp, in which he proposed to surround the oil vessel with a number of capillary tubes. Then it struck him, that if he cut off the middle of the tubes, or made holes in metal plates, placed at a distance from each other, equal to the length of the tubes, the air would get in better, and the effect in preventing the communication of explosion would be the same.

He was encouraged to persevere in the completion of his safety-lamp by the occurrence of several fatal accidents about this time in the Killingworth pit. On the 9th of November a boy was killed by a blast in the *A* pit, at the very place where Stephenson had made the experiments with his first lamp; and, when told of the accident, he observed that if the boy had been provided with his lamp, his life would have been saved. On the 20th of November he went over to Newcastle to order his third lamp from Mr. Watson, a plumber in that town. Mr. Watson referred him to his clerk, Henry Smith, whom Stephenson invited to join him at a neighbouring public-house, where they might quietly talk over the matter together, and the plan of the new lamp could be finally settled. They adjourned to the " Newcastle Arms," near the present High Level Bridge, where they had some ale, and a design of the lamp was drawn in pencil upon a half-sheet of foolscap, with a rough specification subjoined. The sketch, when shown to us by Robert Stephenson some years since, still bore the marks of the ale. It was a very rude design, but sufficient to work from. It was immediately placed in the hands of the workmen, finished in the course

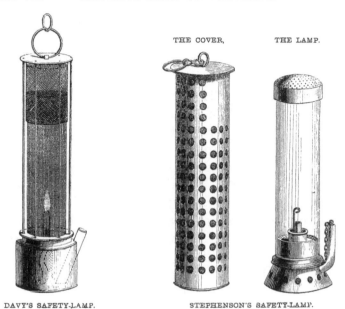

THE COVER.   THE LAMP.

DAVY'S SAFETY-LAMP.   STEPHENSON'S SAFETY-LAMP.

of a few days, and experimentally tested in the Killingworth pit like the previous lamps, on the 30th of
November, by which time neither Stephenson nor
Wood had heard of Sir Humphry Davy's experiments,
nor of the lamp which that gentleman proposed to
construct.

An angry controversy afterwards took place as to
the respective merits of George Stephenson and Sir
Humphry Davy in respect of the invention of the safety-
lamp. A committee was formed on both sides, and
the facts were stated in various ways. It is perfectly
clear, however, that Stephenson had ascertained *the fact*
that flame will not pass through tubes of a certain diameter—the principle on which the safety-lamp is constructed—before Sir Humphry Davy had formed any
definite idea on the subject, or invented the model lamp
afterwards exhibited by him before the Royal Society.
Mr. Stephenson had actually constructed a lamp on
such a principle, and proved its safety, before Sir
Humphry had communicated his views to any indi-

vidual on the subject; and by the time that the first public intimation had been given of his discovery, Stephenson's second lamp had been constructed and tested in like manner in the Killingworth pit. The *first* was tried on the 21st of October, 1815; the *second* was tried on the 4th of November; but it was not until the 9th of November that Sir Humphry Davy presented his first lamp to the public. And by the 30th of the same month, as we have seen, Stephenson had constructed and tested his *third* safety-lamp.

Stephenson's theory of the "burnt air" and the "draught" was no doubt wrong; but his lamp was right, and that was the great fact which mainly concerned him. Torricelli did not know the rationale of his tube, nor Otto Gürike that of his air-pump; yet no one thinks of denying them the merit of their inventions on that account. The discoveries of Volta and Galvani were in like manner independent of theory; the greatest discoveries consisting in bringing to light certain grand facts, on which theories are afterwards framed. Our inventor had been pursuing the Baconian method, though he did not think of that, but of inventing a safe lamp, which he knew could only be done through the process of repeated experiment. He experimented upon the fire-damp at the blowers in the mine, and also by means of the apparatus which was blown up in his cottage, as above described by himself. By experiment he distinctly ascertained that the explosion of fire-damp could not pass through small tubes; and he also did what had not before been done by any inventor—he constructed a lamp on this principle, and repeatedly proved its safety at the risk of his life. At the same time, there is no doubt that it was to Sir Humphry Davy that the merit belonged of having pointed out the true law on which the safety-lamp is constructed.

The subject of this important invention excited so much interest in the northern mining districts, and

Mr. Stephenson's numerous friends considered his lamp
so completely successful—having stood the test of repeated experiments—that they urged him to bring his
invention before the Philosophical and Literary Society
of Newcastle, of some of whose apparatus he had

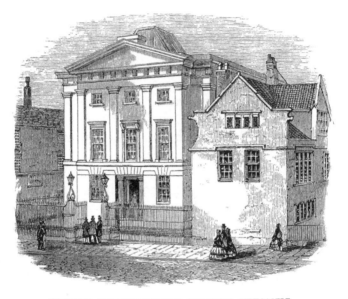

LITERARY AND PHILOSOPHICAL INSTITUTE, NEWCASTLE.

availed himself in the course of his experiments on fire-
damp.   After much persuasion he consented to do so,
and a meeting was appointed for the purpose of receiving his explanations, on the evening of the 5th of
December, 1815.   Mr. Stephenson was at that time so
diffident in manner and unpractised in speech, that he
took with him his friend Mr. Nicholas Wood, to act as
his interpreter and expositor on the occasion.   From
eighty to a hundred of the most intelligent members of
the Society were present at the meeting, when Mr.
Wood stood forward to expound the principles on
which the lamp had been formed, and to describe the
details of its construction.   Several questions were put,

to which Mr. Wood proceeded to give replies to .the best of his knowledge. But Stephenson, who up to that time had stood behind Wood, screened from notice, observing that the explanations given were not quite correct, could no longer control his reserve, and, standing forward, he proceeded in his strong Northumbrian dialect, to describe the lamp, down to its minutest details. He then produced several bladders full of carburetted hydrogen, which he had collected from the blowers in the Killingworth mine, and proved the safety of his lamp by numerous experiments with the gas, repeated in various ways; his earnest and impressive manner exciting in the minds of his auditors the liveliest interest both in the inventor and his invention.

Shortly after, Sir H. Davy's model lamp was received and exhibited to the coal-miners at Newcastle, on which occasion the observation was made by several gentlemen, "Why, it is the same as Stephenson's!"

Notwithstanding Stephenson's claim to be regarded as the first inventor of the Tube Safety-lamp, his merits do not seem to have been generally recognised. Sir Humphry Davy carried off the *éclat* which attached to the discovery. What chance had the unknown workman of Killingworth with so distinguished a competitor? The one was as yet but a colliery engine-wright, scarce raised above the manual-labour class, without chemical knowledge or literary culture, pursuing his experiments in obscurity, with a view only to usefulness; the other was the scientific prodigy of his day, the pet of the Royal Society, the favourite of princes, the most brilliant of lecturers, and the most popular of philosophers.

No small indignation was expressed by the friends of Sir Humphry Davy at this "presumption" on Stephenson's part. The scientific class united to ignore him entirely in the matter. In 1831, Dr. Paris, in his 'Life of Sir Humphry Davy,' thus spoke of Stephenson, in connexion with his claims as the inventor of the safety-lamp :—" It will hereafter be scarcely believed that an invention so eminently scientific, and which could never have been derived but from the sterling treasury of science, should have been claimed on behalf of an engine-wright of Killingworth, of the name of Stephenson—a person not even possessing a knowledge of the elements of chemistry."

But Stephenson was really far above claiming for himself any invention not his own. He had already accomplished a far greater thing even than the making of a safety-lamp—he had constructed a successful locomotive, which was to be seen daily at work upon the Killingworth railway. By the improvements he had made in the engine, he might almost be said to have *invented* it ; but no one—not even the philosophers—detected the significance of that wonderful machine. It excited no scientific interest, called forth no leading articles in the newspapers or the reviews, and formed the subject of no eloquent lectures at the Royal Society ; for railways were still comparatively unknown, and the might which slumbered in the locomotive was scarcely even dreamt of. What railways were to become, rested in a great measure with that " engine-wright of Killingworth, of the name of Stephenson," though he was scarcely known as yet beyond the bounds of his own district.

As to the value of the invention of the safety-lamp, there could be no doubt ; and the colliery owners of Durham and Northumberland, to testify their sense of its importance, determined to present a testimonial to its inventor. The friends of Sir H. Davy met in August 1816 to take steps for raising a subscription for the purpose. The advertised object of the meeting was to present him with a reward for " the invention of *his* safety-lamp." To this no objection could be taken ; for though the principle on which the safety-lamps of

Stephenson and Davy were constructed was the same;
and although Stephenson's lamp was, unquestionably, the
first successful lamp that had been constructed on such
principle, and proved to be efficient,—yet Sir H. Davy
did invent a safety-lamp, no doubt quite independent of
all that Stephenson had done; and having directed his
careful attention to the subject, and elucidated the true
theory of explosion of carburetted hydrogen, he was
entitled to all praise and reward for his labours.   But
when the meeting of coal-owners proposed to raise a
subscription for the purpose of presenting Sir H. Davy
with a reward for " his invention of *the* safety-lamp,"
the case was entirely altered; and Mr. Stephenson's
friends then proceeded to assert his claims to be regarded
as its first inventor.

Many meetings took place on the subject, and much
discussion ensued, the result of which was that a sum of
2000*l.* was presented to Sir Humphry Davy as " the inventor of the safety-lamp;" but, at the same time, a
purse of 100 guineas was voted to George Stephenson,
in consideration of what he had done in the same direction.   This result was, however, very unsatisfactory to
Stephenson, as well as to his friends; and Mr. Brandling, of Gosforth, suggested to him that, the subject
being now fairly before the public, he should publish a
statement of the facts on which his claim was founded.

This was not at all in George Stephenson's line.   He
had never appeared in print before; and it seemed to
him to be a more formidable thing to write a letter for
publication in " the papers" than even to invent a
safety-lamp or design a locomotive.   However, he called
to his aid his son Robert, set him down before a sheet of
foolscap, and when all was ready, told him to " put down
there just what I tell you."   The composition of this
letter, as we were informed by the writer of it, occupied
more evenings than one; and when it was at length
finished, after many corrections, and fairly copied out,

the father and son set out—the latter dressed in his Sunday's round jacket—to lay the joint production before Mr. Brandling, at Gosforth House. Glancing over the letter, Mr. Brandling said, " George, this will not do." " It is all true, sir," was the reply. " That may be ; but it is badly written." Robert blushed, for he thought it was the penmanship that was called in question, and he had written his very best. Mr. Brandling then requested his visitors to sit down while he put the letter in a more polished form, which he did, and it was shortly after published in the local journals.

As the controversy continued for some time longer to be carried on in the Newcastle papers, Mr. Stephenson, in the year 1817, consented to publish the detailed plans, with descriptions, of the several safety-lamps which he had contrived for use in the Killingworth colliery. The whole forms a pamphlet of only sixteen pages of letterpress.[1]

His friends, being fully satisfied of his claims to priority as the inventor of the safety-lamp used in the Killingworth and other collieries, proceeded to hold a public meeting for the purpose of presenting him with a reward " for the valuable service he had thus rendered to mankind." Charles J. Brandling, Esq., occupied the chair ; and a series of resolutions were passed, of which the first and most important was as follows :—" That it is the opinion of this meeting that Mr. George Stephenson, having *discovered the fact* that explosion of hydrogen gas will not pass through tubes and apertures of small dimensions, and having been *the first to apply that principle in the construction of a safety-lamp*, is entitled to a public reward."

A subscription was immediately commenced with this object, and a committee was formed, consisting of the Earl of Strathmore, C. J. Brandling, and others. The

---

[1] ' A Description of the Safety-Lamp, invented by George Stephenson, and now in use in the Killingworth Colliery.' London, 1817.

subscription list was headed by Lord Ravensworth, one of the partners in the Killingworth colliery, who showed his appreciation of the merits of Stephenson by giving 100 guineas.   C. J. Brandling and partners gave a like sum, and Matthew Bell and partners, and John Brandling and partners, gave fifty guineas each.

When the resolutions appeared in the newspapers, the scientific friends of Sir Humphry Davy in London met, and passed a series of counter-resolutions, which they published, declaring their opinion that Mr. Stephenson was *not* the author of the discovery of the fact that explosion of hydrogen will not pass through tubes and apertures of small dimensions, and that he was *not* the first to apply that principle to the construction of a safety-lamp.   To these counter-resolutions were attached the well-known names of Sir Joseph Banks, P.R.S., William Thomas Brande, Charles Hatchett, W. H. Wollaston, and Thomas Young.

Mr. Stephenson's friends then, to make assurance doubly sure, and with a view to set the question at rest, determined to take evidence in detail as to the date of discovery by George Stephenson of the fact in question, and its practical application by him in the formation and actual trial of his safety-lamp.   The witnesses examined were, George Stephenson himself, Mr. Nicholas Wood, and John Moodie, who had been present at the first trial of the lamp; the several tinmen who made the lamps; the secretary and other members of the Literary and Philosophical Society of Newcastle, who were present at the exhibition of the third lamp; and some of the workmen at Killingworth colliery, who had been witnesses of Mr. Stephenson's experiments on fire-damp, made with the lamps at various periods, before Sir Humphry Davy's investigations had been heard of.   This evidence was quite conclusive to the minds of the gentlemen who investigated the subject, and they published it in 1817, together with their Report, in which they declared that, "after a careful inquiry into the merits of the case, conducted, as they trust, in a spirit of fairness and moderation, they can perceive no satisfactory reason for changing their opinion."

The Stephenson subscription, when collected, amounted to 1000*l.* Part of the money was devoted to the purchase of a silver tankard, which was presented to the inventor, together with the balance of the subscription, at a public dinner given in the Assembly Rooms at Newcastle.[1] But what gave Stephenson even greater pleasure than the silver tankard and purse of sovereigns was the gift of a silver watch, purchased by small subscriptions collected amongst the colliers themselves, and presented to him by them as a token of their esteem and regard for him as a man, as well as of their gratitude for the perseverance and skill with which he had prosecuted his valuable and life-saving invention to a successful issue. To the last day of his life he spoke with pride of this gift as amongst the most valuable which he had ever received.

However great the merits of Mr. Stephenson in connection with the invention of the tube safety-lamp, they cannot be regarded as detracting from the reputation of Sir Humphry Davy. His inquiries into the explosive properties of carburetted hydrogen gas were quite original; and his discovery of the fact that explosion will not pass through tubes of a certain diameter was made independently of all that Stephenson had done in verification of the same fact. It even appears that Mr. Smithson Tennant and Dr. Wollaston had observed the

---

[1] The tankard bore the following inscription:—" This piece of plate, purchased with a part of the sum of 1000*l.*, a subscription raised for the remuneration of Mr. GEORGE STEPHENSON for having discovered the fact that inflamed fire-damp will not pass through tubes and apertures of small dimensions, and having been *the* *first* to apply that principle in the construction of a safety-lamp calculated for the preservation of human life in situations formerly of the greatest danger, was presented to him at a general meeting of the subscribers, Charles John Brandling, Esq., in the Chair. January 12th, 1818."

same fact several years before, though neither Stephenson nor Davy knew it while they were prosecuting their experiments. Sir Humphry Davy's subsequent modification of the tube-lamp, by which, while diminishing the diameter, he in the same ratio shortened the tubes without danger, and in the form of wire-gauze enveloped the safety-lamp by a multiplicity of tubes, was a beautiful application of the true theory which he had formed upon the subject.

The increased number of accidents which have occurred from explosions in coal mines since the general introduction of the Davy lamp, have led to considerable doubts as to its safety, and to inquiries as to the means by which it may be further improved; for experience has shown that, under certain circumstances, the Davy lamp is *not* safe. Mr. Stephenson was of opinion that the modification of his own and Sir Humphry Davy's lamp, combining the glass cylinder with the wire-gauze, was the most secure; at the same time it must be admitted that the Davy and the Geordy lamps alike failed to stand the severe tests to which they were submitted by Dr. Pereira, before the Committee on Accidents in Mines. Indeed, Dr. Pereira did not hesitate to say, that when exposed to a current of explosive gas the Davy lamp is " decidedly unsafe," and that the experiments by which its safety had been " demonstrated " in the lecture-room had proved entirely " fallacious."

It is worthy of remark, that under circumstances in which the wire-gauze of the Davy lamp becomes red-hot from the high explosiveness of the gas, the Geordy lamp is extinguished; and we cannot but think that this fact testifies to the decidedly superior safety of the Geordy. An accident occurred in the Oaks Colliery Pit at Barnsley, on the 20th of August, 1857, which strikingly exemplified the respective qualities of the lamps. A sudden outburst of gas took place from the floor of the mine, along a distance of fifty yards. For-

tunately the men working in the pit at the time
were all supplied with safety-lamps—the hewers with
Stephenson's, and the hurriers with Davy's. Upon this
occasion, the whole of the Stephenson's lamps, over a
space of five hundred yards, were extinguished almost
instantaneously; whereas the Davy lamps were filled
with fire, and became red-hot—so much so, that several
of the men using them had their hands burned by the
gauze. Had a strong current of air been blowing
through the gallery at the time, an explosion would most
probably have taken place—an accident which, it will
be observed, could not, under such circumstances, occur
from the use of the Geordy, which is immediately extinguished as soon as the air becomes explosive.[1]

---

[1] The accident above referred to was described in the 'Barnsley Times,' a copy of which, containing the account, Robert Stephenson forwarded to the author, with the observation that "it is evidently written by a practical miner, and is, I think, worthy of record in my father's Life." The superiority of the Stephenson lamp has since formed the subject of a lengthy communication which appeared in the 'Times' of December 24th, 1860, signed John Brown, C.E., of Barnsley, an able mining engineer, in reply to a previous communication urging the sufficiency of ventilation for keeping mines clear of explosive gas.

"I am well acquainted with collieries," says Mr. Brown, "that are liable to yield, without a moment's warning, such large quantities of explosive gas that I am quite sure no amount of ventilation that can practically, and not upon paper, be passed through the workings, would dilute the enormous quantities of this suddenly issuing gas sufficiently to prevent it igniting at the first naked light with which it came in contact.

"I have known in this district gas to issue from beneath the coal with such violence as to rip up the floor, which was almost as hard as stone, producing great fissures several feet in depth and many yards in length, the gas issuing therefrom with a noise like that produced by high-pressure steam escaping from a safety-valve.

"At the period of this occurrence we had two kinds of safety-lamps in use in this pit—viz., 'Davy' and 'Stephenson,' and the gas in going off to the upcast shaft had to pass great numbers of men, who were at work with both kinds of lamps. The whole of the 'Davy's' became red-hot almost instantaneously from the rapid ignition of the gas within the gauze; the 'Stephenson's' were as instantly self-extinguished from the same cause, it being the prominent qualification of these lamps that, in addition to affording a somewhat better light than the 'Davy' lamp, they are suddenly extinguished when placed within a highly explosive atmosphere, so that no person can remain working and run the risk of his lamp becoming red-hot—which, under such circumstances, would be the result with the 'Davy' lamp.

"The red-hot lamps were, most fortunately, all safely put out, although the men in many cases had their hands severely burned by the gauze; but from that time I fully resolved to adopt the exclusive use of the 'Stephenson' lamps, and not expose men to the fearful risk they must

Nicholas Wood, a good judge, has said of the two inventions, " Priority has been claimed for each of them —I believe the inventions to be parallel. By different roads they both arrived at the same result. Stephenson's is the superior lamp. Davy's is safe—Stephenson's is safer."

When the question of priority was under discussion at Mr. Lough's studio, in 1857, Sir Matthew White Ridley asked Robert Stephenson, who was present, for his opinion on the subject. His answer was, " I am not exactly the person to give an unbiassed opinion; but, as you ask me frankly, I will as frankly say, that if George Stephenson had never lived, Sir Humphry Davy could and most probably would have invented the safety-lamp; but again, if Sir Humphry Davy had never lived, George Stephenson certainly would have invented the safety-lamp, as I believe he did, independent of all that Sir Humphry Davy had ever done in the matter."

To this day, the Geordy lamp continues in regular use in the Killingworth Collieries; and the Killingworth pitmen have expressed to the writer their decided preference for it compared with the Davy. It is certainly a ·strong testimony in its favour, that no accident is known to have arisen from its use, since it was generally introduced into the Killingworth pits.

---

run from working with ' Davy ' lamps during the probable recurrence of a similar event.

"I may remark that the ' Stephenson' lamp was originally invented by the great George Stephenson, and in its present shape combines the merits of his discovery with that of Sir Humphry Davy—constituting, to my mind, the safest lamp at present known, and I speak from the long use of many hundreds daily in various collieries."

CHAPTER VIII.

George Stephenson's further Improvements in the Locomotive — The Hetton Railway — Robert Stephenson
as Viewer's Apprentice and Student.

Mr. Stephenson's experiments on fire-damp, and his labours in connexion with the invention of the safety-lamp, occupied but a small portion of his time, which was necessarily devoted for the most part to the ordinary business of the colliery. From the day of his appointment as engine-wright, one of the subjects which particularly occupied his attention was the best practical method of winning and raising the coal. His friend, Nicholas Wood, has said of him that he was one of the first to introduce steam machinery underground with the latter object. Indeed, the Killingworth mines came to be regarded as the models of the district; and when Mr. Robert Bald, the celebrated Scotch mining engineer, was requested by Dr. (afterwards Sir David) Brewster, to prepare the article 'Mine' for the 'Edinburgh Encyclopædia,' he proceeded to Killingworth principally for the purpose of examining Stephenson's underground machinery. Mr. Bald has favoured us with an account of his visit made with this object in 1818, and he states that he was much struck with the novelty, as well as the remarkable efficiency of Stephenson's arrangements, especially in regard to what is called the underdip working. " I found," he says, " that a mine had been commenced near the main pit bottom, and carried forward down the dip or slope of the coal, the rate of dip being about one in twelve; and the coals were drawn from the dip to the pit-bottom by the steam machinery in a very rapid manner. The water which oozed from the upper winning was disposed of at the pit-bottom in a barrel or trunk, and was drawn up by the power of the engine which worked the other machinery. The dip at the time of my visit was nearly a mile in length, but has since been greatly extended. As I was considerably tired by my wanderings in the galleries, when I arrived at the forehead of the dip, Mr. Stephenson said to me, ' You may very speedily be carried up to the rise, by laying yourself flat upon the coal-baskets,' which were laden and ready to be taken up the incline. This I at once did, and was straightway wafted on the wings of fire to the bottom of the pit, from whence I was borne swiftly up to the light by the steam machinery on the pit-head." The whole of the working arrangements seemed to Mr. Bald to be conducted in the most skilful and efficient manner, and reflected the highest credit on the colliery engineer.

Besides attending to the underground arrangements, the improved transit of the coals aboveground from the pit-head to the shipping-place, demanded an increasing share of his attention. Every day's experience convinced him that the locomotive constructed by him after his patent of the year 1815, was far from perfect; though he continued to entertain confident hopes of its complete eventual success. He even went so far as to say that the locomotive would yet supersede every other traction-power for drawing heavy loads. Many still regarded his travelling engine as little better than a curious toy; and some, shaking their heads, predicted for it " a terrible blow-up some day " Nevertheless, it was daily performing its work with regularity, dragging the coal-waggons between the colliery and the staiths, and saving the labour of many men and horses. There was not, however, so marked a saving in the expense of haulage as to induce the northern colliery masters to adopt locomotive power generally as a substitute for horses. How it could be improved and rendered more efficient as well as economical, was never out of Stephenson's mind. He was fully conscious of the imperfections both in the road and the engine; and gave himself no rest until he had brought the efficiency of both up to a higher point. Thus he worked his way inch by inch, slowly but surely; and every step gained was made good as a basis for further improvements.

At an early period of his labours, or about the time when he had completed his second locomotive, he began to direct his particular attention to the state of the road; as he perceived that the extended use of the locomotive must necessarily depend in a great measure upon the perfection, solidity, continuity, and smoothness of the way along which the engine travelled. Even at that early period, he was in the habit of regarding the road and the locomotive as one machine, speaking of the rail and the wheel as "man and wife."

All railways were at that time laid in a careless and loose manner, and great inequalities of level were allowed to occur without much attention being paid to repairs. The consequence was a great loss of power, as well as much wear and tear of the machinery, by the frequent jolts and blows of the wheels against the rails. His first object therefore was, to remove the inequalities produced by the imperfect junction between rail and rail. At that time (1816) the rails were made of cast iron, each rail being about three feet long; and sufficient care was not taken to maintain the points of junction on the same level. The chairs, or cast-iron pedestals into which the rails were inserted, were flat at the bottom; so that, whenever any disturbance took place in the stone blocks or sleepers supporting them, the flat base of the chair upon which the rails rested being tilted by unequal subsidence, the end of one rail became depressed, whilst that of the other was elevated. Hence constant jolts and shocks, the reaction of which

K 2

very often caused the fracture of the rails, and occasionally threw the engine off the road.

To remedy this imperfection, Mr. Stephenson devised
a new chair, with an entirely new mode of fixing
the rails therein.   Instead of adopting the *butt-joint*
which had hitherto been used in all cast-iron rails, he
adopted the *half-lap joint*, by which means the rails
extended a certain distance over each other at the ends,
like a scarf-joint.   These ends, instead of resting upon
the flat chair, were
made to rest upon the
apex of a curve form-
ing the bottom of the
chair.   The supports
were also extended
from three feet to

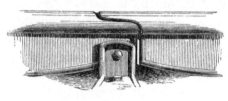

HALF-LAP JOINT.

three feet nine inches or four feet apart.   These rails were
accordingly substituted for the old cast-iron plates on the
Killingworth Colliery Railway, and they were found
to be a very great improvement upon the previous
system, adding both to the efficiency of the horse-power
(still used on the railway) and to the smooth action of
the locomotive engine, but more particularly increasing
the efficiency of the latter.

This improved form of the rail and chair was embodied in a patent taken out in the joint names of
Mr. Losh, of Newcastle, iron-founder, and of Mr.
Stephenson, bearing date the 30th of September, 1816.
Mr. Losh being a wealthy, enterprising iron-manufacturer, and having confidence in George Stephenson and
his improvements, found the money for the purpose of
taking out the patent, which, in those days, was a very
costly as well as troublesome affair.

The specification of the same patent also described
various important improvements in the locomotive itself.
The wheels of the engine were improved, being altered

from cast to malleable iron, in whole or in part, by which they were made lighter as well as more durable and safe. But the most ingenious and original contrivance embodied in this patent was the substitute for springs which Mr. Stephenson invented. He contrived that the steam generated in the boiler should perform this important office. The method by which this was effected displayed such genuine mechanical genius, that we would particularly call the reader's attention to the device, which was the more remarkable, as it was contrived long before the possibility of steam locomotion had become an object of parliamentary inquiry or even of public interest.

It has already been observed that up to, and indeed for some time after, the period of which we speak, there was no such class of skilled mechanics, nor were there any such machinery and tools in use, as are now at the disposal of all inventors and manufacturers. Although skilled workmen were in course of gradual training in a few of the larger manufacturing towns, they did not, at the date of Stephenson's patent, exist in any considerable numbers, nor was there then any class of mechanics capable of constructing springs of sufficient strength and elasticity to support locomotive engines of ten tons weight.

In order to avoid the dangers arising from the inequalities of the road, Mr. Stephenson so arranged the boiler of his new patent locomotive that it was supported upon the frame of the engine by four cylinders, which opened into the interior of the boiler. These cylinders were occupied by pistons with rods, which passed downwards and pressed upon the upper side of the axles. The cylinders opening into the interior of the boiler, allowed the pressure of steam to be applied to the upper side of the piston; and the pressure being nearly equivalent to one-fourth of the weight of the engine, each axle, whatever might be its position, had at all times nearly the same amount of weight to bear, and consequently the entire weight was pretty equally distributed amongst the four wheels of the locomotive. Thus the four floating pistons were ingeniously made to serve the purpose of springs in equalising the weight, and in softening the jerks of the machine; the weight of which, it must also be observed, had been increased, on a road originally calculated to bear a considerably lighter description of carriage. This mode of supporting the engine remained in use until the progress of spring-making had so far advanced that steel springs could be manufactured of sufficient strength to bear the weight of locomotive engines.

The result of the actual working of the new locomotive on the improved road amply justified the promises held forth in the specification. The traffic was conducted with greater regularity and economy, and the superiority of the engine, as compared with horse traction, became still more marked. And it is a fact worthy of notice, that the identical engines constructed by Mr. Stephenson in 1816 are to this day to be seen in regular useful work upon the Killingworth Railway, conveying heavy coal-trains at the speed of between five and six miles an hour, probably as economically as any of the more perfect locomotives now in use.

Mr. Stephenson's endeavours having been attended with such marked success in the adaptation of locomotive power to railways, his attention was called by many of his friends, about the year 1818, to the application of steam to travelling on common roads. It was from this point, indeed, that the locomotive had first started, Trevithick's first engine having been constructed with this special object. Stephenson's friends having observed how far behind he had left the original projector of the locomotive in its application to railroads, perhaps naturally inferred that he would be equally successful in applying it to the purpose for which Trevithick and

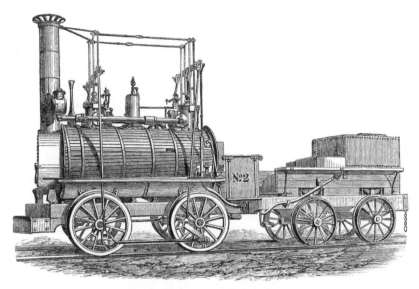

OLD KILLINGWORTH LOCOMOTIVE, STILL IN USE.

Vivian had intended their first engine. But the accuracy
with which he estimated the resistance to which loads were
exposed on railways, arising from friction and gravity,
led him at a very early stage to reject the idea of ever
applying steam power economically to common road
travelling.   In October, 1818, he made a series of
careful experiments in conjunction with Mr. Nicholas
Wood, on the resistance to which carriages were exposed on railways, testing the results by means of a
dynamometer of his own construction.   The series of
practical observations made by means of this instrument were interesting, as the first systematic attempt
to determine the precise amount of resistance to carriages moving along railways.   It was then for the
first time ascertained by experiment that the friction
was a constant quantity at all velocities.   Although
this theory had long before been developed by Vince
and Coulomb, and was well known to scientific men as an
established truth, yet at the time when Mr. Stephenson
made his experiments, the deductions of philosophers

on the subject were neither believed in nor acted upon by practical engineers.

He ascertained that the resistances to traction were mainly three; the first being upon the axles of the carriages, the second, or rolling resistance, being between the circumference of the wheel and the surface of the rail, and the third being the resistance of gravity. The amount of friction and gravity he could accurately ascertain; but the rolling resistance was a matter of greater difficulty, being subject to much variation. But he satisfied himself that it was so great when the surface presented to the wheel was of a rough character, that the idea of working steam carriages economically on common roads was dismissed by him as entirely out of the question. Taking it as 10 lbs. to a ton weight on a level railway, it became obvious to him that so small a rise as 1 in 100 would diminish the useful effort of a locomotive by upwards of 50 per cent. This was demonstrated by repeated experiments, and the important fact, thus rooted in his mind, was never lost sight of in the course of his future railway career.

It was owing in a great measure to these painstaking experiments that he early became convinced of the vital importance, in an economical point of view, of reducing the country through which a railway was intended to pass as nearly as possible to a level. Where, as in the first coal railways of Northumberland and Durham, the load was nearly all one way,—that is, from the colliery to the shipping-place,—it was an advantage to have an inclination in that direction. The strain on the powers of the locomotive was thus diminished, and it was an easy matter for it to haul the empty waggons back to the colliery up even a pretty steep incline. But when the loads were both ways, it was obvious to him that the railroad must be constructed as nearly as possible on a level.[1]

---

[1] This subject will be found further discussed in Robert Stephenson's 'Memoir on the Invention of the Railway Locomotive,' appended to this volume.

These views, thus early entertained, originated in Mr. Stephenson's mind the peculiar character of railroad works as distinguished from all other roads; for, in railroads, he early contended that large sums would be wisely expended in perforating barriers of hills with long tunnels, and in raising the lower levels with the excess cut down from the adjacent high ground. In proportion as these views forced themselves upon his mind and were corroborated by his daily experience, he became more and more convinced of the hopelessness of applying steam locomotion to common roads; for every argument in favour of a level railway was, in his view, an argument against the rough and hilly course of a common road.

At this day it is difficult to understand how the sagacious and strong common-sense views of Stephenson on this subject failed to force themselves sooner upon the minds of those who were persisting in their vain though ingenious attempts to apply locomotive power to ordinary roads. For a long time they continued to hold with obstinate perseverance to the belief that for steam purposes a soft road was better than a hard one— a road easily crushed better than one incapable of being crushed; and they held to this after it had been demonstrated in all parts of the mining districts, that iron tramways were better than paved roads. But the fallacy that iron was incapable of adhesion upon iron continued to prevail, and the projectors of steam-travelling on common roads only shared in the common belief. They still considered that roughness of surface was essential to produce " bite," especially in surmounting acclivities; the truth being, that they confounded roughness of surface with tenacity of surface and contact of parts; not perceiving that a yielding surface which would adapt itself to the tread of the wheel, could never become an unyielding surface to form a fulcrum for its progression.

Although Stephenson's locomotive engines were in daily use for many years on the Killingworth Railway, they excited comparatively little interest. They were no longer experimental, but had become an established tractive power. The experience of years had proved that they worked more steadily, drew heavier loads, and were, on the whole, considerably more economical than horses. Nevertheless eight years passed before another locomotive railway was constructed and opened for the purposes of coal or other traffic.

It is difficult to account for this early indifference on the part of the public to the merits of the greatest mechanical invention of the age. Steam carriages were exciting much interest, and numerous and repeated experiments were made with them. The improvements effected by M'Adam in the mode of constructing turnpike-roads were the subject of frequent discussions in the legislature, on the grants of public money being proposed, which were from time to time made to him. Yet here at Killingworth, without the aid of a farthing of government money, a system of road locomotion had been in existence since 1814, which was destined, before many years, to revolutionise the internal communications of England and of the world, but of which the English public and the English government as yet knew nothing.

Mr. Stephenson had no means of bringing his important invention prominently under the notice of the public. He himself knew well its importance, and he already anticipated its eventual general adoption; but being an unlettered man, he could not give utterance to the thoughts which brooded within him on the subject. Killingworth Colliery lay far from London, the centre of scientific life in England. It was visited by no savans nor literary men, who might have succeeded in introducing to notice the wonderful machine of Stephenson. Even the local chroniclers seem to have taken no notice of the Killingworth Railway. The "Puffing Billy" was doing its daily quota of hard work, and had long ceased to be a curiosity in the neighbourhood. Blenkinsop's clumsier and less successful engine—which has long since been disused, while Stephenson's Killingworth engines continue working to this day—excited far more interest; partly, perhaps, because it was close to the large town of Leeds, and used to be visited by strangers as one of the few objects of interest in that place. Blenkinsop was also an educated man, and was in communication with some of the most distinguished personages of his day upon the subject of his locomotive, which thus obtained considerable notoriety.

The first engine constructed by Mr. Stephenson to order, after the Killingworth model, was made for the Duke of Portland in 1817, for use upon his tramroad, about ten miles long, extending from Kilmarnock to Troon, in Ayrshire. It was employed to haul the coals from the Duke's collieries along the line to Troon harbour. Its use was however discontinued in consequence of the frequent breakages of the cast-iron rails, by which the working of the line was interrupted, and accordingly horses were again employed as before.[1] There seemed, indeed, to be so small a prospect of introducing the locomotive into general use, that Mr. Stephenson,— perhaps feeling the capabilities within him,—again recurred to his old idea of emigrating to the United States. Before entering as sleeping partner in a small foundry at Forth Banks, Newcastle, managed by Mr. John Burrell, he had thrown out the suggestion to the latter that it would be a good speculation for them to emigrate to North America, and introduce steamboats upon the great inland lakes there. The first steamers were then

---

[1] The iron wheels of this engine were afterwards removed, and replaced with wooden wheels, when it was again placed upon the road, and continued working until the year 1848. Its original cost was 750*l.* It was broken up, and the materials were sold, realizing only 13*l.*

plying upon the Tyne before his eyes; and he saw in
them the germ of a great revolution in navigation. It
occurred to him that North America presented the finest
field for trying their wonderful powers. He was an
engineer, and Mr. Burrell was an iron-founder; and
between them, he thought they might strike out a path
to fortune in the mighty West. Fortunately, this idea
remained a mere speculation so far as Mr. Stephenson
was concerned; and it was left to others to do what he
had dreamt of achieving. After all his patient waiting,
his skill, industry, and perseverance were at length
about to bear fruit.

In 1819, the owners of the Hetton Colliery, in the
county of Durham, determined to have their waggon-
way altered to a locomotive railroad. The result of the
working of the Killingworth Railway had been so satisfactory, that they resolved to adopt the same system.
One reason why an experiment so long continued and so
successful as that at Killingworth should have been so
slow in producing results, perhaps was, that to lay down
a railway and furnish it with locomotives, or fixed
engines where necessary, required a very large capital,
beyond the means of ordinary coal-owners; whilst the
small amount of interest felt in railways by the general
public, and the supposed impracticability of working
them to a profit, as yet prevented the ordinary capitalists
from venturing their money in the promotion of such
undertakings. The Hetton Coal Company were, however, possessed of adequate means; and the local reputation of the Killingworth engine-wright pointed him
out as the man best calculated to lay out their line, and
superintend their works. They accordingly invited him
to act as the engineer of the proposed railway. Being
in the service of the Killingworth Company, Mr. Stephenson felt it necessary to obtain their permission to
enter upon this new work. This was at once granted.
The best feeling existed between him and his employers;

and they regarded it as a compliment that their colliery
engineer should be selected for a work so important as
the laying down of the Hetton Railway, which was to
be the longest locomotive line that had, up to that time,
been constructed in the neighbourhood. Mr. Stephenson
accepted the appointment, his brother Robert acting as
resident engineer and personally superintending the
execution of the works.

The Hetton Railway extended from the Hetton Colliery, situated about two miles south of Houghton-le-
Spring, in the county of Durham, to the shipping-places
on the banks of the Wear, near Sunderland. Its length
was about eight miles ; and in its course it crossed
Warden Law, one of the highest hills in the district.
The character of the country forbade the construction of
a flat line, or one of comparatively easy gradients,
except by the expenditure of a much larger capital than
was placed at Mr. Stephenson's command. Heavy
works could not be executed ; it was, therefore, necessary to form the line with but little deviation from the
natural conformation of the district which it traversed,
and also to adapt the mechanical methods employed for
its working to the character of the gradients, which in
some places were necessarily heavy.

Although Mr. Stephenson had, with every step made
towards its increased utility, become more and more
identified with the success of the locomotive engine, he
did not allow his enthusiasm to carry him away into
costly mistakes. He carefully drew the line between
the cases in which the locomotive could be usefully employed, and those in which stationary engines were
calculated to be more economical. This led him, as in
the instance of the Hetton Railway, to execute lines
through and over rough countries, where gradients
within the powers of the locomotive engine of that day
could not be secured, employing in their stead stationary
engines where locomotives were not practicable. In the

present case, this course was adopted by him most successfully.   On the original Hetton line, there were five
self-acting inclines,—the full waggons drawing the empty
ones up,—and two inclines worked by fixed reciprocating
engines of sixty-horse power each.   The locomotive
travelling engine, or " the iron horse " as the people of
the neighbourhood then styled it, did the rest.   On the
day of the opening of the Hetton Railway, the 18th of
November, 1822, crowds of spectators assembled from
all parts to witness the first operations of this ingenious
and powerful machinery, which was entirely successful.
On that day five of Stephenson's locomotives were at
work upon the railway, under the direction of his brother
Robert; and the first shipment of coal was then made
by the Hetton Company, at their new staiths on the
Wear.   The speed at which the locomotives travelled
was about four miles an hour, and each engine dragged
after it a train of seventeen waggons, weighing about
sixty-four tons.

While thus advancing step by step,—attending to the
business of the Killingworth Colliery, and laying out
railways in the neighbourhood,—he was carefully watching over the education of his son.   We have already
seen that Robert was sent to school at Newcastle, and
that he left it about the year 1818.   He was then put
apprentice to Mr. Nicholas Wood, the head viewer at
Killingworth, to learn the business of the colliery; and
he served in that capacity for about three years, during
which time he became familiar with most departments
of underground work.   The occupation was not unattended with peril, as the following incident will show.
Though the use of the Geordy lamp had become general in the Killingworth pits, and the workmen were
bound, under a penalty of half-a-crown, not to use a
naked candle, yet it was difficult to enforce the rule,
and even the masters themselves occasionally broke it.
One day, Nicholas Wood the head viewer, Moodie the

under viewer, and Robert Stephenson, were proceeding along one of the galleries, Wood with a naked candle in his hand, and Robert following him with a lamp.   They came to a place where a fall of stones from the roof had taken place, on which Wood, who was first, proceeded to clamber over the stones, holding high the naked candle.   He had nearly reached the summit of the heap, when the fire-damp, which had accumulated in the hollow of the roof, exploded, and instantly the whole party were blown down, and the lights extinguished.   They were a mile from the shaft, and quite in the dark.   There was a rush of the workpeople from all quarters towards the shaft, for it was feared that the fire might extend to more dangerous parts of the pit, where if the gas had exploded, every soul in the mine must inevitably have perished. Robert Stephenson and Moodie, on the first impulse, ran back at full speed along the dark gallery leading to the shaft, coming into collision, on their way, with the hind quarters of a horse stunned by the explosion.   When they had gone half-way, Moodie halted, and bethought him of Nicholas Wood.   "Stop, laddie!" said he to Robert, "stop; we maun gang back, and seek the maister."   So they retraced their steps.   Happily, no further explosion had taken place.   They found the master lying on the heap of stones, stunned and bruised, with his hands severely burnt.   They then led him to the bottom of the shaft; and he afterwards took care not to venture into the dangerous parts of the mine without the protection of a Geordy lamp.

The time that Robert spent at Killingworth as viewer's apprentice was of advantage both to his father and himself.   The evenings were generally devoted to reading and study, the two from this time working together as friends and co-labourers.   One who used to drop in at the cottage of an evening, well remembers the animated and eager discussions which on some occasions took place, more especially with reference to the growing powers of the locomotive engine. The son was even more enthusiastic than the father on this subject. Robert would suggest alterations and improvements in this, that, and the other detail of the machine. His father, on the contrary, would offer every possible objection, defending the existing arrangements,—proud, nevertheless, of his son's suggestions, and often warmed and excited by his brilliant anticipations of the ultimate triumph of the locomotive.

These discussions probably had considerable influence in inducing Mr. Stephenson to take the next important step in the education of his son. Although Robert, who was only nineteen years of age, was doing well, and was certain at the expiration of his apprenticeship to rise to a higher position, his father was not satisfied with the amount of instruction which he had as yet given him. Remembering the disadvantages under which he had himself laboured in consequence of his ignorance of practical chemistry during his investigations connected with the safety-lamp, more especially with reference to the properties of gas, as well as in the course of his experiments with the object of improving the locomotive engine, he determined to furnish his son with as complete a scientific culture as his means would afford. He was also of opinion that a proper training in technical science was almost indispensable to success in the higher walks of the engineer's profession; and he determined to give to his son that kind and degree of education which he so much desired for himself. He would thus, he knew, secure a hearty and generous co-worker in the elaboration of the great ideas now looming before him, and with their united practical and scientific knowledge he probably felt that they would be equal to any enterprise.

He accordingly took Robert from his labours as underviewer in the West Moor Pit, and, in the year 1820,

sent him to the Edinburgh University, there being then
no college in England accessible to persons of moderate
means, for purposes of scientific culture. Robert was
furnished with letters of introduction to several men of
scientific eminence in Edinburgh; his father's reputation
in connexion with the safety-lamp being of service to
him in this respect. He lodged in Drummond Street,
in the immediate vicinity of the college, and attended
the Chemical Lectures of Dr. Hope, the Natural Philosophy Lectures of Sir John Leslie, and the Natural
History Class of Professor Jameson. He also devoted
several evenings in each week to the study of practical
Chemistry under Dr. John Murray, himself one of the
numerous designers of a safety-lamp. The young student
entered upon his studies with so keen a zest and interest,
his mind was so ripe for the pursuit and reception of
knowledge, and he prosecuted his labours with such
laborious zeal, that it is not too much to say that in the
six months' study to which his college career was
limited, he acquired more real knowledge than the
average of students do during their entire course. He
took careful notes of all the lectures, which he copied
out at night before he went to bed; so that, when he
returned to Killingworth, he might read them over to
his father. He afterwards had the notes bound up, and
placed in his library. Long years after, when conversing
with Thomas Harrison, C.E., at his house in Gloucester
Square, Mr. Stephenson rose from his seat and took
down a volume from the shelves. Mr. Harrison observed that the book was in MS., neatly written out.
" What have we here? " he asked. The answer was—
" When I went to college, I knew the difficulty my father
had in collecting the funds to send me there. Before
going I studied short-hand; while at Edinburgh, I took
down verbatim every lecture; and in the evenings, before
I went to bed, I transcribed those lectures word for
word. You see the result in that range of books."

Robert was not without the pleasures of social intercourse either, during his stay at Edinburgh. Among
the letters of introduction which he took with him was
one to Robert Bald. the mining engineer, which proved
of much service to him. "I remember Mr. Bald very
well," he said on one occasion, when recounting his
reminiscences of his Edinburgh college life. "He
introduced me to Dr. Hope, Dr. Murray, and several
of the distinguished men of the north. Bald was the
Buddle of Scotland. He knew my father from having
visited the pits at Killingworth, with the object of
describing the system of working them, in his article
intended for the ' Edinburgh Encyclopædia.' A strange
adventure befel that article before it appeared in print.
Bald was living at Alloa when he wrote it; and when
finished he sent it to Edinburgh by the hands of young
Maxton, one of his nephews, whom he enjoined to take
special care of it and deliver it safely into the hands of
the editor. He took passage for Newhaven by one of
the little steamers which then plied upon the Forth; but
on the voyage down the Frith, she struck upon a rock
nearly opposite Queensferry, and soon sunk. When the
accident happened, Maxton's whole concern was about
his uncle's article. He durst not return to Alloa if he
lost it, and he must not go on to Edinburgh without it.
So he desperately clung to the chimney chains, with the
paper parcel under his arm, while most of the other passengers were washed away and drowned. And there he
continued to cling, until rescued by some boatmen, parcel
and all; after which he made his way to Edinburgh,
and the article duly appeared."

Returning to the subject of his life in Edinburgh,
Robert continued : "Besides taking me with him to the
meetings of the Royal and other Societies, Mr. Bald introduced me to a very agreeable family, relatives of his own,
at whose house I spent many pleasant evenings. It was
there I met Jeannie M——. She was a bonnie lass,

and I, being young and susceptible, fairly fell in love with
her. But, like most very early attachments, mine proved
evanescent. Years passed, and I had all but forgotten
Jeannie, when one day I received a letter from her,
from which it appeared that she was in great distress
through the ruin of her relatives. I sent her a sum of
money, and continued to do so for several years; but
the last remittance not being acknowledged, I directed
Sanderson, my solicitor, to make inquiries. I afterwards
found that the money had reached her at Portobello just
as she was dying, and so, poor thing! she had been
unable to acknowledge it."

One of the practical sciences in the study of which
Robert Stephenson took special interest while at Edinburgh was that of geology. The situation of the city,
in the midst of a district of highly interesting geological
formation, easily accessible to pedestrians, is indeed most
favourable to the pursuit of such a study; and it was
the practice of Professor Jameson frequently to head a
band of his pupils, armed with hammers, chisels, and
clinometers, and take them with him on a long ramble
into the country, for the purpose of teaching them habits
of observation and reading to them from the open book
of Nature itself. The professor was habitually grave
and taciturn, but on such occasions he would relax and
even become genial. For his own special science he had
an almost engrossing enthusiasm, which on such occasions he did not fail to inspire into his pupils; who thus
not only got their knowledge in the pleasantest possible
way, but also fresh air and exercise in the midst of
glorious scenery and in joyous company. At the close
of this session, the professor took with him a select
body of his pupils on an excursion along the Great
Glen of the Highlands, in the line of the Caledonian
Canal, and Robert formed one of the party. They
passed under the shadow of Ben Nevis, examined the
famous old sea-margins known as the " parallel roads of

Glen Roy," and extended their journey as far as Inverness; the professor teaching the young men as they travelled how to observe in a mountain country. Not long before his death, Robert Stephenson spoke in glowing terms of the great pleasure and benefit which he had derived from that interesting excursion. " I have travelled far, and enjoyed much," he said; " but that delightful botanical and geological tour I shall never forget; and I am just about to start in the *Titania* for a trip round the east coast of Scotland, returning south through the Caledonian Canal, to refresh myself with the recollection of that first and brightest tour of my life."

Towards the end of the summer the young student returned to Killingworth to re-enter upon the active business of life. The six months' study had cost his father 80*l.*, a considerable sum to him in those days; but he was amply repaid by the sound scientific culture which his son had acquired, and the evidence of ability and industry which he was enabled to exhibit in the prize for mathematics which he had won at the University.

WEST MOOR COLLIERY, KILLINGWORTH.

CHAPTER IX.

George Stephenson Engineer of the Stockton and Darlington Railway.

The district lying to the west of Darlington, in the county of Durham, is one of the richest mineral fields of the North. Vast stores of coal underlie the Bishop Auckland Valley; and from an early period it was felt to be an exceedingly desirable object to open up new communications to enable the article to be sent to market. But as yet it remained almost a closed field, the cost of transport of the coal in carts, or on horses' or donkeys' backs, greatly limiting the sale. Long ago, in the days of canal formations, Brindley was consulted about a canal; afterwards, in 1812, a tramroad was surveyed by Rennie; and eventually, in 1817, a railway was projected from Witton Colliery, a few miles above Darlington, to Stockton-on-Tees.

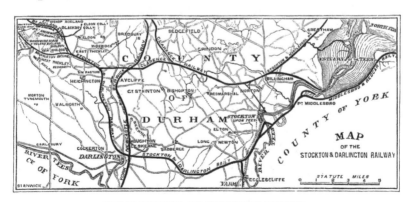

MAP OF STOCKTON AND DARLINGTON RAILWAY.

Of this railway Edward Pease was the projector. A thoughtful and sagacious man, ready in resources, possessed of indomitable energy and perseverance, he was

eminently qualified to undertake what appeared to
many the desperate enterprise of obtaining an Act of
Parliament to construct a railway through a rather
unpromising district.   One who knew him in 1818
said, " he was a man who could see a hundred years
ahead."   When the writer last saw him, in the autumn
of 1854, Mr. Pease was in his eighty-eighth year; yet
he still possessed the hopefulness and mental vigour of
a man in his prime.   Hale and hearty, and full of reminiscences of the past, he continued to take an active
interest in all measures calculated to render the lives of
men happier and better.   Still sound in health, his eye
had not lost its brilliancy, nor his cheek its colour; and
there was an elasticity in his step which younger men
might have envied.[1]

In getting up a company for the purpose of surveying
and forming a railway, Mr. Pease had great difficulties
to encounter.   The people of the neighbourhood spoke
of it as a ridiculous undertaking, and predicted that
it would be the ruin of all who had to do with it.
Even those who were most interested in the opening
out of new markets for the vend of their coal, were
indifferent, if not actually hostile.   The Stockton merchants and shipowners, whom the formation of a
railway was calculated so greatly to benefit, gave the
project no support; and not twenty shares were subscribed for in the whole town.   Mr. Pease nevertheless
persevered with the formation of a company; and he
induced many of his friends and relations to subscribe
for shares.   The Richardsons and Backhouses, members,
like himself, of the Society of Friends, influenced by his
persuasion, united themselves with him; and so many
of the same denomination (having great confidence in
those influential Darlington names) followed their
example and subscribed for shares, that the railway sub-

---

[1] Mr. Pease died at Darlington, on the 31st of July, 1858, aged ninety-two.

sequently obtained the designation, which it still enjoys, of " The Quakers' Line."

The engineer first employed to make a survey of the line was a Mr. Overton, who had had considerable experience in the formation of similar roads in Wales. The necessary preliminary steps were taken in the year 1818 to apply for an Act to authorise the construction of a tramroad from Witton to Stockton.   The measure was however, strongly opposed by the Duke of Cleveland, because the proposed line passed near to one of his fox covers; and, having considerable parliamentary influence, he succeeded in throwing out the bill by a majority of only thirteen,—above one hundred members voting in support of the measure.   A nobleman said, when he heard of the division, " Well, if the Quakers in these times, when nobody knows anything about railways, can raise up such a phalanx in their support, I should recommend the county gentlemen to be very wary how they oppose them in future."

A new survey was then made, avoiding the Duke's fox cover ; and in 1819 a renewed application was made to Parliament for an Act.   But George III. dying in January, 1820, while Parliament was still sitting, there was a dissolution, and the Bill was necessarily suspended. The promoters, however, did not lose sight of their project.   They had now spent a considerable sum of money in surveys and legal and parliamentary expenses, and were determined to proceed, though they were still unable to enlist the active support of the inhabitants of the district proposed to be served by the railway.

The energy of Edward Pease, backed by the support of his Quaker friends, enabled him to hold the company together, to raise the requisite preliminary funds from time to time for the purpose of prosecuting the undertaking, and eventually to overcome the opposition raised against the measure in Parliament.   The bill at length passed ; and the royal assent was given to the first Stockton and Darlington Railway Act on the 19th of April, 1821.

The preamble of this Act recites, that " the making and maintaining of a Railway or Tramroad, for the passage of waggons and other carriages " from Stockton to Witton Park Colliery (by Darlington), "will be of great public utility, by facilitating the conveyance of coal, iron, lime, corn, and other commodities " between the places mentioned.   The projectors of the line did not originally contemplate the employment of locomotives; for in the Act they provide for the making and maintaining of the tramroads for the passage upon them " of waggons and other carriages" "*with men and horses* or otherwise,*" and a further clause made provision as to the damages which might be done in the course of traffic by the " waggoners."   The public were to be free " to use, with horses, cattle and carriages," the roads formed by the company, on payment of the authorised rates, " between the hours of seven in the morning and six in the evening," during the winter months; " between six in the morning and eight in the evening," in two of the spring and autumn months each; and " between five in the morning and ten in the evening," in the high summer months of May, June, July, and August.   From this it will be obvious that the projectors of the line had themselves at first no very large conceptions as to the scope of their project.

Some time elapsed before any active steps were taken to proceed with the construction of the railway.   Doubts had been raised whether the line was the best that could be adopted for the district; and the subscribers generally were not so sanguine about the undertaking as to induce them to press it forward.

One day, about the end of the year 1821, two strangers knocked at the door of Mr. Pease's house in Darlington; and the message was brought to him that some persons from Killingworth wanted to speak with

him.   They were invited in, on which one of the
visitors introduced himself as Nicholas Wood, viewer at
Killingworth, and then, turning to his companion, he
introduced him to Mr. Pease as George Stephenson, of
the same place.   Mr. Stephenson came forward and
handed to Mr. Pease a letter from Mr. Lambert, the
manager at Killingworth, in which it was stated that
the bearer was the engine-wright at the pits, that
he had had experience in the laying out of railways
and had given satisfaction to his employers, and that
he would therefore recommend him to the notice of
Mr. Pease if he stood in need of the services of such
a person.

Mr. Pease entered into conversation with his visitors,
and soon ascertained the object of their errand.   Stephenson had heard of the passing of the Stockton and Darlington Act, and desiring to increase his railway experience,
and also to employ in some larger field the practical
knowledge he had already gained, he determined to
visit Mr. Pease, the known projector of the undertaking,
with the view of being employed to carry it out.   He
had brought with him his friend Nicholas Wood, for
the purpose at the same time of relieving his diffidence,
and supporting his application.

Mr. Pease liked the appearance of his visitor.   "There
was," as he afterwards remarked, in speaking of Stephenson, " such an honest, sensible look about him, and
he seemed so modest and unpretending.   He spoke in
the strong Northumbrian dialect of his district, and
described himself as 'only the engine-wright at Killingworth ; that's what he was.' "

Mr. Pease soon saw that our engineer was the very
man for his purpose.   The whole plans of the railway
being still in an undetermined state, Mr. Pease was glad
to have the opportunity of gathering from George Stephenson the results of his experience.   The latter strongly
recommended a *railway* in preference to a tramroad, in

which Mr. Pease was disposed to concur with him. The conversation next turned on the tractive power which the company intended to employ, and Mr. Pease said that they had based their whole calculations on the employment of *horse* power. " I was so satisfied," said he afterwards, " that a horse upon an iron road would draw ten tons for one ton on a common road, that I felt sure that before long the railway would become the King's highway."

But Mr. Pease was scarcely prepared for the bold assertion made by his visitor, that the locomotive engine with which he had been working the Killingworth Railway for many years past was worth fifty horses, and that engines made after a similar plan would yet entirely supersede all horse power upon railroads. Mr. Stephenson was daily becoming more positive as to the superiority of his locomotive; and on this, as on all subsequent occasions, he strongly urged Mr. Pease to adopt it. " Come over to Killingworth," said he, " and see what my engines can do; seeing is believing, sir." And Mr. Pease promised that on some early day he would go over to Killingworth with his friend Thomas Richardson, and take a look at the wonderful machine that was to supersede horses.

On Mr. Pease referring to the difficulties and the opposition which the projectors of the railway had had to encounter, and the obstacles which still lay in their way, Stephenson said to him, " I think, sir, I have some knowledge of craniology, and from what I see of your head, I feel sure that if you will fairly *buckle* to this railway, you are the man successfully to carry it through." " I think so, too," rejoined Mr. Pease; " and I may observe to thee, that if thou succeed in making this a good railway, thou may consider thy fortune as good as made." He added that all they would require at present was an estimate of the cost of re-surveying the line, with the direction of which the company were not quite satisfied;

and as they had already paid away several hundred pounds, and found themselves very little advanced, Mr. Pease asked that this new survey should be done at as little expense as possible.   This Stephenson readily assented to; and after Mr. Pease had pledged himself to bring his application for the appointment of engineer before the Directors on an early day, and to support it with his influence, the two visitors prepared to take their leave, informing Mr. Pease that they intended to return as they had come, " by nip;" that is, they expected to get a smuggled lift on the stage-coach, by tipping Jehu, —for in those days the stage-coachmen were wont to regard all casual roadside passengers as their special per-quisite.   They had, however, been so much engrossed by their interesting conversation, that the lapse of time was forgotten, and when Stephenson and his friend Wood left Mr. Pease's house to make enquiries about the return coach, they found the last had left; and they were there-fore under the necessity of walking the eighteen miles to Durham on their way back to Newcastle.[1]

Mr. Pease having made further inquiries respecting Stephenson's character and qualifications, and having received from John Grimshaw — also a Friend, the inventor of endless spinning—a very strong recommen-dation of him as the right man for the intended work,

---

[1] Mr. Nicholas Wood has given the following account of this remark-able day's proceedings:—" It was my good fortune to have accompanied Mr. Stephenson—as you will have seen recorded in his Life—on his visit to Darlington, to communicate with Mr. Pease on the establishment of the Darlington and Stockton Railway. It was rather a heavy day for us, as we first of all started from Killing-worth, and rode six miles; we then went upon a coach thirty miles or more to Stockton; then we had a walk of twelve miles through the fields over the line of the proposed railway; then had an interview with Mr. Pease; and lastly, we walked from Darlington to Durham, eighteen miles further. Unfortunately for me, I broke down, about three miles from Durham, at the " Traveller's Rest." I hoped I might get accommodation there; but unfortunately I was told there was no room in the house, and had to go the remaining three miles. That was a joke, and a very satisfac-tory one, with Mr. Stephenson against me, during the whole of his life. This was only one instance of the very great energy which he displayed in accomplishing objects he had under-taken."—Speech at Newcastle, 26th October, 1858.

he brought the subject of his application before the
directors of the Stockton and Darlington Company.
They resolved to adopt his recommendation that a railway be formed instead of a tramroad; and they further
requested Mr. Pease to write to Mr. Stephenson, which
he accordingly did, requesting him to report as to the
practicability, or otherwise, of the line laid out by
Mr. Overton, and to state his suggestions as to any
deviations or improvements in its course, together with
estimates of comparative expenses. " In short," said Mr.
Pease, " we wish thee to proceed in all thy levels, estimates, and calculations, with that care and economy
which would influence thee if the whole of the work
were thy own."

A man was despatched on a horse with the letter,
and when he reached Killingworth he made diligent
enquiry after the person named upon the address,
" George Stephenson, Esquire, Engineer." No such
person was known in the village. It is said that the
man was on the point of giving up all further search,
when the happy thought struck some of the colliers'
wives who had gathered about him, that it must be
" Geordie " the man was in search of; and to Geordie's
cottage he accordingly went, found him at home, and
delivered the letter. In his reply, Mr. Stephenson informed Mr. Pease that the re-survey of the line would
occupy at least four weeks, and that his charge would
include all necessary assistance for the accomplishment
of the survey, estimates of the expense of cuts and
batteries (since called cuttings and embankments) on
the different projected lines, together with all remarks,
reports, &c., on the same; also the comparative cost of
malleable and cast iron rails, laying the same, winning
and preparing the blocks of stone, and all other materials
wanted to complete the line. " I could not do this,"
said he, " for less than 140l., allowing me to be moderately paid. Such a survey would of course have to be

made before the work could be begun, as it is impossible
to form any idea of contracting for the cuts and batteries
by the former one; and I assure you I shall, in completing the undertaking, act with that economy which
would influence me if the whole of the work was my
own."

About the end of September Mr. Stephenson again
went carefully over the line of the proposed railway, for
the purpose of suggesting such improvements and deviations as he might consider desirable. He was accompanied by an assistant and a chainman,—his son Robert
entering the figures while his father took the sights.
After being engaged in the work at intervals for
about six weeks, Stephenson reported the result of his
survey to the Board of Directors, and showed that by
certain deviations, a line shorter by about three miles
might be constructed at a considerable saving in expense,
while at the same time more favourable gradients—an
important consideration—would be secured.

The directors of the company, being satisfied that the
improvements suggested in the line, and the saving
which would be effected in mileage and in money,
fully warranted them in incurring the trouble, delay,
and expense of making a further application to Parliament for an amended Act, took the requisite steps with
this object. And in the mean time they directed Mr.
Stephenson to prepare the specifications for the rails
and chairs, and make arrangements to enter into contracts for the supply of the stone and wooden blocks on
which the rails and chairs were to be laid. It was
determined in the first place to proceed with the works
at those parts of the line where no deviation was proposed; and the first rail of the Stockton and Darlington
Railway was laid with considerable ceremony, by Thomas
Meynell, Esq., of Yarm, at a point near St. John's Well,
Stockton, on the 23rd of May, 1822.

It is worthy of note that Stephenson, in making

his first estimate of the cost of forming the railway according to the instructions of the directors, set down, as part of the cost, 6200*l.* for stationary engines, not mentioning locomotives at all. The directors as yet confined their views to the employment only of horses for the haulage of the coals, and of fixed engines and ropes where horse-power was not applicable. The whole question of steam locomotive power was, in the estimation of the public, as well as of practical and scientific men, as yet in doubt. The confident anticipations of George Stephenson, as to the eventual success of locomotive engines, were regarded as mere speculations; and when he gave utterance to his views, as he frequently took the opportunity of doing, it had the effect of shaking the confidence of some of his friends in the solidity of his judgment and his practical qualities as an engineer.

When Mr. Pease discussed the question with Stephenson, his remark was, " Come over and see my engines at Killingworth, and satisfy yourself as to the efficiency of the locomotive. I will show you the colliery books, that you may ascertain for yourself the actual cost of working. And I must tell you that the economy of the locomotive engine is no longer a matter of theory, but a matter of fact." So confident was the tone in which Stephenson spoke of the success of his engines, and so important were the consequences involved in arriving at a correct conclusion on the subject, that Mr. Pease at length resolved upon paying a visit to Killingworth; and he proceeded thither accordingly, in the summer of 1822, in company with his friend Mr. Thomas Richardson,[1] a considerable subscriber to the Stockton and Darlington project.

When Mr. Pease arrived at Killingworth village, he inquired for George Stephenson, and was told that he

---

[1] Mr. Richardson was the founder of the celebrated discount house of Richardson, Overend, and Gurney, in Lombard Street.

must go over to the West Moor, and seek for a cottage by the roadside, with a dial over the door—that was where George Stephenson lived. They soon found the house with the dial; and on knocking, the door was opened by Mrs. Stephenson—his second wife (Elizabeth Hindmarsh), the daughter of a farmer at Black Callerton, whom he had married in 1819.[1] Her husband, she said, was not in the house at present, but she would send for him to the colliery. And in a short time Stephenson appeared before them in his working dress, just as he had come out of the pit.

He very soon had his locomotive brought up to the crossing close by the end of the cottage,—made the gentlemen mount it, and showed them its paces. Harnessing it to a train of loaded waggons, he ran it along the railroad, and so thoroughly satisfied his visitors of its powers and capabilities, that from that day Edward Pease was a declared supporter of the locomotive engine.

---

[1] The story has been told that George was a former suitor of Miss Hindmarsh, while occupying the position of a humble workman at Black Callerton, but that having been rejected by her, he proceeded to make love to her servant, whom he married; and that after her death, when he had become a comparatively thriving man, and rode a galloway, he again made up to Miss Hindmarsh, and was on the second occasion accepted. The story is, however, without any foundation, as George's first wife was never a servant in the Hindmarsh family, nor had he ever exchanged a word with Miss Hindmarsh until the year 1818, when he was introduced to her at his own desire by Thomas Hindmarsh, her brother, the author's informant as to the facts. It may be observed in passing, that the writer of the article "George Stephenson," in the eighth edition of the 'Encyclopædia Britannica,' while objecting to the accuracy in certain respects of the 'Life of George Stephenson,' as written by the author of this book, points to his own "Biography in Brief," as published in the interesting little book, entitled 'Our Coal and Our Coal-Pits.' On turning to the book itself, it will be found that the "Biography in Brief" is substantially taken from a sketch of George Stephenson's life which appeared in 'Eliza Cook's Journal,' of June the 2nd, 1849. Among the errors contained in that article, is the statement that Stephenson was sent into a coal-pit, to work as a "trapper," when between six and seven years old; and also the above anecdote of his having first courted the mistress, and then descended to the maid—both of which are adopted almost *verbatim* in 'Our Coal and Our Coal-Pits.' The author of this book has the less hesitation in stating these to be errors, as the article in 'Eliza Cook's Journal,' where they originally appeared, was written by himself, on imperfect information, and before he had the opportunity of thoroughly sifting, as he has since done, the facts of George Stephenson's early life.

In preparing the Amended Stockton and Darlington Act of 1823, at Stephenson's urgent request Mr. Pease had a clause inserted, taking power to work the railway by means of locomotive engines, and to employ them for the haulage of passengers as well as of merchandise.[1]

The second Stockton and Darlington Act was obtained in the session of 1823, not, however, without opposition, the Duke of Cleveland and the road trustees still appearing as the determined opponents of the bill. Nevertheless, the measure passed into law; Stephenson was appointed the company's engineer at a salary of 300*l.* per annum ; and it was determined that the line should be constructed and opened for traffic as soon as practicable.

He at once proceeded with the working survey of the improved line of the Stockton and Darlington Railway, laying out every foot of the ground himself, accompanied by his assistants. Railway surveying was as yet in its infancy, and was very slow and deliberate work. It afterwards became a separate branch of railway business, and was left to a special staff of surveyors. Indeed on no subsequent line did George Stephenson take the sights through the spirit level with his own hands and eyes as he did on this railway. He started very early in the morning, and surveyed until dusk. John Dixon, who assisted in the survey, mentions that he remembers on one occasion, after a long day's work near Aycliffe, when the light had completely failed them, the party separated—some to walk to Darlington, four miles off, Stephenson himself to the Simpasture farmhouse, where he had arranged to stay for the night; and his last stringent injunction was, that they must all be on the ground to resume levelling as soon as there was light enough for the purpose. "You must not," he said, " set off from Darlington by daybreak, for then we

---

[1] The first clause in any railway act, empowering the employment of locomotive engines for the working of passenger traffic.

shall lose an hour; but you must be *here*, ready to begin work as soon as it is daylight."

Stephenson performed the survey in top-boots and breeches—a usual dress at the time. He was not at any time particular as to his living; and during the survey, he took his chance of getting a drink of milk and a bit of bread at some cottager's house along the line, or occasionally joined in a homely dinner at some neigh- bouring farmhouse. The country people were accus- tomed to give him a hearty welcome when he appeared at their door; for he was always full of cheery and homely talk, and, when there were children about the house, he had plenty of humorous chat for them as well as for their seniors.

After the day's work was over, George would drop in at Mr. Pease's, to talk over with him the progress of the survey, and discuss various matters connected with the railway. Mr. Pease's daughters were usually present; and on one occasion, finding the young ladies learning the art of embroidery, he volunteered to instruct them.[1] " I know all about it," said he; " and you will wonder how I learnt it. I will tell you. When I was a brakes- man at Killingworth, I learnt the art of embroidery while working the pitman's button-holes by the engine fire at nights." He was never ashamed, but on the con- trary rather proud, of reminding his friends of these humble pursuits of his early life. Mr. Pease's family were greatly pleased with his conversation, which was always amusing and instructive; full of all sorts of experience, gathered in the oddest and most out-of- the-way places. Even at that early period, before he mixed in the society of educated persons, there was a dash of speculativeness in his remarks, which gave a high degree of originality to his conversation; and

---

[1] This incident, communicated to the author by the late Edward Pease, has since been made the subject of a fine picture by Mr. A. Rankley, A.R.A., exhibited at the Royal Academy Ex- hibition of 1861.

he would sometimes, in a casual remark, throw a flash of light upon a subject, which called up a whole train of pregnant suggestions.

One of the most important subjects of discussion at these meetings with Mr. Pease, was the establishment of a manufactory at Newcastle for the building of locomotive engines.   Up to this time all the locomotives constructed after Stephenson's designs, had been made by ordinary mechanics working amongst the collieries in the North of England.   But he had long felt that the accuracy and style of their workmanship admitted of great improvement, and that upon this the more perfect action of the locomotive engine, and its general adoption as the tractive power on railways, in a great measure depended.   One great object that he had in view in establishing the proposed factory was, to concentrate a number of good workmen for the purpose of carrying out the improvements in detail which he was constantly making in his engine.   He felt hampered by the want of efficient helpers in the shape of skilled mechanics, who could work out in a practical form the ideas of which his busy mind was always so prolific. Doubtless, too, he believed that the locomotive manufactory would prove a remunerative investment, and that, on the general adoption of the railway system, which he now anticipated, he would derive solid advantages from the fact of his manufactory being the only establishment of the kind for the special construction of railway locomotives.

He still believed in the eventual success of railways, though it might be slow.   Much, he believed, would depend upon the issue of this great experiment at Darlington ; and as Mr. Pease was a man on whose sound judgment he could rely, he determined upon consulting him about his proposed locomotive factory. Mr. Pease approved of his design, and strongly recommended him to carry it into effect.   But there was the question of means; and Stephenson did not think he had capital enough for the purpose. He told Mr. Pease that he could advance a thousand pounds—the amount of the testimonial presented by the coal-owners for his safety-lamp invention, which he had still left untouched; but he did not think this sufficient for the purpose, and he thought that he should require at least another thousand pounds. Mr. Pease had been very much struck with the successful performances of the Killingworth engine; and being an accurate judge of character, he was not slow to perceive that he could not go far wrong in linking a portion of his fortune with the energy and industry of George Stephenson. He consulted his friend Thomas Richardson in the matter; and the two consented to advance 500*l.* each for the purpose of establishing the engine factory at Newcastle. A piece of land was accordingly purchased in Forth Street, in August, 1823, on which a small building was erected—the nucleus of the gigantic establishment which was afterwards formed around it; and active operations commenced early in 1824.

While the Stockton and Darlington Railway works were in progress, Mr. Stephenson held many interesting discussions with Mr. Pease, on points connected with its construction and working, the determination of which in a great measure affected the formation and working of all future railways. The most important points were these: 1. The comparative merits of cast and wrought iron rails. 2. The gauge of the railway. 3. The employment of horse or engine power in working it, when ready for traffic.

The kind of rails to be laid down to form the permanent road was a matter of considerable importance. A wooden tramroad had been contemplated when the first Act was applied for; but Stephenson having advised that an iron road should be laid down, he was instructed to draw up a specification of the rails. He

went before the directors to discuss with them the kind
of material to be specified. He was himself interested in the patent for cast-iron rails, which he had
taken out in conjunction with Mr. Losh in 1816; and,
of course, it was to his interest that his articles should
be used. But when requested to give his opinion on
the subject, he frankly said to the directors, " Well,
gentlemen, to tell you the truth, although it would put
500*l.* in my pocket to specify my own patent rails, 1
cannot do so after the experience I have had. If you
take my advice, you will not lay down a single cast-iron
rail." " Why?" asked the directors. " Because they
will not stand the weight, and you will be at no end
of expense for repairs and relays." " What kind of
road, then," he was asked, " would you recommend?"
" Malleable rails, certainly," said he; " and I can recommend them with the more confidence from the fact that
at Killingworth we have had some Swedish bars laid
down—nailed to wooden sleepers—for a period of fourteen years, the waggons passing over them daily; and
there they are, in use yet, whereas the cast rails are
constantly giving way." [1]

The price of malleable rails was, however, so high—
being then worth about 12*l.* per ton as compared with
cast-iron rails at about 5*l.* 10*s.*—and the saving of expense was so important a consideration with the subscribers to the railway, that Mr. Stephenson was directed
to provide, in the specification drawn by him, that only

---

[1] The most suitable kind of iron for
rails had formed the subject of frequent conversations between George
Stephenson and his son in their cottage at Killingworth many years before; and they had both come to the
conclusion that malleable iron only
should be used for the purpose.
While Robert Stephenson was attending college at Edinburgh, he wrote
a letter on the subject to Richard
Scorton, a gentleman interested in the
formation of a railway near Durham,
and his opinion on the point was clear
and explicit. Robert was only eighteen
years old at the time, but his letter
was full of practical information on
the then little known subject of railways, indicating habits of careful observation, and the action of a vigorous
and well-disciplined intellect. The
letter was published in the ' Mining
Journal' of April 5th, 1862.

one-half of the quantity of the rails required—or 800 tons—should be of malleable iron, the remainder being of cast-iron. The malleable rails were of the kind called "fish-bellied," and weighed only 28 lbs. to the yard, being 2¼ inches broad at the top, with the upper flange ¾ inch thick. They were only 2 inches in depth at the points at which they rested on the chairs, and 3¼ inches in the middle or bellied part.

When forming the road, the proper gauge had also to be determined. What width was this to be? The gauge of the first tramroad laid down had virtually settled the point. The gauge of wheels of the common vehicles of the country—of the carts and waggons employed on common roads, which were first used on the tramroads—was about 4 feet 8½ inches. And so the first tramroads were laid down of this gauge. The tools and machinery for constructing coal-waggons and locomotives were formed with this gauge in view. The Wylam waggon-way, afterwards the Wylam plate-way, the Killingworth railroad, and the Hetton railroad, were as nearly as possible on the same gauge. Some of the earth-waggons used to form the Stockton and Darlington road were brought from the Hetton railway; and others which were specially constructed were formed of the same dimensions, these being intended to be afterwards employed in the working of the traffic.

As the period drew near for the opening of the line, the question of the tractive power to be employed was anxiously discussed. At the Brusselton incline, fixed engines must necessarily be made use of; but with respect to the mode of working the railway generally, it was decided that horses were to be largely employed, and arrangements were made for their purchase. The influence of Mr. Pease also secured that a fair trial should be given to the experiment of working the traffic by locomotive power; and three engines were ordered from the firm of Stephenson and Co., Newcastle, which

were put in hand forthwith, in anticipation of the
opening of the railway. These were constructed after
Mr. Stephenson's most matured designs, and embodied
all the improvements in the locomotive which he had contrived up to that time. No. I. engine, the " Locomotion,"
which was first delivered upon the line, weighed about
eight tons. It had one large flue or tube through the
boiler, by which the heated air passed direct from the
furnace at one end, lined with fire-bricks, to the chimney
at the other. The combustion in the furnace was quickened by the adoption of the steam-blast in the chimney.
The heat raised was sometimes so great, and it was so
imperfectly abstracted by the surrounding water, that
the chimney became almost red-hot. Such engines,
when put to the top of their speed, were found capable
of running at the rate of from twelve to sixteen miles an
hour ; but they were better adapted for the heavy work
of hauling coal-trains at low speeds—for which, indeed,
they were specially constructed—than for running at
the higher speeds afterwards adopted. Nor was it contemplated by the directors as possible, at the time when
they were ordered, that locomotives could be made
available for the purposes of passenger travelling. Besides, the Stockton and Darlington Railway did not run
through a district in which passengers were supposed to
be likely to constitute any considerable portion of the
expected traffic.

We may easily imagine the anxiety felt by Mr.
Stephenson during the progress of the works towards
completion, and his mingled hopes and doubts (though
his doubts were but few) as to the issue of this great
experiment. When the formation of the line near
Stockton was well advanced, Mr. Stephenson one day,
accompanied by his son Robert and John Dixon, made a
journey of inspection of the works. The party reached
Stockton, and proceeded to dine at one of the inns there.
After dinner, Mr. Stephenson ventured on the very

unusual measure of ordering in a bottle of wine, to drink success to the railway.   John Dixon remembers and relates with pride the utterance of the master on the occasion.   " Now, lads," said he to the two young men, " I will tell you that I think you will live to see the day, though I may not live so long, when railways will come to supersede almost all other methods of conveyance in this country—when mail-coaches will go by railway, and railroads will become the great highway for the king and all his subjects.   The time is coming when it will be cheaper for a working man to travel upon a railway than to walk on foot.   I know there are great and almost insurmountable difficulties that will have to be encountered ; but what I have said will come to pass as sure as you live.   I only wish I may live to see the day, though that I can scarcely hope for, as I know how slow all human progress is, and with what difficulty I have been able to get the locomotive adopted, notwithstanding my more than ten years' successful experiment at Killingworth."   The result, however, outstripped even the most sanguine anticipations of Stephenson ; and his son Robert, shortly after his return from America in 1827, saw his father's locomotive generally adopted as the tractive power on railways.

The Stockton and Darlington line was opened for traffic on the 27th of September, 1825.   An immense concourse of people assembled from all parts to witness the ceremony of opening this first public railway.   The powerful opposition which the project had encountered, the threats which were still uttered against the company by the road-trustees and others, who declared that they would yet prevent the line being worked, and perhaps the general unbelief as to its success which still prevailed, tended to excite the curiosity of the public as to the result.   Some went to rejoice at the opening, some to see the " bubble burst ;" and there were many prophets of evil who would not miss the blowing up of the boasted travelling engine. The opening was, however, auspicious. The proceedings commenced at Brusselton Incline, about nine miles above Darlington, when the fixed engine drew a train of loaded waggons up the incline from the west, and lowered them on the east

PROCESSION AT THE OPENING OF THE STOCKTON AND DARLINGTON RAILWAY.

[Fac-simile of a local lithograph.]

side. At the foot of the incline a locomotive was in readiness to receive them, Mr. Stephenson himself driving the engine. The train consisted of six waggons loaded with coals and flour; after these was the passenger-coach, filled with the directors and their friends, and then twenty-one waggons fitted up with temporary seats for passengers; and lastly came six waggon-loads of coals, making in all a train of thirty-eight vehicles. The local chronicler of the day went almost out of breath in describing the extraordinary event:—" The signal being given," he says, " the engine started off with this immense train of carriages; and such was its velocity, that in some parts the speed was frequently 12 miles an hour." By the time the train reached Stockton there were about 600 persons in the train or hanging on to the waggons, which must have gone at a safe and steady pace of from four to six miles an hour from Darlington. " The arrival at Stockton," it is added, " excited a deep interest and admiration."

The working of the line then commenced, and the results were such as to surprise even the most sanguine of its projectors. The traffic upon which they had formed their estimates of profit proved to be small in comparison with the traffic which flowed in upon them that had never been taken into account. Thus, what the company had principally relied upon for their profit was the carriage of coals for land sale at the stations along the line, whereas the haulage of coals to the seaports for exportation to the London market was not contemplated as possible. When the bill was before Parliament, Mr. Lambton (afterwards Earl of Durham) succeeded in getting a clause inserted, limiting the charge for the haulage of all coal to Stockton-on-Tees for the purpose of shipment, to one halfpenny per ton per mile ; whereas a rate of fourpence per ton was allowed to be taken for all coals led upon the railway for land sale. Mr. Lambton's object in enforcing the low rate of one halfpenny was to protect his own trade in coal exported from Sunderland and the northern ports. He believed, in common with everybody else, that the halfpenny rate would effectually secure him against any competition on the part of the Stockton and Darlington Company ; for it was not considered possible for coals to be led at that low price, and the proprietors of the railway themselves considered that to carry coals at such a rate would be utterly ruinous. The projectors never contemplated sending more than 10,000 tons a year to Stockton, and those only for shipment as ballast ; they looked for their profits almost exclusively to the land sale. The result, however, was as surprising to them as it must have been to Mr. Lambton. The halfpenny rate which was forced upon them, instead of being

ruinous, proved the vital element in the success of the
railway.   In the course of a few years, the annual shipment of coal, led by the Stockton and Darlington Railway to Stockton and Middlesborough, exceeded five
hundred thousand tons; and it has since far exceeded
this amount.   Instead of being, as anticipated, a subordinate branch of traffic, it proved, in fact, the main
traffic, while the land sale was merely subsidiary.

The anticipations of the company as to passenger
traffic were in like manner more than realised.   At
first, passengers were not thought of; and it was only
while the works were in progress that the starting of
a passenger coach was seriously contemplated.   The
number of persons travelling between the two towns
was very small; and it was not known whether these
would risk their persons upon the iron road.   It was
determined, however, to make the trial of a railway
coach; and Mr. Stephenson was authorised by the
directors to have one built to his order at Newcastle, at
the cost of the company.   This was done accordingly;
and the first railway passenger carriage was built after
our engineer's plans.   It was, however, a very modest,
and indeed a somewhat uncouth machine, more resembling the caravans still to be seen at country fairs containing the "Giant and the Dwarf" and other wonders
of the world, than a passenger coach of any extant
form.   A row of seats ran along each side of the interior, and a long deal table was fixed in the centre; the
access being by means of a door at the back end, in the
manner of an omnibus.   This coach arrived from Newcastle the day before the opening, and formed part of
the railway procession above described.   Mr. Stephenson was consulted as to the name of the coach, and he
at once suggested "The Experiment;" and by this name
it was called.   The Company's arms were afterwards
painted on her side, with the motto "Periculum pri-
vatum utilitas publica."   Such was the sole passenger-

THE FIRST RAILWAY COACH.

carrying stock of the Stockton and Darlington Company
in the year 1825. But the "Experiment" proved the forerunner of a mighty traffic: and long time did not elapse
before it was displaced, not only by improved coaches
(still drawn by horses), but afterwards by long trains of
passenger-carriages drawn by locomotive engines.

No sooner did the coal and merchandise trains begin
to run regularly upon the line, than new business relations sprang up between Stockton and Darlington, and
there were many more persons who found occasion to
travel between the two towns,—merchandise and mineral
traffic invariably stimulating, if not calling into existence, an entirely new traffic in passengers. Before the
construction of the line, the attempt had been made to
run a coach between Stockton, Darlington, and Barnard
Castle three times a week; but it was starved off the road
for want of support. Now, however, that there were
numbers of people desiring to travel, the stage-coach by
the common road was revived and prospered, and many
other persons connected with the new traffic got a
" lift " by the railway waggons, which were even more
popular than the stage-coach.

" The Experiment" was fairly started as a passenger coach on the 10th of October, 1825, a fortnight after the opening of the line.   It was drawn by one horse, and performed a journey daily each way between the two towns, accomplishing the distance of twelve miles in about two hours.   The fare charged was a shilling, without distinction of class; and each passenger was allowed fourteen pounds of luggage free.   The "Experiment" was not, however, worked by the company, but was let to Messrs. Pickersgill and Harland, carriers on the railway, under an arrangement with them as to the payment of tolls for the use of the line, rent of booking-cabins, &c.

The speculation answered so well, that several coaching companies were shortly after got up by innkeepers at Darlington and Stockton, for the purpose of running other coaches upon the railroad; and an active competition for passenger traffic sprang up.[1]   " The Experiment" being found too heavy for one horse to draw between Stockton and Darlington, besides being found an uncomfortable machine, was banished to the coal district, and ran for a time between Darlington and Shildon.   Its place on the line between Stockton and Darlington was supplied by other and better vehicles,—though they were no other than old stage-coach bodies, purchased by the company, and each mounted upon an underframe with flange-wheels.   These were let on hire to the coaching com-

---

[1] The coaches were not allowed to be run upon the line without considerable opposition. We find Edward Pease writing to Joseph Sanders, of Liverpool, on the 18th January, 1827:—" Our railway coach proprietors have individually received notices of a process in the Exchequer for various fines, to the amount of 150l., in penalties of 20l. each, for neglecting to have the plates, with the numbers of their licenses, on the coach doors, agreeably to the provisions of the Act 95 George IV. In looking into the nature of this proceeding and its consequences, it is clear, if the Court shall confirm it by conviction, that we are undone as to the conveyance of passengers." Mr. Pease incidentally mentions the names of the several coach proprietors at the time —" Pickersgill and Co., Richard Scott, and Martha Hewson." The proceeding was eventually defeated, it being decided that the penalties only applied to coaches travelling on common or turnpike-roads.

panies, who horsed and managed them under an arrangement as to tolls, in like manner as the "Experiment" had been worked. Now began the distinction of inside and outside passenger, equivalent to first and second class, paying different fares. The competition with each other upon the railway, and with the ordinary stage-coaches upon the road, soon brought up the speed, which was increased to ten miles an hour—the mail-coach rate of travelling in those days, and considered very fast.

Mr. Clephan, a native of the district, has described some of the curious features of the competition between the rival coach companies :—" There were two separate coach companies in Stockton, and amusing collisions sometimes occurred between the drivers—who found on the rail a novel element for contention. Coaches cannot pass each other on the rail as on the road ; and, as the line was single, with four sidings in the mile, when two coaches met, or two trains, or coach and train, the question arose which of the drivers must go back ? This was not always settled in silence. As to trains, it came to be a sort of understanding that light waggons should give way to loaded ; and as to trains and coaches, that the passengers should have preference over coals ; while coaches, when they met, must quarrel it out. At length, midway between sidings, a post was erected, and a rule was laid down that he who had passed the pillar must go on, and the 'coming man' go back. At the Goose Pool and Early Nook, it was common for these coaches to stop ; and there, as Jonathan would say, passengers and coachmen 'liquored.' One coach, introduced by an innkeeper, was a compound of two mourning-coaches,—an approximation to the real railway coach, which still adheres, with multiplying exceptions, to the stage-coach type. One Dixon, who drove the 'Experiment' between Darlington and Shildon, is the inventor of carriage-lighting on the rail. On a dark winter night, having compassion on his passengers, he would buy a penny candle, and place it lighted amongst them on the table of the 'Experiment'—the first railway coach (which, by the way, ended its days at Shildon as a railway cabin), being also the first coach on the rail (first, second, and third class jammed all into one) that indulged its customers with light in darkness."

The traffic of all sorts increased so steadily and so rapidly that considerable difficulty was experienced in working it satisfactorily. It had been provided by the first Stockton and Darlington Act that the line should be free to all parties who chose to use it at certain prescribed rates, and that any person might put horses and waggons on the railway, and carry for himself. But this arrangement led to increasing confusion and difficulty, and could not continue in the face of a large and rapidly-increasing traffic. The goods trains got so long that the carriers found it necessary to call in the aid of the locomotive engine to help them on their way. Then mixed trains of passengers and merchandise began to run; and the result was that the railway company found it necessary to take the entire charge and working of the traffic. In course of time new passenger carriages were specially built for the better accommodation of the public, until at length regular passenger trains were run, drawn by the locomotive engine,—though this was not until after the Liverpool and Manchester Company had established these as a distinct branch of their traffic.

The three Stephenson locomotives were from the first regularly employed to work the coal trains; and their proved efficiency for this purpose led to the gradual increase of the locomotive power. The speed of the engines—slow though it seems now—was in those days regarded as something marvellous. A race actually came off between No. I. engine, the "Locomotion," and one of the stage-coaches travelling from Darlington to Stockton by the ordinary road; and it was regarded as a great triumph of mechanical skill that the locomotive reached Stockton first, beating the stage-coach by about a hundred yards!   The same engine continued in good working order in the year 1846, when it headed the railway procession on the opening of the Middlesborough and Redcar Railway, travelling at the rate of about fourteen miles an hour.   This engine, the first that travelled upon the first public railway, has recently been placed upon a pedestal in front of the railway station at Darlington.

THE NO. I. ENGINE AT DARLINGTON.

For some years, however, the principal haulage of the line was performed by horses.   The inclination of the gradients being towards the sea, this was perhaps the cheapest mode of traction, so long as the traffic was not very large.   The horse drew the train along the level road, until, on reaching a descending gradient, down which the train ran by its own gravity, the animal was unharnessed, and, when loose, he wheeled round to the other end of the waggons, to which a "dandy-cart" was attached, its bottom being only a few inches from

the rail.   Bringing his step into unison with the speed
of the train, the horse learnt to leap nimbly into his
place in this waggon, which was usually fitted with a
well-filled hay-rack.

The details of the working were gradually perfected
by experience, the projectors of the line being scarcely
conscious at first of the importance and significance of
the work which they had taken in hand, and little
thinking that they were laying the foundations of a
system which was yet to revolutionise the internal
communications of the world, and confer the greatest
blessings on mankind.   It is important to note that the
commercial results of the enterprise were considered
satisfactory from the opening of the railway.   Besides
conferring a great public benefit upon the inhabitants
of the district and throwing open entirely new markets
for the almost boundless stores of coal found in the
Bishop Auckland district, the profits derived from the
traffic created by the railway enabled increasing dividends
to be paid to those who had risked their capital in the
undertaking, and thus held forth an encouragement to
the projectors of railways generally, which was not
without an important effect in stimulating the projection
of similar enterprises in other districts.   These results,
as displayed in the annual dividends, must have been
eminently encouraging to the astute commercial men of
Liverpool and Manchester, who were then engaged in
the prosecution of their railway.   Indeed, the commercial success of the Stockton and Darlington Company
may be justly characterised as the turning-point of the
railway system.   With that practical illustration daily
in sight of the public, it was no longer possible for
Parliament to have prevented its eventual extension.

Before leaving the subject of the Stockton and Darlington Railway, we cannot avoid alluding to one of its
most remarkable and direct results—the creation of the
town of Middlesborough-on-Tees.   When the railway

was opened in 1825, the site of this future metropolis of Cleveland was occupied by one solitary farm-house and its outbuildings. All round was pasture-land or mud-banks; scarcely another house was within sight. The corporation of the town of Stockton being unwilling or unable to provide accommodation for the rapidly increasing coal traffic, Mr. Edward Pease, in 1829, joined by a few of his Quaker friends, bought about 500 or 600 acres of land, five miles lower down the river—the site of the modern Middlesborough—for the purpose of there forming a new seaport for the shipment of coals brought to the Tees by the railway. The line was accordingly extended thither; docks were excavated; a town sprang up; churches, chapels, and schools were built, with a custom-house, mechanics' institute, banks, shipbuilding yards, and iron-factories; and in a few years the port of Middlesborough became one of the most thriving on the north-east coast of England. In ten years a busy population of some 6000 persons (since swelled to about 20,000) occupied the site of the original farmhouse. More recently, the discovery of vast stores of ironstone in the Cleveland Hills, close adjoining Middlesborough, has tended still more rapidly to augment the population and increase the commercial importance of the place.

It is pleasing to relate, in connexion with this great work—the Stockton and Darlington Railway, projected by Edward Pease and executed by George Stephenson —that when Mr. Stephenson became a prosperous and a celebrated man, he did not forget the friend who had taken him by the hand, and helped him on in his early days. He continued to remember Mr. Pease with gratitude and affection, and that gentleman, to the close of his life, was proud to exhibit a handsome gold watch, received as a gift from his celebrated *protégé*, bearing these words:—" Esteem and gratitude : from George Stephenson to Edward Pease."

CHAPTER X.

THE LIVERPOOL AND MANCHESTER RAILWAY PROJECTED.

THE rapid growth of the trade and manufactures of South Lancashire gave rise, about the year 1821, to the project of a tramroad for the conveyance of goods between Liverpool and Manchester. Since the construction of the Bridgewater Canal by Brindley, some fifty years before, the increase in the business transacted between the two towns had become quite marvellous. The steam-engine, the spinning-jenny, and the canal, working together, had accumulated in one focus a vast aggregate of population, manufactures, and trade.

The Duke's Canal, when first made, furnished a cheap and ready means of conveyance between the seaport and the manufacturing towns, for the raw cotton in the one direction and the manufactured produce in the other. During the first thirty years of its existence the traffic was small and easily managed. About the end of last century, for instance, it was considered satisfactory if one cotton-flat a day reached Manchester by canal from Liverpool. But such was the expansion of business caused by the inventions to which we have referred that, before the lapse of many more years, the navigation was found altogether inadequate to accommodate the traffic, which completely outgrew all the Canal Companies' appliances of wharves, boats, and horses. Cotton lay at Liverpool for weeks together, waiting to be removed; and it occupied a longer time to transport the cargoes from Liverpool to Manchester than it had done to bring them across the Atlantic from the United States to England. Carts and waggons were tried, but proved altogether insufficient. Sometimes manufacturing operations had to be suspended altogether, and during a frost, when the canals were frozen up, the communication was entirely stopped. The consequences were often disastrous, alike to operatives, merchants, and manufacturers. The same difficulty was experienced in the conveyance of manufactured goods from Manchester to Liverpool for export. Mr. Huskisson, in the House of Commons, referring to these ruinous delays, observed that "cotton was sometimes detained a fortnight at Liverpool, while the Manchester manufacturers were obliged to suspend their labours; and goods manufactured at Manchester for foreign markets could not be transmitted in time, in consequence of the tardy conveyance."

Expostulation with the Canal Companies was of no use. They were overcrowded with business at their own prices, and disposed to be very dictatorial. When the Duke first constructed his canal, it will be remembered that he had to encounter the fierce opposition of the Irwell and Mersey Navigation, whose monopoly his new line of water conveyance threatened to interfere with.[1] But the innovation of one generation often becomes the obstruction of the next. The Duke's agents would scarcely listen to the expostulations of the Liverpool merchants and Manchester manufacturers, and the Bridgewater Canal was accordingly, in its turn, denounced as a monopoly.

Under these circumstances any new mode of transit between the two towns which offered a reasonable prospect of relief was certain to receive a cordial welcome. The scheme of a tramroad was, however, so new and comparatively untried, that it is not surprising that the parties interested should have hesitated before committing themselves to it. Mr. Sandars, an influential

---

[1] Lives of the Engineers, vol. i. p. 371.

Liverpool merchant, was amongst the first to broach the subject. He himself had suffered in his business, in common with so many others, from the insufficiency of the existing modes of communication, and was ready to give due consideration to any plan presenting elements of practical efficiency which proposed a remedy for the generally admitted grievance. Having caused inquiry to be made as to the success which had attended the haulage of heavy coal-trains by locomotive power on the northern railways, he was led to form the opinion that the same means might be equally efficient in conducting the increasing traffic in merchandise between Liverpool and Manchester. He ventilated the subject amongst his friends, and about the beginning of 1821 a committee was formed for the purpose of bringing the scheme of a railroad before the public.

The novel project having become noised abroad, attracted the attention of the friends of railways in other quarters. Tramroads were by no means new expedients for the transit of heavy articles. The Croydon and Wandsworth Railway, laid down by William Jessop as early as the year 1801, had been regularly used for the conveyance of lime and stone in waggons hauled by mules or donkeys from Merstham to London.[1] The sight of this humble railroad in 1813 led Sir Richard Phillips to throw out the following thoughtful observations in his ‘ Morning Walk to Kew ’:—“ I found delight,” said he, “ in witnessing at Wandsworth the

---

[1] This line was purchased by the London and Brighton Railway Company, and has long since been disused, though the traveller to Brighton can still discern the marks of the old tramroad along the hill-side, a little to the south of Croydon. “ The *genius loci*,” says Charles Knight, “ must look with wonder on the gigantic offspring of the little railway, which has swallowed up its own sire. Lean mules no longer crawl leisurely along the little rails with trucks of stone through Croydon, once perchance during the day, but the whistle and the rush of the locomotive are now heard all day long. Not a few loads of lime, but all London and its contents, by comparison — men, women, children, horses, dogs, oxen, sheep, pigs, carriages, merchandise, food—would seem to be now-a-days passing Croydon; for day after day, more than 100 journeys are made by the great railroads which pass the place.”

economy of horse labour on the iron railway. Yet a heavy sigh escaped me as I thought of the inconceivable millions of money which had been spent about Malta, four or five of which might have been the means of extending double lines of iron railway from London to Edinburgh, Glasgow, Holyhead, Milford, Falmouth, Yarmouth, Dover, and Portsmouth. A reward of a single thousand would have supplied coaches and other vehicles, of various degrees of speed, with the best tackle for readily turning out; and we might, ere this, have witnessed our mail coaches running at the rate of ten miles an hour drawn by a single horse, or impelled fifteen miles an hour by Blenkinsop's steam-engine. Such would have been a legitimate motive for over-stepping the income of a nation, and the completion of so great and useful a work would have afforded rational ground for public triumph in general jubilee."

In the same year we find Mr. Lovell Edgworth, who had for fifty years been advocating the superiority of tram or railroads over common roads, writing to James Watt (7th August, 1813) : " I have always thought that steam would become the universal lord, and that we should in time scorn post-horses; an iron railroad would be a cheaper thing than a road upon the common construction." Thomas Gray, of Nottingham, was another speculator on the same subject. Though he was no mechanic nor inventor, he had an enthusiastic belief in the powers of the railroad system. Being a native of Leeds, he had, when a boy, seen Blenkinsop's locomotive at work on the Middleton cogged railroad, and from an early period he seems to have entertained almost as sanguine views on the subject as Sir Richard Phillips himself. It would appear that Gray was residing in Brussels in 1816, when the project of a canal from Charleroi, for the purpose of connecting Holland with the mining districts of Belgium, was the subject of discussion ; and, in conversation with Mr. John Cockerill

and others, he took the opportunity of advocating the
superior advantages of a railway.   He occupied himself
for some time with the preparation of a pamphlet on
the subject.   He shut himself up in his room, secluded
from his wife and relations, declining to give them any
information on the subject of his mysterious studies,
beyond the assurance that his scheme "would revolu-
tionise the whole face of the material world and of
society."   In 1820 Mr. Gray published the result of his
studies in his 'Observations on a General Iron Railway,'[1] in which, with great cogency, he urged the
superiority of a locomotive railway over common roads
and canals, pointing out, at the same time, the advantages to all classes of the community of this mode of
conveyance for merchandise and persons.   In this book
Mr. Gray suggested the propriety of making a railway
between Manchester and Liverpool, "which," he observed, "would employ many thousands of the distressed
population of Lancashire."   The treatise seems to have
met with a ready sale, for we find that, two years later,
it had already passed into a fourth edition.   In 1822,
Mr. Gray added a diagram to the book, showing a
number of suggested lines of railway connecting the
principal towns of England, and another in like manner
connecting the principal towns of Ireland.

The publication of this essay had the effect of bringing
the subject of railway extension prominently under the
notice of the public.   Although little able to afford it,
Gray also pressed his favourite project of a general iron
road on the attention of public men—mayors, members
of Parliament, and prime ministers.   He sent memorials
to Lord Sidmouth in 1820, and to the Lord Mayor and

---

[1] 'Observations on a General Iron
Railway (with Plates and Map illustrative of the plan); showing its great
superiority, by the general introduction of mechanic power, over all the
present methods of conveyance by
turnpike-roads and canals; and claiming the particular attention of merchants, manufacturers, farmers, and
indeed every class of society.' London: Baldwin, Cradock, and Joy,
1820.

Corporation of London in 1821. In 1822, he addressed the Earl of Liverpool, Sir Robert Peel, and others, urging the great national importance of his plan. In the year following, he petitioned the ministers of state to the same effect. He was so pertinacious that public men pronounced him to be a "bore," and in the town of Nottingham, where he then lived, those who knew him declared him to be "cracked." William Howitt, who frequently met Gray at that time, has published a lively portraiture of this indefatigable and enthusiastic projector, who seized all men by the button, and would not let them go until he had unravelled to them his wonderful scheme. With Thomas Gray, says he, "begin where you would, on whatever subject—the weather, the news, the political movement or event of the day—it would not be many minutes before you would be enveloped with steam, and listening to an harangue on the practicability and immense advantages, to the nation and to every man in it, of 'a general iron railway.'"

These speculations show that the subject of railways was gradually becoming familiar to the public mind, and that thoughtful men were anticipating with confidence the adoption of steam-power for the purposes of railway traction. At the same time, a still more profitable class of labourers was at work — first, men like Stephenson, who were engaged in improving the locomotive and making it a practicable and economical working power, and next, those like Edward Pease of Darlington, and Joseph Sandars of Liverpool, who were organizing the means of laying down the railways. Mr. William James, of West Bromwich, belonged to the active class of projectors. He was a man of considerable social influence, of an active temperament, and had from an early period taken a warm interest in the formation of tramroads. Acting as land-agent for gentlemen of property in the mining districts, he had laid down several lines in the neighbourhood of Birmingham,

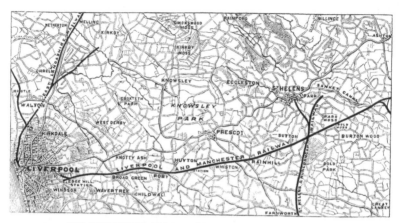

MAP OF LIVERPOOL AND MANCHESTER RAILWAY. (Western Part.)

Gloucester, and Bristol; and he published many pamphlets urging their formation in other places. At one period of his life he was a large iron-manufacturer, for some time acting as Chairman of the Staffordshire ironmasters, The times, however, went against him. It was thought he was too bold, some considered him even reckless, in his speculations; and he lost almost his entire fortune.. He continued to follow the business of a land-agent, and it was while engaged in making a survey for one of his clients in the neighbourhood of Liverpool early in 1821, that he first heard of Mr. Sandars' project of a railway between that town and Manchester. He at once called upon Mr. Sandars, and offered his services as surveyor of the proposed line. After conferring with his friend Mr. Moss, Mr. Sandars authorized James to proceed, and agreed to pay him for the survey at the rate of 10l. a mile, or 300l. for the entire survey.

The trial survey was then proceeded with, but it was conducted with great difficulty, the inhabitants of the district entertaining the most violent prejudices against the formation of the proposed railway. In some places Mr. James and his surveying party even encountered

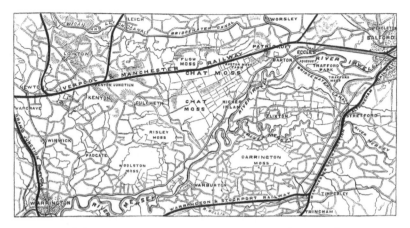

MAP OF LIVERPOOL AND MANCHESTER RAILWAY. (Eastern Part.)

personal violence. Near Newton-in-the-Willows the farmers stationed men at the field-gates with pitchforks, and sometimes with guns, to drive the surveyors back. At St. Helen's, one of the chainmen was laid hold of by a mob of colliers, and threatened to be hurled down a coal-pit. A number of men, women, and children, collected and ran after the surveyors wherever they made their appearance, bawling nicknames and throwing stones at them. As one of the chainmen was climbing over a gate one day, a labourer made at him with a pitchfork, and ran it through his clothes into his back; other watchers running up, the chainman, who was more stunned than hurt, took to his heels and fled. But that mysterious-looking instrument — the theodolite — most excited the fury of the natives, who concentrated on the man who carried it their fiercest execrations and most offensive nicknames.

A powerful fellow, a noted bruiser, was hired by the surveyors to carry the instrument, with a view to its protection against all assailants; but one day an equally powerful fellow, a St. Helen's collier, cock of the walk in his neighbourhood, made up to the theodolite bearer to wrest it from him by sheer force. A battle took place, the collier was soundly pummelled, the natives poured in volleys of stones upon the surveyors and their instruments, and the theodolite was smashed to pieces.

From a letter before us, written by Mr. James to Mr. Sandars, on the 21st October, 1821, it appears that an outline-survey had then been made, and the notices were published of the intended application to Parliament. Mr. James there states that he is "going to Newcastle principally to get a certificate from Stephenson of the operations of his engine. Until a deputation goes down, it may serve to prevent the existence and spread of doubts, which are so mortifying to honourable intentions." Mr. James accordingly proceeded to Killingworth, and his son, who accompanied him, has informed us of the result of the visit. Mr. James was not so fortunate as to meet Mr. Stephenson on the occasion; but he examined the locomotive at work, and was very much struck by its power and efficiency. He saw at a glance the magnificent uses to which it might be applied. "Here," said he, "is an engine that will, before long, effect a complete revolution in society." Returning to Moreton-in-the-Marsh, he wrote to Mr. Losh (Stephenson's partner in the patent) expressing his admiration of the Killingworth engine. "It is," said he, "the greatest wonder of the age, and the forerunner, as I firmly believe, of the most important changes in the internal communications of the kingdom." Mr. Losh invited him again to visit Killingworth, for the purpose of having an interview with Mr. Stephenson on the subject of his locomotive. Accordingly, in September of the same year, Mr. James, accompanied by his two sons, made a second journey to Killingworth, where he met both Losh and Stephenson. The visitors were at once taken to where the locomotive was working, and invited to mount it. The uncouth and extraordinary appearance of the machine, as it came snorting along, was somewhat alarming to the youths, who expressed their fears lest it should burst; and they were with some difficulty induced to mount.

The engine went through its usual performances, dragging a heavy load of coal-waggons at about six miles an hour, with apparent ease, at which Mr. James expressed his extreme satisfaction, and declared to Mr. Losh his opinion that Stephenson "was the greatest practical genius of the age," and that, "if he developed the full powers of that engine (the locomotive), his fame in the world would rank equal to that of Watt." Mr. James informed Stephenson and Losh of his survey of the proposed tramroad between Liverpool and Manchester, and did not hesitate to state that he would thenceforward advocate the adoption of a locomotive railroad instead of the tramroad which had originally been proposed.

Stephenson and Losh were naturally desirous of enlisting James's good services on behalf of their patent locomotive, for as yet it had proved comparatively unproductive. They believed that he might be able so to advocate it in influential quarters as to ensure its more extensive adoption, and with this object they proposed to give him an interest in their patent. Accordingly they assigned him one-fourth of the profits derived from the use of their patent locomotive on any lines which might be constructed south of a line drawn across England from Liverpool to Hull. The arrangement, however, led to no beneficial results. Mr. James endeavoured to introduce the engine on the Moreton-on-Marsh Railway; but it was opposed by the engineer of the line, and the attempt failed. He next urged that a locomotive should be sent for trial upon the Merstham tramroad; but, anxious though Stephenson was respecting its extended employment, he was too cautious to risk an experiment which might only bring discredit upon the engine; and the Merstham road being only laid with cast-iron plates, which would not bear its weight, the invitation was declined.

It turned out that the first survey of the Liverpool and Manchester line was very imperfect, and it was determined to have a second and more complete one made in the following year. Robert Stephenson was sent over by his father to Liverpool to assist in this survey. He was present with Mr. James on the occasion on which he tried to lay out the line across Chat Moss,—a proceeding which was not only difficult but dangerous. The Moss was very wet at the time, and only its edges could be ventured on. Mr. James was a heavy, thick-set man ; and one day, when endeavouring to obtain a stand for his theodolite, he felt himself suddenly sinking. He immediately threw himself down, and rolled over and over until he reached firm ground again, in a sad mess. Other attempts which he subsequently made to enter upon the Moss for the same purpose, were abandoned for the same reason—the want of a solid stand for the theodolite.

On the 4th October, 1822, we find Mr. James writing to Mr. Sandars, " I came last night to send my aid, Robert Stephenson, to his father, and to-morrow I shall pay off Evans and Hamilton, two other assistants. I have now only Messrs. Padley and Clarke to finish the copy of plans for Parliament, which will be done in about a week or nine days' time." It would appear however, that, notwithstanding all his exertions, Mr. James was unable to complete his plans and estimates in time for the ensuing Session of Parliament; and another year was thus lost. The Railroad Committee became impatient at the delay. Mr. James's financial embarrassments reached their climax ;[1]

---

[1] In 'The Two Jameses and the Two Stephensons' (London, 1861), the following letter is given, from Robert Stephenson (then not quite twenty years of age) to William James, dated Newcastle, August 29th, 1823 :—" Dear Sir,—It gives rise to feelings of true regret when I reflect

and, what with illness and what with debt, he was no longer in a position to fulfil his promises to the Committee. They were, therefore, under the necessity of calling to their aid some other engineer.

Mr. Sandars had by this time visited George Stephenson at Killingworth, and, like all who came within reach of his personal influence, was charmed with him at first sight. The energy which he had displayed in carrying on the works of the Stockton and Darlington Railway, now approaching completion; his readiness to face difficulties, and his practical ability in overcoming them; the enthusiasm which he displayed on the subject of railways and railway locomotion,—concurred in satisfying Mr. Sandars that he was, of all men, the best calculated to help forward the Liverpool undertaking at this juncture. On his return he stated this opinion to the Committee, who approved his recommendation, and George Stephenson was unanimously appointed engineer of the projected railway. On the 25th May, 1824, Mr. Sandars writes to Mr. James,—" I think it right to inform you that the Committee have engaged your friend Mr. George Stephenson. We expect him here in a few days. The subscription list for 300,000l. is filled, and the Manchester gentlemen have conceded to us the entire management. I very much regret that, by delay and promises, you have forfeited the confidence of the sub-

---

on your situation; but yet a consolation arises when I consider your persevering spirit will for ever bear you up in the arms of triumph, instances of which I have witnessed of too forcible a character to be easily effaced from my memory.. It is these thoughts, and these alone, that could banish from my soul feelings of despair for one, the respect I have for whom can be easier conceived than described. Can I forget the advice you have afforded me in your letters? and what a heavenly inducement you pointed before me at the close, when you said that attention and obedience to my dear father would afford me music at midnight. Ah, and so it has already. . . . . My father and I set off for London on Monday next, on our way to Cork. Our return will probably be about the time you wish me to be at Liverpool. If all be right, we may possibly call and see what is going on. That line [the Liverpool and Manchester] is the finest project in England. Hoping to see you and Mr. Padley in a few days, believe me, &c. &c., ROBERT STEPHENSON."

scribers. I cannot help it. I fear now that you will
only have the fame of being connected with the commencement of the undertaking."[1]

It will be observed that Mr. Sandars had held to his
original purpose with great determination and perseverance, and he gradually succeeded in enlisting on
his side an increasing number of influential merchants
and manufacturers both at Liverpool and Manchester.
Early in 1824 he published a pamphlet, in which he
strongly urged the great losses and interruptions to the
trade of the district by the delays in the forwarding of
merchandise; and in the same year he had a Public De-
claration drawn up, and signed by upwards of 150 of the
principal merchants of Liverpool, setting forth that they
considered " the present establishments for the transport
of goods quite inadequate, and that a new line of conveyance has become absolutely necessary to conduct the
increasing trade of the country with speed, certainty,
and economy."

A public meeting was then held to consider the best
plan to be adopted, and resolutions were passed in favour
of a railroad. A committee was appointed to take the
necessary measures; but, as if reluctant to enter upon
their arduous struggle with the " vested interests," they
first waited on Mr. Bradshaw, the Duke of Bridgewater's canal agent, in the hope of persuading him to
increase the means of conveyance, as well as to reduce
the charges; but they were met by an unqualified
refusal. They suggested the expediency of a railway,

---

[1] In 1858 Mr. Robert Stephenson
sent the author a large bundle of
letters, which had been forwarded to
him by Mr. Sandars, "descriptive of
the birth and progress of the Liverpool and Manchester Railway." In
the letter accompanying them Mr.
Stephenson said, " there is a bundle
of James's, which characterise the
man very clearly as a ready, dashing writer, but no thinker at all on
the practical part of the subject he
had taken up. It was the same with
everything he touched. He never
succeeded in anything, and yet possessed a great deal of taking talent.
His fluency of conversation I never
heard equalled, and so you would
judge from his letters."

and invited Mr. Bradshaw to become a proprietor of
shares in it. But his reply was—" All or none !" The
canal proprietors, confident in their imagined security,
ridiculed the proposed railway as a chimera. It had
been spoken about years before, and nothing had come
of it then : it would be the same now.

In order to form a better opinion as to the practicability of the railroad, a deputation of gentlemen interested in the project proceeded to Killingworth, to inspect the engines which had been so long in use there.
They first went to Darlington, where they found the
works of the Stockton line in progress, though still unfinished. Proceeding next to Killingworth with Mr.
Stephenson, they there witnessed the performances of
his locomotive engines. The result of their visit was,
on the whole, so satisfactory, that on their report being
delivered to the committee at Liverpool, it was finally
determined to form a company of proprietors for the
construction of a double line of railway between Liverpool and Manchester.

The first prospectus of the scheme was dated the 29th
of October, 1824, and had attached to it the names of
the leading merchants of Liverpool and Manchester. It
was a modest document, very unlike the inflated balloons
which were sent up by railway speculators in succeeding
years. It set forth as its main object the establishment
of a safe and cheap mode of transit for merchandise, by
which the conveyance of goods between the two towns
would be effected in five or six hours (instead of thirty-
six hours, as by the canal), whilst the charges would be
reduced one-third. On looking at the prospectus now,
it is curious to note that, while the advantages anticipated from the carriage of merchandise were strongly
insisted upon, the conveyance of passengers—which
proved to be the chief source of profit—was only very
cautiously referred to. " As a cheap and expeditious
means of conveyance for travellers," says the prospectus

in conclusion, " the railway holds out the fair prospect
of a public accommodation, the magnitude and importance of which cannot be immediately ascertained." The
estimated expense of forming the line was set down at
400,000l.,—a sum which was eventually found to be
quite inadequate. The subscription list when opened
was filled up without difficulty.

While the project was still under discussion, its promoters, desirous of removing the doubts which existed
as to the employment of steam power on the proposed
railway, sent a second deputation to Killingworth
for the purpose of again observing the action of Mr.
Stephenson's engines. The deputation was on this
occasion accompanied by Mr. Sylvester, an ingenious
mechanic and engineer, who afterwards presented an
able report on the subject to the committee. Mr. Sylvester showed that the high-pressure engines employed
by Mr. Stephenson were both safe and economical in
their working. With respect to the speed of the engines,
he said :—" Although it would be practicable to go at
any speed limited by the means of creating steam, the
size of the wheels, and the number of strokes in the
engine, it would not be safe to go at a greater rate than
nine or ten miles an hour."

Satisfactory though the calculations and statements of
Mr. Sylvester were, the cautious projectors of the railway were not yet quite satisfied ; and a third journey
was made to Killingworth, in January, 1825, by several
gentlemen of the committee, accompanied by practical
engineers, for the purpose of being personal eye-wit-
nesses of what steam-carriages were able to perform
upon a railway. There they saw a train, consisting of
a locomotive and loaded waggons, weighing in all fifty-
four tons, travelling at the average rate of about seven
miles an hour, the greatest speed being about nine and
a half miles an hour. But when the engine was run by
itself, with only one waggon attached containing twenty

gentlemen, five of whom were engineers, the speed attained was from ten to twelve miles an hour.

In the mean time the survey was proceeded with, in the face of great opposition on the part of the proprietors of the lands through which the railway was intended to pass. The prejudices of the farming and labouring classes were strongly excited against the persons employed upon the ground, and it was with the greatest difficulty that the levels could be taken. This opposition was especially manifested when the attempt was made to survey the line through the properties of Lords Derby and Sefton, and also where it crossed the Duke of Bridgewater's canal. At Knowsley, Mr. Stephenson was driven off the ground by the keepers, and threatened with rough handling if found there again. Lord Derby's farmers also turned out their men to watch the surveying party, and prevent them entering upon any lands where they had the power of driving them off. Afterwards, Mr. Stephenson suddenly and unexpectedly went upon the ground with a body of surveyors and their assistants, who out-numbered Lord Derby's keepers and farmers, hastily collected to resist them; and this time they were only threatened with the legal consequences of their trespass. The same sort of resistance was offered by Lord Sefton's keepers and farmers, with whom the following ruse was adopted. A minute was concocted, purporting to be a resolution of the Old Quay Canal Company to oppose the projected railroad by every possible means, and calling upon landowners and others to afford every facility for making such a survey of the intended line as should enable the opponents to detect errors in the scheme of the promoters, and thereby ensure its defeat. A copy of this minute, without any signature, was exhibited by the surveyors who went upon the ground, and the farmers, believing them to have the sanction of the landlords, permitted them to proceed with the hasty completion of their survey.[1]

The principal opposition, however, was experienced from Mr. Bradshaw, the manager of the Duke of Bridgewater's canal property, who offered a vigorous and protracted resistance to the survey in all its stages. The Duke's farmers obstinately refused permission to enter upon their fields, although Mr. Stephenson offered to pay for any damage that might be done. Mr. Bradshaw positively refused his sanction in any case ; and being a strict preserver of game, with a large staff of keepers in his pay, he declared that he would order them to shoot or apprehend any persons attempting a survey over his property. But one moonlight night a survey was obtained by the following ruse. Some men, under the orders of the surveying party, were set to fire off guns in a particular quarter ; on which all the gamekeepers on the watch made off in that direction, and they were drawn away to such a distance in pursuit of the supposed poachers, as to enable a rapid survey to be made during their absence.

Mr. Stephenson, afterwards describing before Parliament the difficulties which he encountered in making the survey, said :—" I was threatened to be ducked in the pond if I proceeded, and, of course, we had a great deal of the survey to take by stealth, at the time when the people were at dinner. We could not get it done by night : indeed, we were watched day and night, and guns were discharged over the grounds

---

[1] Mr. Sandars, when forwarding to Robert Stephenson the original of this document (amongst the bundle of documents referred to in a previous note), added to it—"The foregoing was written by me, and given to Mr. Oliver, one of the surveyors of the railway intended to pass through Lord Derby and Lord Sefton's property. Lord Sefton never spoke to me afterwards when he found out the ruse that had been practised. I little thought then that railways would in the end overwhelm me." Mr. Sandars died at Taplow, Bucks, a few years since, unhappily in very reduced circumstances.

belonging to Captain Bradshaw to prevent us. I can state further that I was myself twice turned off Mr. Bradshaw's grounds by his men; and they said if I did not go instantly, they would take me up and carry me off to Worsley."

When the canal companies found that the Liverpool merchants were determined to proceed with their scheme —that they had completed their survey, and were ready to apply to Parliament for an Act to enable them to form the railway—they at last reluctantly, and with a bad grace, made overtures of conciliation. They promised to employ steam-vessels both on the Mersey and on the Canal. One of the companies offered to reduce its length by three miles, at a considerable outlay. At the same time they made a show of lowering their rates. But it was all too late; for the project of the railway had now gone so far that the promoters (who might have been conciliated by such overtures at an earlier period) felt they were fully committed to it, and that now they could not well draw back. Besides, the remedies offered by the canal companies could only have had the effect of staving off the difficulty for a brief season,—the absolute necessity of forming a new line of communication between Liverpool and Manchester becoming more urgent from year to year. Arrangements were therefore made for proceeding with the bill in the parliamentary session of 1825.

On this becoming known, the canal companies prepared to resist the measure tooth and nail. The public were appealed to on the subject; pamphlets were written and newspapers were hired to revile the railway. It was declared that its formation would prevent cows grazing and hens laying. The poisoned air from the locomotives would kill birds as they flew over them, and render the preservation of pheasants and foxes no longer possible. Householders adjoining the projected

line were told that their houses would be burnt up by
the fire thrown from the engine-chimneys; while the
air around would be polluted by clouds of smoke. There
would no longer be any use for horses; and if railways
extended, the species would become extinguished, and
oats and hay be rendered unsaleable commodities. Travelling by rail would be highly dangerous, and country inns
would be ruined. Boilers would burst and blow passengers to atoms. But there was always this consolation to wind up with—that the weight of the locomotive
would completely prevent its moving, and that railways, even if made, could *never* be worked by steam-
power.

Nevertheless, the canal companies of Leeds, Liverpool, and Birmingham, called upon every navigation
company in the kingdom to oppose railways wherever
they were projected, but more especially the Liverpool
and Manchester scheme, the battle with which they
evidently regarded as their Armageddon. A Birmingham journal invited a combined opposition to the measure, and a public subscription was entered into for the
purpose of making it effectual. The newspapers generally spoke of the project as a mere speculation; some
wishing it success, although greatly doubting; others
ridiculing it as a delusion, similar to the many other
absurd projects of that madly-speculative period. It
was a time when balloon companies proposed to work
passenger traffic through the air at forty miles an hour,
and when coaching companies projected carriages to run
on turnpikes at twelve miles an hour, with relays of
bottled gas instead of horses. There were companies for
the working of American gold and silver mines,—companies for cutting ship canals through Panama and
Nicaragua,—milk companies, burying companies, fish
companies, and steam companies of all sorts; and many,
less speculatively disposed than their neighbours, were

ready to set down the projected railways of 1825 as mere bubbles of a similarly delusive character.[1]

Among the most sagacious newspaper articles of the day, calling attention to the application of the locomotive engine to the purposes of rapid steam-travelling on railroads, was a series which appeared in 1824, in the *Scotsman* newspaper, then edited by Mr. Charles Maclaren. In those publications the wonderful powers of the locomotive were logically demonstrated, and the writer, arguing from the experiments on friction made more than half a century before by Vince and Coulomb, which scientific men seemed to have altogether lost sight of, clearly showed that, by the use of steam-power on railroads, the more rapid, as well as cheaper, transit of persons and merchandise might be confidently anticipated.

Not many years passed before the anticipations of the writer, sanguine and speculative though they were regarded at the time, were amply realised. Even Mr. Nicholas Wood, in 1825, speaking of the powers of the locomotive, and referring doubtless to the speculations of the *Scotsman* as well as of his equally sanguine friend Stephenson, observed —" It is far from my wish to promulgate to the world that the ridiculous expectations, or rather professions, of the enthusiastic speculist will be realised, and that we shall see engines travelling at the rate of twelve, sixteen, eighteen, or twenty miles an hour. Nothing could do more harm towards their general adoption and improvement than the promulgation of such nonsense." [2]

---

[1] " Many years ago I met in a public library with a bulky volume, consisting of the prospectuses of various projects bound up together, and labelled, ' Some of the bubbles of 1825.' Among the projects thus described, was one that has since been productive of the greatest and most rapid advance in the social condition of mankind effected since the first dawn of civilisation : it was the plan of the Company for constructing a railway between Liverpool and Manchester."—W. B. Hodge, in ' Journal of the Institute of Actuaries,' No. 40, July, 1860.

[2] Wood on Railroads. Ed. 1825, p. 290.

Indeed, when Mr. Stephenson, at the interviews with counsel, held previous to the Liverpool and Manchester bill going into Committee of the House of Commons, confidently stated his expectation of being able to impel his locomotive at the rate of twenty miles an hour, Mr. William Brougham, who was retained by the promoters to conduct their case, frankly told him that if he did not moderate his views, and bring his engine within a *reasonable* speed, he would " inevitably damn the whole thing, and be himself regarded as a maniac fit only for Bedlam."

Amongst the papers left by Mr. Sandars we find a letter addressed to him by Sir John Barrow .of the Admiralty, as to the proper mode of conducting the case in Parliament, which pretty accurately represents the state of public opinion as to the practicability of locomotive travelling on railroads, at the time at which it was written, the 10th of January, 1825. Sir John strongly urged Mr. Sandars to keep the locomotive altogether in the background,—to rely upon the proved inability of the canals and common roads to accommodate the existing traffic,—and to be satisfied with proving the absolute necessity of a new line of conveyance ; above all, he recommended him not even to hint at the intention of carrying passengers. " My objection to great speed being attended with danger," said he, " applies only to the conveyance of passengers, and not to vehicles appended to the extremity of a long string of waggons, in which, however, I still think you will not get many who will suffer themselves to be conveyed even at the rate of eight miles an hour, amidst the hissing noise and the dense smoke of their own and other passing engines. . . I think it would be wise, for the present at least, to give up the passengers, for it is *there* you will fail, if you persevere. You will at once raise a host of enemies in the proprietors of coaches, post-chaises, innkeepers, &c., whose interests will be attacked, and who, I have no doubt, will be strongly supported, and for what? Some thousands of passengers, *you* say—but a few hundreds *I* should say—in the year." He accordingly urged that *passengers* as well as *speed* should be kept entirely out of the act; but if the latter were insisted on, then he recommended that it should be kept as low as possible—say at five miles an hour.

The idea thrown out by Stephenson, of travelling at a rate of speed double that of the fastest mail-coach, appeared at the time so preposterous that he was unable to find any engineer who would risk his reputation in supporting such "absurd views." Speaking of his isolation at the time, he subsequently observed, at a public meeting of railway men in Manchester: "He remembered the time when he had very few supporters in bringing out the railway system—when he sought England over for an engineer to support him in his evidence before Parliament, and could find only one man, James Walker, but was afraid to call that gentleman, because be knew nothing about railways. He had then no one to tell his tale to but Mr. Sandars, of Liverpool, who did listen to him, and kept his spirits up; and his schemes had at length been carried out only by dint of sheer perseverance."

George Stephenson's idea was at that time regarded as but the dream of a chimerical projector. It stood before the public friendless, struggling hard to gain a footing, and scarcely daring to lift itself into notice for fear of ridicule. The civil engineers generally rejected the notion of a Locomotive Railway; and when no leading man of the day could be found to stand forward in support of the Killingworth mechanic, its chances of success must indeed have been pronounced but small.

When such was the hostility of the civil engineers, no wonder the reviewers were puzzled. The 'Quarterly,' in an able article in support of the projected Liverpool and Manchester Railway,—while admitting its *absolute*

necessity, and insisting that there was no choice left but
a railroad, on which the journey between Liverpool and
Manchester, whether performed by horses or engines,
would always be accomplished " within the day,"—
nevertheless scouted the idea of travelling at a greater
speed than eight or nine miles an hour.   Adverting to
a project for forming a railway to Woolwich, by which
passengers were to be drawn by locomotive engines,
moving with twice the velocity of ordinary coaches, the
reviewer observed :—" What can be more palpably
absurd and ridiculous than the prospect held out of
locomotives travelling *twice as fast* as stage-coaches !
We would as soon expect the people of Woolwich to
suffer themselves to be fired off upon one of Congreve's
ricochet rockets, as trust themselves to the mercy of such
a machine going at such a rate.   We will back old
Father Thames against the Woolwich Railway for any
sum.   We trust that Parliament will, in all railways it
may sanction, limit the speed to *eight or nine miles an
hour*, which we entirely agree with Mr. Sylvester is as
great as can be ventured on with safety."

SURVEYING ON CHAT MOSS.

CHAPTER XI.

PARLIAMENTARY CONTEST ON THE LIVERPOOL AND MANCHESTER
BILL.

THE Liverpool and Manchester Bill went into Committee
of the House of Commons on the 21st of March, 1825.
There was an extraordinary array of legal talent on the
occasion, but especially on the side of the opponents to
the measure. Their wealth and influence enabled them
to retain the ablest counsel at the bar; Mr. (afterwards
Baron) Alderson, Mr. Stephenson, Mr. (afterwards Baron)
Parke, Mr. Rose, Mr. Macdonnell, Mr. Harrison, Mr. Erle,
and Mr. Cullen, made common cause with each other in
their opposition to the bill; the case for which was conducted by Mr. Adam, Mr. Serjeant Spankie, Mr. William
Brougham, and Mr. Joy.

Evidence was taken at great length as to the difficulties
and delays in forwarding raw goods of all kinds from
Liverpool to Manchester, as also in the conveyance of
manufactured articles from Manchester to Liverpool.
The evidence adduced in support of the bill on these
grounds was overwhelming. The utter inadequacy of
the existing modes of conveyance to carry on satisfactorily
the large and rapidly-growing trade between the two
towns was fully proved. But then came the gist of the
promoters' case—the evidence to prove the practicability
of a railroad to be worked by locomotive power. Mr.
Adam, in his opening speech, referred to the cases of
the Hetton and the Killingworth railroads, where heavy
goods were safely and economically transported by means
of locomotive engines. "None of the tremendous consequences," he observed, "have ensued from the use of
steam in land carriage that have been stated. The

horses have not started, nor the cows ceased to give their milk, nor have ladies miscarried at the sight of these things going forward at the rate of four miles and a half an hour." Notwithstanding the petition of two ladies alleging the great danger to be apprehended from the bursting of the locomotive boilers, he urged the safety of the high-pressure engine when the boilers were constructed of wrought-iron; and as to the rate at which they could travel, he expressed his full conviction that such engines "could supply force to drive a carriage at the rate of five or six miles an hour."

The taking of the evidence as to the impediments thrown in the way of trade and commerce by the existing system extended over a month, and it was the 21st of April before the Committee went into the engineering evidence, which was the vital part of the question.

On the 25th, George Stephenson was called into the witness-box. It was his first appearance before a Committee of the House of Commons, and he well knew what he had to expect. He was aware that the whole force of the opposition was to be directed against him; and if they could break down his evidence, the canal monopoly might yet be upheld for a time. Many years afterwards, when looking back at his position on this trying occasion, he said :—" When I went to Liverpool to plan a line from thence to Manchester, I pledged myself to the directors to attain a speed of ten miles an hour. I said I had no doubt the locomotive might be made to go much faster, but that we had better be moderate at the beginning. The directors said I was quite right; for that if, when they went to Parliament, I talked of going at a greater rate than ten miles an hour, I should put a cross upon the concern. It was not an easy task for me to keep the engine down to ten miles an hour, but it must be done, and I did my best. I had to place myself in that most unpleasant of all positions— the witness-box of a Parliamentary Committee. I was not long in it, before I began to wish for a hole to creep out at! I could not find words to satisfy either the Committee or myself. I was subjected to the cross-examination of eight or ten barristers, purposely, as far as possible, to bewilder me. Some member of the Committee asked if I was a foreigner, and another hinted that I was mad. But I put up with every rebuff, and went on with my plans, determined not to be put down."

Mr. Stephenson stood before the Committee to prove what the public opinion of that day held to be impossible. The self-taught mechanic had to demonstrate the practicability of accomplishing that which the most distinguished engineers of the time regarded as impracticable. Clear though the subject was to himself, and familiar as he was with the powers of the locomotive, it was no easy task for him to bring home his convictions, or even to convey his meaning, to the less informed minds of his hearers. In his strong Northumbrian dialect, he struggled for utterance, in the face of the sneers, interruptions, and ridicule of the opponents of the measure, and even of the Committee, some of whom shook their heads and whispered doubts as to his sanity, when he energetically avowed that he could make the locomotive go at the rate of twelve miles an hour! It was so grossly in the teeth of all the experience of honourable members, that the man "must certainly be labouring under a delusion !"

And yet his large experience of railways and locomotives, as described by himself to the Committee, entitled this "untaught, inarticulate genius," as he has so well been styled, to speak with confidence on such a subject. Beginning with his experience as a brakesman at Killingworth in 1803, he went on to state that he was appointed to take the entire charge of the steam-engines in 1813, and had superintended the railroads connected with the numerous collieries of the Grand Allies

from that time downwards. He had laid down or
superintended the railways at Burradon, Mount Moor,
Springwell, Bedlington, Hetton, and Darlington, besides
improving those at Killingworth, South Moor, and
Derwent Crook. He had constructed fifty-five steam-
engines, of which sixteen were locomotives. Some of
these had been sent to France. The engines constructed
by him for the working of the Killingworth Railroad,
eleven years before, had continued steadily at work ever
since, and fulfilled his most sanguine expectations. He
was prepared to prove the safety of working high-
pressure locomotives on a railroad, and the superiority
of this mode of transporting goods over all others. As
to speed, he said he had recommended eight miles an
hour with twenty tons, and four miles an hour with
forty tons; but he was quite confident that much more
might be done. Indeed, he had no doubt they might
go at the rate of twelve miles. As to the charge that
locomotives on a railroad would so terrify the horses in
the neighbourhood, that to travel on horseback or to
plough the adjoining fields would be rendered highly
dangerous, the witness said that horses learnt to take no
notice of them, though there *were* horses that would shy
at a wheelbarrow. A mail-coach was likely to be more
shied at by horses than a locomotive. In the neighbourhood of Killingworth, the cattle in the fields went
on grazing while the engines passed them, and the
farmers made no complaints.

Mr. Alderson, who had carefully studied the subject,
and was well skilled in practical science, subjected the
witness to a protracted and severe cross-examination as
to the speed and power of the locomotive, the stroke of
the piston, the slipping of the wheels upon the rails,
and various other points of detail. Mr. Stephenson
insisted that no slipping took place, as attempted to be
extorted from him by the counsel. He said; " It is
impossible for slipping to take place so long as the

adhesive weight of the wheel upon the rail is greater than the weight to be dragged after it." There was a good deal of interruption to the witness's answers by Mr. Alderson, to which Mr. Joy more than once objected. As to accidents, Mr. Stephenson knew of none that had occurred with his engines. There had been one, he was told, at the Middleton Colliery, near Leeds, with a Blenkinsop engine. The driver had been in liquor, and put a considerable load on the safety-valve, so that upon going forward the engine blew up and the man was killed. But he added, if proper precautions had been used with that boiler, the accident could not have happened. The following cross-examination occurred in reference to the question of speed :—

" Of course," he was asked, " when a body is moving upon a road, the greater the velocity the greater the momentum that is generated?" " Certainly."—" What would be the momentum of forty tons moving at the rate of twelve miles an hour?" " It would be very great."—" Have you seen a railroad that would stand that?" " Yes."—" Where?" " Any railroad that would bear going four miles an hour : I mean to say, that if it would bear the weight at four miles an hour, it would bear it at twelve."—" Taking it at four miles an hour, do you mean to say that it would not require a stronger railway to carry the same weight twelve miles an hour?" " I will give an answer to that. I dare say every person has been over ice when skating, or seen persons go over, and they know that it would bear them better at a greater velocity than it would if they went slower; when they go quick, the weight in a measure ceases."—" Is not that upon the hypothesis that the railroad is perfect?" " It is; and I mean to make it perfect."

It is not necessary to state that to have passed such an ordeal scatheless, needed no small amount of courage, intelligence, and ready shrewdness on the part of the witness.   Nicholas Wood, who was present on the occasion, has since stated that the point on which Stephenson was hardest pressed was that of speed. " I believe," he says, " that it would have lost the Company their Bill if he had gone beyond eight or nine miles an hour.   If he had stated his intention of going twelve or fifteen miles an hour, not a single person would have believed it to be practicable."   Mr. Alderson, had, indeed, so pressed the point of "twelve miles an hour," and the promoters were so alarmed lest it should appear in evidence that they contemplated any such extravagant rate of speed, that immediately on Mr. Alderson sitting down, Mr. Joy proceeded to re-examine Mr. Stephenson, with the view of removing from the minds of the Committee an impression so unfavourable, and, as was supposed, so damaging to their case.   " With regard," asked Mr. Joy, " to all those hypothetical questions of my learned friend, they have been all put on the sup-position of going twelve miles an hour : now that is not the rate at which, I believe, any of the engines of which you have spoken have travelled ?"   " No," replied Mr. Stephenson, "except as an experiment for a short distance."—" But what they have gone has been three, five, or six miles an hour ? "   " Yes."—" So that those hypothetical cases of twelve miles an hour do not fall within your general experience ? " · " They do not."

The Committee also seem to have entertained some alarm as to the high rate of speed which had been spoken of, and proceeded to examine the witness further on the subject.   They supposed the case of the engine being upset when going at nine miles an hour, and asked what, in such a case, would become of the cargo astern.   To which the witness replied that it would not be upset.   One of the members of the Committee pressed the witness a little further.   He put the following case :—" Suppose, now, one of these engines to be going along a railroad at the rate of nine or ten miles an hour,

and that a cow were to stray upon the line and get in
the way of the engine; would not that, think you, be a
very awkward circumstance?" "Yes," replied the
witness, with a twinkle in his eye, "very awkward
—*for the coo!*" The honourable member did not proceed further with his cross-examination; to use a
railway phrase, he was "shunted." Another asked if
animals would not be very much frightened by the
engine passing at night, especially by the glare of the
red-hot chimney? "But how would they know that it
was'nt painted?" said the witness.

On the following day (the 26th April), Mr. Stephenson
was subjected to a very severe examination. On that
part of the scheme with which he was most practically
conversant, his evidence was clear and conclusive. Now,
he had to give evidence on the plans made by his
surveyors, and the estimates which had been founded on
such plans. So long as he was confined to locomotive
engines and iron railroads, with the minutest details of
which he was more familiar than any man living, he
felt at home, and in his element. But when the designs
of bridges and the cost of constructing them had to be
gone into, the subject being in a great measure new to
him, his evidence was much less satisfactory.

Mr. Alderson cross-examined him at great length on
the plans of the bridges, the tunnels, the crossings of
the roads and streets, and the details of the survey,
which, it soon clearly appeared, were in some respects
seriously at fault. It seems that, after the plans had
been deposited, Mr. Stephenson found that a much more
favourable line might be made; and he made his estimates accordingly, supposing that Parliament would not
confine the Company to the precise plan which had been
deposited. This was felt to be a serious blot in the parliamentary case, and one very difficult to be got over.

For three entire days was Mr. Stephenson subjected
to cross-examination by Mr. Alderson, Mr. Cullen, and

the other leading counsel for the opposition. He held
his ground bravely, and defended the plans and estimates with remarkable ability and skill; but it was
clear they were imperfect, and the result was on the
whole damaging to the bill. Mr. (afterwards Sir
William) Cubitt was called by the promoters,—Mr.
Adam stating that he proposed by this witness to correct
some of the levels as given by Mr. Stephenson. It
seems a singular course to have been taken by the
promoters of the measure; for Mr. Cubitt's evidence
went to upset the statements made by Mr. Stephenson as to the survey. This adverse evidence was,
of course, made the most of by the opponents of the
scheme.

Mr. Serjeant Spankie then summed up for the bill, on
the 2nd of May, in a speech of great length; and the
case of the opponents was next gone into, Mr. Harrison
opening with a long and eloquent speech on behalf of
his clients, Mrs. Atherton and others. He indulged
in strong vituperation against the witnesses for the
bill, and especially dwelt upon the manner in which
Mr. Cubitt, for the promoters, had proved that Mr.
Stephenson's levels were wrong. " They got a person,"
said he, "whose character and skill I do not dispute,
though I do not exactly know that I should have gone
to the inventor of the treadmill as the fittest man to
take the levels of Knowsley Moss and Chat Moss,
which shook almost as much as a treadmill, as you
recollect, for he (Mr. Cubitt) said Chat Moss trembled
so much under his feet that he could not take his
observations accurately. . . . . . In fact, Mr. Cubitt did
not go on to Chat Moss, because he knew that it was
an immense mass of pulp, and nothing else. It actually
rises in height, from the rain swelling it like a sponge,
and sinks again in dry weather; and if a boring instrument is put into it, it sinks immediately by its own
weight. The making of an embankment out of this

pulpy, wet moss, is no very easy task.   Who but Mr.
Stephenson would have thought of entering into Chat
Moss, carrying it out almost like wet dung?   It is
ignorance almost inconceivable.   It is perfect madness,
in a person called upon to speak on a scientific subject,
to propose such a plan. . . . . . Every part of the scheme
shows that this man has applied himself to a subject of
which he has no knowledge, and to which he has no
science to apply."   Then adverting to the proposal to
work the intended line by means of locomotives, the
learned gentleman proceeded : " When we set out with
the original prospectus, we were to gallop, I know not
at what rate ;—I believe it was at the rate of twelve
miles an hour.   My learned friend, Mr. Adam, contemplated—possibly alluding to Ireland—that some of
the Irish members would arrive in the waggons to a
division.   My learned friend says that they would go at
the rate of twelve miles an hour with the aid of the
devil in the form of a locomotive, sitting as postilion on
the fore horse, and an honourable member sitting behind
him to stir up the fire, and keep it at full speed.   But
the speed at which these locomotive engines are to go
has slackened : Mr. Adam does not go faster now than
five miles an hour   The learned serjeant (Spankie) says
he should like to have seven, but he would be content
to go six.   I will show he cannot go six ; and probably,
for any practical purposes, I may be able to show that
I can keep up with him *by the canal* . . . . . Locomotive
engines are liable to be operated upon by the weather.
You are told they are affected by rain, and an attempt
has been made to cover them ; but the wind will affect
them ; and any gale of wind which would affect the traffic
on the Mersey would render it *impossible* to set off a locomotive engine, either by poking of the fire, or keeping
up the pressure of the steam till the boiler was ready to
burst."   How amusing it now is to read these extraordinary views as to the formation of a railway over

Chat Moss, and the impossibility of starting a locomotive engine in the face of a gale of wind!

Evidence was called to show that the house property passed by the proposed railway would be greatly deteriorated — in some places almost destroyed; that the locomotive engines would be terrible nuisances, in consequence of the fire and smoke vomited forth by them; and that the value of land in the neighbourhood of Manchester alone would be deteriorated by no less than 20,000*l.*! Evidence was also given at great length showing the utter impossibility of forming a road of any kind upon Chat Moss. A Manchester builder, who was examined, could not imagine the feat possible, unless by arching it across in the manner of a viaduct from one side to the other. It was the old story of " nothing like leather." But the opposition mainly relied upon the evidence of the leading engineers—not, like Mr. Stephenson, self-taught men, but regular professionals. Mr. Francis Giles, C.E., was their great card. He had been twenty-two years an engineer, and could speak with some authority. His testimony was mainly directed to the utter impossibility of forming a railway over Chat Moss. " *No engineer in his senses,*" said he, " would go through Chat Moss if he wanted to make a railroad from Liverpool to Manchester. . . . . In my judgment *a railroad certainly cannot be safely made over Chat Moss without going to the bottom of the Moss.* The soil ought all to be taken out, undoubtedly; in doing which, it will not be practicable to approach each end of the cutting, as you make it, with the carriages. No carriages would stand upon the Moss short of the bottom. My estimate for the whole cutting and embankment over Chat Moss is 270,000*l.* nearly, at those quantities and those prices which are decidedly correct. . . . . It will be necessary to take this Moss completely out at the bottom, in order to make a solid road."

Mr. H. R. Palmer, C.E., gave evidence to prove that

resistance to a moving body going under four and a quarter miles an hour was *less* upon a canal than upon a railroad; and that, when going against a strong wind, the progress of a locomotive was retarded " very much." Mr. George Leather, C.E., the engineer of the Croydon and Wandsworth Railway, on which he said the waggons went at from two and a half to three miles an hour, also testified against the practicability of Mr. Stephenson's plan. He considered his estimate a " very wild " one. He himself had no confidence in locomotive power. The Weardale Railway, of which he was engineer, had given up the use of locomotive engines. He supposed that, when used, they travelled at three and a half to four miles an hour, because they were considered to be then more effective than at a higher speed.

When these distinguished engineers had given their evidence, Mr. Alderson summed up in a speech which extended over two days. He declared Mr. Stephenson's plan to be " the most absurd scheme that ever entered into the head of man to conceive. My learned friends," said he, " almost endeavoured to stop my examination ; they wished me to put in the plan, but I had rather have the exhibition of Mr. Stephenson in that box. I say he never had a plan—I believe he never had one— I do not believe he is capable of making one. His is a mind perpetually fluctuating between opposite difficulties : he neither knows whether he is to make bridges over roads or rivers, of one size or of another ; or to make embankments, or cuttings, or inclined planes, or in what way the thing is to be carried into effect. Whenever a difficulty is pressed, as in the case of a tunnel, he gets out of it at one end, and when you try to catch him at that, he gets out at the other." Mr. Alderson proceeded to declaim against the gross ignorance of this so-called engineer, who proposed to make " impossible ditches by the side of an impossible railway " through Chat Moss ; and he contrasted with his evidence that given " by that most respectable gentleman we have called before you, I mean Mr. Giles, who has executed a vast number of works," &c. Then Mr. Giles's evidence as to the impossibility of making any railway over the Moss that would stand short of the bottom, was emphatically dwelt upon; and Mr. Alderson proceeded to say,—"Having now, sir, gone through Chat Moss, and having shown that Mr. Giles is right in his principle when he adopts a solid railway,—and I care not whether Mr. Giles is right or wrong in his estimate, for whether it be effected by means of piers raised up all the way for four miles through Chat Moss, whether they are to support it on beams of wood or by erecting masonry, or whether Mr. Giles shall put a solid bank of earth through it,— in all these schemes there is not one found like that of Mr. Stephenson's, namely, to cut impossible drains on the side of this road; and it is sufficient for me to suggest and to show, that this scheme of Mr. Stephenson's is impossible or impracticable, and that no other scheme, if they proceed upon this line, can be suggested which will not produce enormous expense. I think that has been irrefragably made out. Every one knows Chat Moss—every one knows that Mr. Giles speaks correctly when he says the iron sinks immediately on its being put upon the surface. I have heard of culverts, which have been put upon the Moss, which, after having been surveyed the day before, have the next morning disappeared; and that a house (a poet's house, who may be supposed in the habit of building castles even in the air), story after story, as fast as one is added, the lower one sinks! There is nothing, it appears, except long sedgy grass, and a little soil, to prevent its sinking into the shades of eternal night. I have now done, sir, with Chat Moss, and there I leave this railroad." Mr. Alderson, of course, called upon the Committee to reject the Bill; and he protested " against the despotism of the Exchange at Liverpool striding across the land of this

country. I do protest," he concluded, " against a measure
like this, supported as it is by such evidence, and founded
upon such calculations."

The case of the other numerous petitioners against
the bill still remained to be gone into. Witnesses were
called to prove the residential injury which would be
caused by the " intolerable nuisance " of the smoke and
fire from the locomotives ; and others to prove that the
price of coals and iron would " infallibly " be greatly
raised throughout the country. This was part of the
case of the Duke of Bridgewater's trustees, whose witnesses " proved " many very extraordinary things. The
Leeds and Liverpool Canal Company were so fortunate
as to pick up a witness from Hetton, who was ready
to furnish some damaging evidence as to the use of
Stephenson's locomotives on that railway. This was
Thomas Wood, one of the Hetton company's clerks,
whose testimony was to the effect that the locomotives,
having been found ineffective, were about to be discontinued in favour of fixed engines. The evidence of
this witness, incompetent though he was to give an
opinion on the subject, and exaggerated as his statements
were afterwards proved to be, was made the most of by
Mr. Harrison, when summing up the case of the canal
companies. " At length," he said, " we have come to
this,—having first set out at twelve miles an hour, the
speed of these locomotives is reduced to six, and now
comes down to two or two and a half. They must be
content to be pulled along by horses and donkeys ; and
all those fine promises of galloping along at the rate of
twelve miles an hour are melted down to a total failure
—the foundation on which their case stood is cut from
under them completely ; for the Act of Parliament, the
Committee will recollect, prohibits any person using
any animal power, of any sort, kind, or description,
except the projectors of the railway themselves ; therefore, I say, that the whole foundation on which this

project exists is gone." After further personal abuse of Mr. Stephenson, whose evidence he spoke of as "trash and confusion," he closed the case of the canal companies on the 30th of May. Mr. Adam replied for the promoters, recapitulating the principal points of their case, and vindicating Mr. Stephenson and the evidence which he had given before the Committee.

The Committee then divided on the preamble, which was carried by a majority of only *one*—thirty-seven voting for it, and thirty-six against it. The clauses were next considered, and on a division the first clause, empowering the Company to make the railway, was lost by a majority of nineteen to thirteen. In like manner, the next clause, empowering the Company to take land, was lost; on which Mr. Adam, on the part of the promoters, withdrew the bill.

Thus ended this memorable contest, which had extended over two months—carried on throughout with great pertinacity and skill, especially on the part of the opposition, who left no stone unturned to defeat the measure. The want of a third line of communication between Liverpool and Manchester had been clearly proved; but the engineering evidence in support of the proposed railway having been thrown almost entirely upon Stephenson, who fought this, the most important part of the battle, single-handed, was not brought out so clearly as it would have been had he secured more efficient engineering assistance—which he was not able to do, as the principal engineers of that day were against the locomotive railway. The obstacles thrown in the way of the survey by the landowners and canal companies, by which the plans were rendered exceedingly imperfect, also tended in a great measure to defeat the bill.

Mr. Gooch says the rejection of the bill was probably the most severe trial George Stephenson underwent in the course of his whole life. The circumstances connected with the defeat of the measure, the errors in the levels, his rigid cross-examination, followed by the fact of his being superseded by another engineer, all told fearfully upon him, and for some time he was as much weighed down as if a personal calamity of the most serious kind had befallen him.   It is also right to add that he was badly served by his surveyors, who were unpractised and incompetent.   On the 27th of September, 1824, we find him writing to Mr. Sandars: " I am quite shocked with Auty's conduct; we must throw him aside as soon as possible.   Indeed, I have begun to fear that he has been fee'd by some of the canal proprietors to make a botch of the job.   I have a letter from Steele,[1] whose views of Auty's conduct quite agree with yours."

The result of this first application to Parliament was so far discouraging.   Mr. Stephenson had been so terribly abused by the leading counsel for the opposition in the course of the proceedings before the Committee —stigmatised by them as an ignoramus, a fool, and a maniac—that even his friends seem for a time to have lost faith in him and in the locomotive system, whose efficiency he nevertheless continued to uphold.   Things never looked blacker for the success of the railway system than at the close of this great parliamentary struggle. And yet it was on the very eve of its triumph.

The Committee of Directors appointed to watch the measure in Parliament were so determined to press on the project of a railway, even though it should have to be worked merely by horse-power, that the bill had scarcely been thrown out ere they met in London to con-

---

[1] Hugh Steele and Elijah Galloway had conducted the survey at one part of the line, and Messrs. Oliver and Blackett at another.   The former couple seem to have made some grievous blunder in the levels on Chat Moss, and the circumstance weighed so heavily on Steele's mind that, shortly after hearing of the re- jection of the Bill, he committed suicide in Stephenson's office at Newcastle. Mr. Gooch informs us that this unhappy affair served to impress upon the minds of Stephenson's other pupils the necessity of ensuring greater accuracy and attention in future, and that the lesson, though sad, was not lost upon them.

sider their next step.   They called their parliamentary
friends together to consult as to future proceedings.
Among those who attended the meeting of gentlemen
with this object, in the Royal Hotel, St. James's Street,
on the 4th of June, were Mr. Huskisson, Mr. Spring
Rice, and General Gascoyne.   Mr. Huskisson urged
the promoters to renew their application to Parliament.
They had secured the first step by the passing of their
preamble ; the measure was of great public importance ;
and whatever temporary opposition it might meet with,
he conceived that Parliament must ultimately give its
sanction to the undertaking.   Similar views were expressed by other speakers; and the deputation went
back to Liverpool determined to renew their application
to Parliament in the ensuing session.

It was not considered desirable to employ Mr. Stephenson in making the new survey.   He had not as yet
established his reputation as an engineer beyond the
boundaries of his own district ; and the promoters of the
bill had doubtless felt the disadvantages of this in the
course of their parliamentary struggle.   They therefore
resolved now to employ engineers of the highest established reputation, as well as the best surveyors that
could be obtained.   In accordance with these views
they engaged Messrs. George and John Rennie to be
the engineers of the railway ; and Mr. Charles Vignolles,
on their behalf, was appointed to prepare the plans and
sections.   The line which was eventually adopted differed somewhat from that surveyed by Mr. Stephenson
—entirely avoiding Lord Sefton's property, and passing
through only a few detached fields of Lord Derby's at a
considerable distance from the Knowsley domain.   The
principal game-preserves of the district were carefully
avoided.   The promoters thus hoped to get rid of the
opposition of the most influential of the resident landowners.   The crossing of certain of the streets of Liverpool was also avoided, and the entrance contrived by

means of a tunnel and an inclined plane. The new line stopped short of the river Irwell at the Manchester end, by which the objections grounded on an illegal interruption to the canal or river traffic were in some measure removed. With reference to the use of the locomotive engine, the promoters, remembering with what effect the objections to it had been urged by the opponents of the bill, intimated, in their second prospectus, that "as a guarantee of their good faith towards the public they will not require any clause empowering them to use it; or they will submit to such restrictions in the employment of it as Parliament may impose, for the satisfaction and ample protection both of proprietors on the line of road and of the public at large."

It was found that the capital required to form the line of railway, as laid out by the Messrs. Rennie, was considerably beyond the amount of Stephenson's estimate, and it became a question with the Committee in what way the new capital should be raised. A proposal was made to the Marquis of Stafford, who was principally interested in the Duke of Bridgewater's Canal, to become a shareholder in the railway. A similar proposal, it will be remembered, had at an earlier period been made to Mr. Bradshaw, the trustee for the property; but his answer was "all or none," and the negotiation was broken off. The Marquis of Stafford, however, now met the projectors of the railway in a more conciliatory spirit; and it was ultimately agreed that he should become a subscriber to the extent of a thousand shares.

The survey of the new line having been completed, the plans were deposited, the standing orders duly complied with, and the bill went before Parliament. The same counsel appeared for the promoters, but the examination of witnesses was not nearly so protracted as on the previous occasion. Mr. Erle and Mr. Harrison led the case of the opposition. The bill went into Com-

mittee on the 6th of March, and on the 16th the
preamble was declared proved by a majority of forty-
three to eighteen.   On the third reading in the House
of Commons, an animated, and what now appears a
very amusing, discussion took place. The Hon. Edward
Stanley moved that the bill be read that day six months;
and in the course of his speech he undertook to prove
that the railway trains would take *ten hours* on the
journey, and that they could only be worked by horses.
Sir Isaac Coffin seconded the motion, and in doing so
denounced the project as a most flagrant imposition.
He would not consent to see widows' premises invaded;
and "What, he would like to know, was to be done
with all these who had advanced money in making and
repairing turnpike-roads?   What with those who may
still wish to travel in their own or hired carriages, after
the fashion of their forefathers?   What was to become
of coach-makers and harness-makers, coach-masters and
coachmen, inn-keepers, horse-breeders, and horse-dealers?
Was the House aware of the smoke and the noise, the
hiss and the whirl, which locomotive engines, passing
at the rate of ten or twelve miles an hour, would occasion?   Neither the cattle ploughing in the fields or
grazing in the meadows could behold them without
dismay.   Iron would be raised in price 100 per cent.,
or more probably exhausted altogether!   It would be
the greatest nuisance, the most complete disturbance of
quiet and comfort in all parts of the kingdom, that the
ingenuity of man could invent!"

Mr. Huskisson and other speakers, though unable to
reply to such arguments as these, strongly supported
the bill; and it was carried on the third reading by a
majority of eighty-eight to forty-one. The bill passed
the House of Lords almost unanimously, its only opponents being the Earl of Derby and his relative the Earl
of Wilton.   The cost of obtaining the Act amounted to
the enormous sum of 27,000*l.*

CHAPTER XII.

CHAT MOSS — CONSTRUCTION OF THE RAILWAY.

THE selection of principal engineer of the railway was taken into consideration at the first meeting of the directors held at Liverpool subsequent to the passing of the Act of incorporation. The magnitude of the proposed works, and the vast consequences involved in the experiment, were deeply impressed upon their minds; and they resolved to secure the services of a resident engineer of proved experience and ability. Their attention was naturally directed to Mr. Stephenson; at the same time they desired to have the benefit of the Messrs. Rennie's professional assistance in superintending the works. Mr. George Rennie had an interview with the board on the subject, at which he proposed to undertake the chief superintendence, making six visits in each year, and stipulating that he should have the appointment of the resident engineer. But the responsibility attaching to the direction in the matter of the efficient carrying on of the works, would not admit of their being influenced by ordinary punctilios on the occasion; and they accordingly declined Mr. Rennie's proposal, and proceeded to appoint Mr. Stephenson their principal engineer at a salary of 1000*l.* per annum.

He at once removed his residence to Liverpool, and made arrangements to commence the works. He began with the "impossible thing"—to do that which the most distinguished engineers of the day had declared that "no man in his senses would undertake to do"— namely, to make the road over Chat Moss! It was indeed a most formidable undertaking; and the project

of carrying a railway along, under, or over such a material as that of which it consisted, would certainly never
have occurred to an ordinary mind.   Michael Drayton
supposed the Moss to have had its origin at the Deluge.
Nothing more impassable could have been imagined
than that dreary waste ; and Mr. Giles only spoke the
popular feeling of the day when he declared that no
carriage could stand on it " short of the bottom."   In
this bog, singular to say, Mr. Roscoe, the accomplished
historian of the Medicis, buried his fortune in the hopeless attempt to cultivate a portion of it which he had
bought.

Chat Moss is an immense peat-bog of about twelve
square miles in extent.   Unlike the bogs or swamps of
Cambridge and Lincolnshire, which consist principally
of soft mud or silt, this bog is a vast mass of spongy
vegetable pulp, the result of the growth and decay of
ages.   The spagni, or bog-mosses, cover the entire area ;
one year's growth rising over another, — the older
growths not entirely decaying, but remaining partially
preserved by the antiseptic properties peculiar to peat.
Hence the remarkable fact that, although a semifluid
mass, the surface of Chat Moss rises above the level of
the surrounding country.   Like a turtle's back, it declines
from the summit in every direction, having from thirty
to forty feet gradual slope to the solid land on all sides.
From the remains of trees, chiefly alder and birch,
which have been dug out of it, and which must have
previously flourished upon the surface of soil now deeply
submerged, it is probable that the sand and clay base
on which the bog rests is saucer-shaped, and so retains
the entire mass in position.   In rainy weather, such is
its capacity for water that it sensibly swells, and rises in
those parts where the moss is the deepest.   This occurs
through the capillary attraction of the fibres of the submerged moss, which is from twenty to thirty feet in
depth, whilst the growing plants effectually check evapo-

ration from the surface. This peculiar character of the
Moss has presented an insuperable difficulty in the way
of reclaiming it by any system of extensive drainage—
such as by sinking shafts in its substance, and pumping
up the water by steam power, as has been proposed by
some engineers. Supposing a shaft of thirty feet deep
to be sunk, it has been calculated that this would only
be effectual for draining a circle of about one hundred
yards, the water running down an incline of about 5 to
1 ; for it was found in the course of draining the bog,
that a ditch three feet deep only served to drain a space
of less than five yards on either side, and two ditches of
this depth, ten feet apart, left a portion of the Moss
between them scarcely affected by the drains.

The three resident engineers selected by Mr. Stephenson to superintend the construction of the line, were
Mr. Joseph Locke, Mr. Allcard, and Mr. John Dixon.
The last was appointed to that portion which included
the proposed road across the Moss, the other two being
by no means desirous of exchanging posts with him.
On Mr. Dixon's arrival, about the month of July, 1826,
Mr. Locke proceeded to show him over the length
he was to take charge of, and to instal him in office.
When they reached Chat Moss, Mr. Dixon found that
the line had already been staked out and the levels
taken in detail by the aid of planks laid upon the bog.
The cutting of the drains along each side of the proposed
road had also been commenced ; but the soft pulpy stuff
had up to this time flowed into the drains and filled
them up as fast as they were cut. Proceeding across
the Moss, on the first day's inspection, the new resident,
when about half-way over, slipped off the plank on
which he walked, and sank to his knees in the bog.
Struggling only sent him the deeper, and he might
have disappeared altogether, but for the workmen,
who hastened to his assistance upon planks, and rescued
him from his perilous position. Much disheartened, he

desired to return, and even for the moment thought of
giving up the job; but Mr. Locke assured him that the
worst part was now past; so the new resident plucked
up heart again, and both floundered on until they
reached the further edge of the Moss, wet and plastered
over with bog sludge.　Mr. Dixon's companions endeavoured to comfort him by the assurance that he might
in future avoid similar perils, by walking upon " pattens,"
or boards fastened to the soles of his feet, as they had
done when taking the levels, and as the workmen did
when engaged in making drains in the softest parts of
the Moss.　Still the resident engineer could not help
being puzzled by the problem of how to construct a
road for heavy locomotives, with trains of passengers
and goods, upon a bog which he had found incapable of
supporting his single individual weight!

Mr. Stephenson's idea was, that such a road might be
made to *float* upon the bog, simply by means of a
sufficient extension of the bearing surface.　As a ship,
or a raft, capable of sustaining heavy loads, floated in
water, so in his opinion, might a light road be floated
upon a bog, which was of considerably greater con-
sistency than water.　Long before the railway was
thought of, Mr. Roscoe had adopted the remarkable
expedient of fitting his plough horses with flat wooden
soles or pattens, to enable them to walk upon the Moss
land which he had brought into cultivation.　These
pattens were fitted on by means of a screw apparatus,
which met in front of the foot and was easily fastened.
The mode by which these pattens served to sustain the
horse is capable of easy explanation, and it will be
observed that the *rationale* alike explains the floating of
a railway train.　The foot of an ordinary farm horse
presents a base of about five inches diameter, but if this
base be enlarged to seven inches—the circles being to
each other as the squares of the diameters—it will be
found that, by this slight enlargement of the base, a

circle of nearly double the area has been secured; and consequently the pressure of the foot upon every unit of ground upon which the horse stands has been reduced one half. In fact, this contrivance has an effect tantamount to setting the horse upon eight feet instead of four.

Apply the same reasoning to the ponderous locomotive, and it will be found, that even such a machine may be made to stand upon a bog, by means of a similar extension of the bearing surface. Suppose the engine to be twenty feet long and five feet wide, thus covering a surface of a hundred square feet, and, provided the bearing has been extended by means of cross sleepers supported upon a matting of heath and branches of trees covered with a few inches of gravel, the pressure of an engine of twenty tons will be only equal to about three pounds per inch over the whole surface on which it stands. Such was George Stephenson's idea in contriving his floating road—something like an elongated raft across the Moss; and we shall see that he steadily kept it in view in carrying the work into execution.

The first thing done was to form a footpath of ling or heather along the proposed road, on which a man might walk without risk of sinking. A single line of temporary railway was then laid down, formed of ordinary cross-bars about three feet long and an inch square, with holes punched through them at the end and nailed down to temporary sleepers. Along this way ran the waggons in which were conveyed the materials requisite to form the permanent road. These waggons carried about a ton each, and they were propelled by boys running behind them along the narrow bar of iron. The boys became so expert that they would run the four miles across at the rate of seven or eight miles an hour without missing a step; if they had done so, they would have sunk in many places up to their middle.[1] The

---

[1] When the Liverpool directors went to inspect the works in progress on the Moss, they were run along the temporary rails in the little three-feet

slight extension of the bearing surface was thus sufficient to enable the bog to bear this temporary line, and the circumstance was a source of increased confidence and hope to our engineer in proceeding with the formation of the permanent road alongside.

The digging of drains had been proceeding for some time along each side of the intended railway; but they filled up almost as soon as dug, the sides flowing in, and the bottom rising up; and it was only in some of the drier parts of the bog that a depth of three or four feet could be reached. The surface-ground between the drains, containing the intertwined roots of heather and long grass, was left untouched, and upon this was spread branches of trees and hedge-cuttings; in the softest places rude gates or hurdles, some eight or nine feet long by four feet wide, interwoven with heather, were laid in double thicknesses, their ends overlapping each other; and upon this floating bed was spread a thin layer of gravel, on which the sleepers, chairs, and rails were laid in the usual manner. Such was the mode in which the road was formed upon the Moss.

It was found, however, after the permanent road had been thus laid, that there was a tendency to sinking at those parts where the bog was the softest. In ordinary cases, where a bank subsides, the sleepers are packed up with ballast or gravel; but in this case the ballast was dug away and removed in order to lighten the road, and the sleepers were packed instead with cakes of dry turf or bundles of heath. By these expedients the subsided parts were again floated up to the level, and an approach was made towards a satisfactory road. But the most formidable difficulties were encountered at the centre

---

gauge-waggons used for forming the road. They were being thus impelled one day at considerable speed, when the waggon suddenly ran off the road, and Mr. Moss, one of the directors, was thrown out in a soft place, from which, however, he was speedily extricated, not without leaving his deep mark. George used afterwards laughingly to refer to the circumstance as "the meeting of the Mosses."

and towards the edges of the Moss; and it required no
small degree of ingenuity and perseverance on the part
of the engineer successfully to overcome them.

The Moss, as has been already observed, was highest
in the centre, and it there presented a sort of hunchback
with a rising and falling gradient. At that point it
was found necessary to cut deeper drains in order to
consolidate the ground between them on which the road
was to be formed. But, as at other parts of the Moss,
the deeper the cutting the more rapid was the flow of
fluid bog into the drain, the bottom rising up almost as
fast as it was removed. To meet this emergency, a
quantity of empty tar-barrels was brought from Liverpool; and as soon as a few yards of drain were dug, the
barrels were laid down end to end, firmly fixed to each
other by strong slabs laid over the joints, and nailed;
they were then covered over with clay, and thus formed
an underground sewer of wood instead of bricks. This
expedient was found to answer the purpose intended,
and the road across the centre of the Moss having thus
been prepared, it was then laid with the permanent
materials.

The greatest difficulty was, however, experienced in
forming an embankment upon the edge of the bog at
the Manchester end. Moss as dry as it could be cut,
was brought up in small waggons, by men and boys,
and emptied so as to form an embankment; but the bank
had scarcely been raised three or four feet in height,
when the stuff broke through the heathery surface of the
bog and sunk overhead. More moss was brought up
and emptied in with no better result; and for many
weeks the filling was continued without any visible
embankment having been made. It was the duty of
the resident engineer to proceed to Liverpool every fortnight to obtain the wages for the workmen employed
under him; and on these occasions he was required to
colour up, on a section drawn to a working scale sus-

pended against the wall of the directors' room, the amount of excavation, embankment, &c., executed from time to time. But on many of these occasions, Mr. Dixon had no progress whatever to show for the money expended upon the Chat Moss embankment. Sometimes, indeed, the visible work done was *less* than it had appeared a fortnight or a month before !

The directors now became seriously alarmed, and feared that the evil prognostications of the eminent engineers were about to be fulfilled. The resident himself was greatly disheartened, and he was even called upon to supply the directors with an estimate of the cost of filling up the Moss with solid stuff from the bottom, as also the cost of piling the roadway, and in effect, constructing a four mile viaduct of timber across the Moss, from twenty to thirty feet high. But the expense appalled the directors, and the question then arose, whether the work was to be proceeded with or *abandoned !*

Mr. Stephenson himself afterwards described the alarming position of affairs at a public dinner given at Birmingham, on the 23rd of December, 1837, on the occasion of a piece of plate being presented to his son, on the completion of the London and Birmingham Railway. He related the anecdote, he said, for the purpose of impressing upon the minds of those who heard him the necessity of perseverance.

" After working for weeks and weeks," said he, " in filling in materials to form the road, there did not yet appear to be the least sign of our being able to raise the solid embankment one single inch ; in short we went on filling in without the slightest apparent effect. Even my assistants began to feel uneasy, and to doubt of the success of the scheme. The directors, too, spoke of it as a hopeless task : and at length they became seriously alarmed, so much so, indeed, that a board meeting was held on Chat Moss to decide whether I should proceed any further. They had previously taken the opinion of other engineers, who reported unfavourably. There was no help for it, however, but to go on. An immense outlay had been incurred; and great loss would have been occasioned had the scheme been then abandoned, and the line taken by another route. So the directors were *compelled* to allow me to go on with my plans, of the ultimate success of which I myself never for one moment doubted."

During the progress of this part of the works, the Worsley and Trafford men, who lived near the Moss, and plumed themselves upon their practical knowledge of bog-work, declared the completion of the road to be utterly impracticable. "If you knew as much about Chat Moss as we do," they said, "you would never have entered on so rash an undertaking; and depend upon it, all you have done and are doing will prove abortive. You must give up altogether the idea of a floating railway, and either fill the Moss up with hard material from the bottom, or else deviate the line so as to avoid it altogether." Such were the conclusions of science and experience.

In the midst of all these alarms and prophecies of failure, Stephenson never lost heart, but held to his purpose. His motto was "Persevere!" "You must go on filling in," he said; "there is no other help for it. The stuff emptied in is doing its work out of sight, and if you will but have patience, it will soon begin to show." And so the filling in went on; several hundreds of men and boys were employed to skin the Moss all round for many thousand yards, by means of sharp spades, called by the turf-cutters "tommy-spades;" and the dried cakes of turf were afterwards used to form the embankment, until at length as the stuff sank and rested upon the bottom, the bank gradually rose above the surface, and slowly advanced onwards, declining in height and consequently in weight, until it became

joined to the floating road already laid upon the Moss.
In the course of forming the embankment, the pressure
of the bog turf tipped out of the waggons caused a
copious stream of bog-water to flow from the end of it,
in colour resembling Barclay's double stout; and when
completed, the bank looked like a long ridge of tightly
pressed tobacco-leaf. The compression of the turf may
be understood from the fact that 670,000 cubic yards of
raw moss formed only 277,000 cubic yards of embankment at the completion of the work.

At the western, or Liverpool end of the Chat Moss,
there was a like embankment; but, as the ground was
there solid, little difficulty was experienced in forming
it, beyond the loss of substance caused by the oozing out
of the water held by the moss-earth.

At another part of the Liverpool and Manchester
line, Parr Moss was crossed by an embankment about a
mile and a half in extent. In the immediate neighbourhood was found a large excess of cutting, which it would
have been necessary to " put out in spoil banks " (according to the technical phrase), but for the convenience of
Parr Moss, into which the surplus clay, stone, and shale,
were tipped, waggon after waggon, until a solid but
concealed embankment, from fifteen to twenty-five feet
high, was formed, although to the eye it appears to be
laid upon the level of the adjoining surface, as at Chat
Moss.

The road across Chat Moss was finished by the 1st of
January, 1830, when the first experimental train of passengers passed over it, drawn by the " Rocket; " and it
turned out that, instead of being the most expensive
part of the line, it was about the cheapest. The total
cost of forming the line over the Moss was 28,000l.,
whereas Mr. Giles's estimate was 270,000l.! It also
proved to be one of the best portions of the railway.
Being a floating road, it was smooth and easy to run
upon, just as Dr. Arnott's water-bed is soft and easy to

lie upon—the pressure being equal at all points.   There was, and still is, a sort of springiness in the road over the Moss, such as is felt when passing along a suspended bridge ; and those who looked along the line as a train passed over it, said they could observe a waviness, such as precedes and follows a skater upon ice.

During the progress of these works the most ridiculous rumours were set afloat.   The drivers of the stage-coaches who feared for their calling, brought the alarming intelligence into Manchester from time to time, that " Chat Moss was blown up ! "   " Hundreds of men and horses had sunk in the bog ; and the works were completely abandoned ! "   The engineer himself was declared to have been swallowed up in the Serbonian bog ; and " railways were at an end for ever ! "

In the construction of the railway, Mr. Stephenson's capacity for organising and directing the labours of a large number of workmen of all kinds eminently displayed itself.   A vast quantity of ballast-waggons had to be constructed for the purposes of the work, and implements and materials had to be collected, before the mass of labour to be employed could be efficiently set in motion at the various points of the line.   There were not at that time, as there are now, large contractors possessed of railway plant, capable of executing earthworks on a large scale.   The first railway engineer had not only to contrive the plant, but to organise the labour, and direct it in person.   The very labourers themselves had to be trained to their work by him ; and it was on the Liverpool and Manchester line that Mr. Stephenson organised the staff of that formidable band of railway navvies, whose handiworks will be the wonder and admiration of succeeding generations.   Looking at their gigantic traces, the men of some future age may be found to declare, of the engineer and of his workmen, that " there were giants in those days."

Although the works of the Liverpool and Manchester

Railway are of a much less formidable character than those of many lines that have since been constructed, they were then regarded as of the most stupendous description. Indeed, the like of them had not before been executed in England. Several of the heaviest and most expensive works were caused by the opposition of Lords Derby and Sefton, whose objections to the line passing near or through their properties forced it more to the south, and thereby involved much tunnelling and heavy stone cutting. It had been our engineer's original intention to carry the railway from the north end of Liverpool, round the red-sandstone ridge on which the upper part of the town is built, and also round the higher rise of the coal formation at Rainhill, by following the natural levels to the north of Knowsley. But the line having been forced to the south, it was rendered necessary to cut through the hills, and go over the high grounds instead of round them. The first consequence of this alteration in the plans was the necessity for constructing a tunnel under the town of Liverpool a mile and a half in length, from the docks at Wapping to the top of Edgehill; the second was the necessity for forming a long and deep cutting through the red-sandstone rock at Olive Mount; and the third and worst of all, was the necessity for ascending and descending the Whiston and Sutton hills by means of inclined planes of 1 in 96. The line was also, by the same forced deviation, prevented passing through the Lancashire coal-field, and the engineer was compelled to carry it across the Sankey valley, at a point where the waters of the brook had dug out an excessively deep channel through the marl-beds of the district.

The principal difficulty was experienced in pushing on the works connected with the formation of the tunnel under Liverpool, 2200 yards in length. The blasting and hewing of the rock were vigorously carried on night and day; and the engineer's practical experience

in the collieries here proved of great use to him.   Many
obstacles had to be encountered and overcome in the
formation of the tunnel, the rock varying in hardness
and texture at different parts.   In some places the
miners were deluged by water, which surged from the
soft blue shale found at the lowest level of the tunnel.
In other places, beds of wet sand were cut through ;
and there careful propping and pinning were necessary
to prevent the roof from tumbling in, until the masonry
to support it could be erected.   On one occasion, while
Mr. Stephenson was absent from Liverpool, a mass of
loose moss-earth and sand fell from the roof, which
had been insufficiently propped.   The miners withdrew
from the work ; and on the engineer's return, he found
them in a refractory state, refusing to re-enter the
tunnel.   He induced them, however, by his example, to
return to their labours ; and when the roof had been
secured, the work went on again as before.   When there
was danger, he was always ready to share it with the
men ;   and gathering confidence from his fearlessness,
they proceeded vigorously with the undertaking, boring
and mining their way towards the light.

The Olive Mount cutting was the first extensive
stone cutting executed on any railway, and to this day
it is one of the most formidable.   It is about two miles
long, and in some parts more than a hundred feet deep.
It is a narrow ravine or defile cut out of the solid rock ;
and not less than four hundred and eighty thousand
cubic yards of stone were removed from it.   Mr. Vignolles, afterwards describing it, said it looked as if it
had been dug out by giants.

The crossing of so many roads and streams involved
the necessity for constructing an unusual number of
bridges.   There were not fewer than sixty-three, under
or over the railway, on the thirty miles between Liverpool and Manchester.   Up to this time, bridges had
been applied generally to high roads, where inclined

approaches were of comparatively small importance, and in determining the rise of his arch the engineer selected any headway he thought proper. Every consideration was indeed made subsidiary to constructing the bridge itself, and the completion of one large structure of this sort was regarded as an epoch in engineering history. Yet here, in the course of a few years, no fewer than

OLIVE MOUNT CUTTING.   [By Percival Skelton.]

sixty-three bridges were constructed on one line of railway! Mr. Stephenson early found that the ordinary arch was inapplicable in certain cases, where the headway was limited, and yet the level of the railway must be preserved. In such cases he employed simple cast-iron beams, by which he safely bridged gaps of moderate

SANKEY VIADUCT. [By Percival Skelton.]

width, economizing headway, and introducing the use
of a new material of the greatest possible value to the
railway engineer.   The bridges of masonry upon the
line were of many kinds ; several of them askew bridges,
and others, such as those at Newton and over the Irwell
at Manchester, were straight and of considerable dimensions.   But the principal piece of masonry on the line
was the Sankey viaduct.

This fine work is principally of brick, with stone
facings.   It consists of nine arches of fifty feet span
each.   The massive piers are supported on two hundred
piles driven deep into the soil ; and they rise to a great
height,—the coping of the parapet being seventy feet
above the level of the valley, in which flow the Sankey
brook and canal.   Its total cost was about 45,000*l.*

By the end of 1828 the directors found they had expended 460,000*l.* on the works, and that they were still far from completion. They looked at the loss of interest on this large investment, and began to grumble at the delay. They desired to see their capital becoming productive; and in the spring of 1829 they urged the engineer to push on the works with increased vigour. Mr. Cropper, one of the directors, who took an active interest in their progress, said to Stephenson one day, " Now, George, thou must get on with the railway, and have it finished without further delay: thou must really have it ready for opening by the first day of January next." " Consider the heavy character of the works, sir, and how much we have been delayed by the want of money, not to speak of the wetness of the weather: it is impossible." " Impossible!" rejoined Cropper; " I wish I could get Napoleon to thee—he would tell thee there is no such word as 'impossible' in the vocabulary." " Tush!" exclaimed Stephenson, with warmth; "don't speak to me about Napoleon! Give me men, money, and materials, and I will do what Napoleon could'nt do —drive a railroad from Liverpool to Manchester over Chat Moss!" And truly the formation of a high road over that bottomless bog was, apparently, a far more difficult task than the hewing even of Napoleon's far-famed road across the Simplon.

The directors had more than once been embarrassed by want of funds to meet the heavy expenditure. The country had scarcely yet recovered from the general panic and crash of 1825: and it was with difficulty that the calls could be raised from the shareholders. A loan of 100,000*l.* was obtained from the Exchequer Loan Commissioners in 1826; and in 1829 an Act was passed enabling the company to raise further capital, to provide working plant for the railway. Two Acts were also obtained during the progress of the works, enabling deviations and alterations to be made ; one to improve the curves and shorten the line near Rainhill, and the other to carry the line across the Irwell into the town of Manchester.   Thanks to the energy of the engineer, the industry of his labourers, and the improved supply of money by the directors, the railway made rapid progress in the course of the year 1829.   Double sets of labourers were employed on Chat Moss and at other points, in carrying on the works by night and day, the night shifts working by torch and fire light ; and at length, the work advancing at all points, the directors saw their way to the satisfactory completion of the undertaking.

It may well be supposed that Mr. Stephenson's time was fully occupied in superintending the extensive, and for the most part novel works, connected with the railway, and that even his extraordinary powers of labour and endurance were taxed to the utmost during the four years that they were in progress.   Almost every detail in the plans was directed and arranged by himself. Every bridge, from the simplest to the most complicated, including the then novel structure of the " skew bridge," iron girders, siphons, fixed engines, and the machinery for working the tunnel at the Liverpool end, had to be thought out by his own head, and reduced to definite plans under his own eyes.   Besides all this, he had to design the working plant in anticipation of the opening of the railway.   He must be prepared with waggons, trucks, and carriages, himself superintending their manufacture.   The permanent road, turntables, switches, and crossings, — in short, the entire structure and machinery of the line, from the turning of the first sod to the running of the first train of carriages upon the railway,—went on under his immediate supervision.   And it was in the midst of this vast accumulation of work and responsibility that the battle of the locomotive engine had to be fought,—a battle, not merely against material difficulties, but against the still more trying obstructions of deeply-rooted mistrust and prejudice on the part of a considerable minority of the directors.

He had no staff of experienced assistants,—not even a staff of draughtsmen in his office,—but only a few pupils learning their business; and he was frequently without even their help. The time of his engineering inspectors was fully occupied in the actual superintendence of the works at different parts of the line; and he took care to direct all their more important operations in person. He had brought three young men from Newcastle with him—fellow-pupils in the workshops there—by name Joseph Locke, Thomas L. Gooch, and William Allcard. These were afterwards joined by John Dixon, and at a later period by Frederick Swanwick. Locke, Allcard, and Dixon, were appointed to superintend the work at different parts of the line; whilst Gooch resided with Mr. Stephenson, and officiated as his sole draughtsman and secretary from the commencement of the works in 1826, until April, 1829, when he proceeded to take charge of another undertaking. " I may say," writes Mr. Gooch, " that the whole of the working and other drawings, as well as the various land-plans for the railway, were drawn by my own hand. They were done at the Company's office in Clayton Square during the day, from instructions supplied in the evenings by Mr. Stephenson, either by word of mouth, or by little rough hand sketches on letter-paper. The evenings were also generally devoted to my duties as secretary, in writing (mostly from his own dictation) his letters and reports, or in making calculations and estimates. The mornings before breakfast were not unfrequently spent by me in visiting and lending a helping hand in the tunnel and other works near Liverpool,—the untiring zeal and perseverance of George

Stephenson never for an instant flagging, and inspiring with a like enthusiasm all who were engaged under him in carrying forward the works."[1]

The usual routine of his life at this time—if routine it might be called—was, to rise early, by sunrise in summer and before it in winter, and thus "break the back of the day's work" by mid-day. While the tunnel under Liverpool was in progress, one of his first duties in a morning before breakfast was to go over the various shafts, clothed in a suitable dress, and inspect the progress of the work at different points; on other days he would visit the extensive workshops at Edgehill, where most of the "plant" for the line was manufactured. Then, returning to his house, in Upper Parliament Street, Windsor, after a hurried breakfast, he would ride along the works to inspect their progress, and push them on with greater energy where needful. On other days he would prepare for the much less congenial engagement of meeting the Board, which was often a cause of great anxiety and pain to him; for it was difficult to satisfy men of all tempers, and some of these not of the most generous sort. On such occasions he might be seen with his right-hand thumb thrust through the topmost button-hole of his coat-breast, vehemently hitching his right shoulder, as was his habit when labouring under any considerable excitement. Occasionally he would take an early ride before breakfast, to inspect the progress of the Sankey viaduct. He had a favourite horse, brought by him from Newcastle, called "Bobby," —so tractable that, with his rider on his back, he would

[1] Mr. Gooch's Letter to the author, December 13th, 1861. Referring to the preparation of the plans and drawings, Mr. Gooch adds, " When we consider the extensive sets of drawings which most engineers have since found it right to adopt in carrying out similar works, it is not the least surprising feature in George Stephenson's early professional career, that he should have been able to confine himself to so limited a number as that which could be supplied by the hands of one person in carrying out the construction of the Liverpool and Manchester Railway; and this may still be said, after full allowance is made for the alteration of system involved by the adoption of the large contract system."

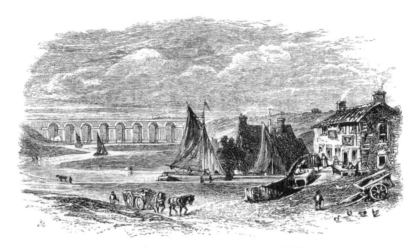

STEPHENSON'S BAITING-PLACE AT SANKEY.

walk up to a locomotive with the steam blowing off, and
put his nose against it without shying. "Bobby,"
saddled and bridled, was brought to Mr. Stephenson's
door betimes in the morning; and mounting him, he
would ride the fifteen miles to Sankey, putting up at a
little public house which then stood upon the banks of
the canal. There he had his breakfast of "crowdie,"
which he made with his own hands. It consisted of
oatmeal stirred into a basin of hot water,—a sort of
porridge,—which was supped with cold sweet milk.
After this frugal breakfast, he would go upon the works,
and remain there, riding from point to point for the
greater part of the day. If he returned home before
mid-day, it would be to examine the pay-sheets in the
different departments, sent in by the assistant engineers,
or by the foremen of the workshops; all this he did
himself, with the greatest care, requiring a full explanation of every item.

After a late dinner, which occupied very short time
and was always of a plain and frugal description,[1] he

---

[1] While at Liverpool, Stephenson
had very little time for " company;"
but on one particular occasion he invited his friend Mr. Sandars to din-

would proceed to dispose of his correspondence, or prepare sketches of drawings, and give instructions as to
their completion. He would occasionally refresh himself
for this evening work by a short doze, which, however,
he would never admit had exceeded the limits of
" winking," to use his own term.   Mr. Frederick
Swanwick, who officiated as his secretary, after the
appointment of Mr. Gooch as Resident Engineer to
the Bolton and Leigh Railway, has informed us that
he then remarked—what in after years he could better
appreciate— the clear, terse and vigorous style of Mr.
Stephenson's dictation ; there was nothing superfluous
in it ; but it was close, direct, and to the point,—in
short, thoroughly business-like.   And if, in passing
through the pen of the amanuensis, his meaning happened in any way to be distorted or modified, it did
not fail to escape his detection, though he was always
tolerant of any liberties taken with his own form of
expression, so long as the words written down conveyed
his real meaning.   His strong natural acumen showed
itself even in such matters as grammar and composition,
—a department of knowledge in which, it might be
supposed, he could scarcely have had either time or
opportunity to acquire much information.   But here, as
in all other things, his shrewd common sense came to
his help ; and his simple, vigorous English might almost
be cited as a model of composition.

His letters and reports written, and his sketches of

---

ner; and as that gentleman was a
connoisseur in port wine, his host
determined to give him a special treat
of that drink.   Stephenson accordingly went to the small merchant
with whom he usually dealt, and
ordered "half a dozen of his very best
port wine," which was promised of
first-rate quality.   After dinner the
wine was produced; and when Mr.
Sandars had sipped a glass, George,
after waiting a little for the expected
eulogium, at length asked, " Well,
Sandars, how d'ye like the port?"
" Poor stuff! " said the guest, " Poor
stuff!"   George was very much
shocked, and with difficulty recovered
his good-humour.   But he lived to be
able to treat Mr. Sandars to a better
article at Tapton House, when he used
to laugh over his first futile attempt
at Liverpool to gain a reputation for
his port.

drawings made and explained, the remainder of the evening was usually devoted to conversation with his wife and those of his pupils who lived under his roof, and constituted, as it were, part of the family. He then delighted to test the knowledge of his young companions, and to question them upon the principles of mechanics. If they were not quite "up to the mark" on any point, there was no escaping detection by evasive or specious explanations on their part. These always met with the verdict of, "Ah! you know nought about it now; but think it over again, and tell me the answer when you understand it." If there were even partial success in the reply, it would at once be acknowledged, and a full explanation given, to which the master would add illustrative examples for the purpose of impressing the principle more deeply upon the pupil's mind.

It was not so much his object and purpose to "cram" the minds of the young men committed to his charge with the *results* of knowledge, as to stimulate them to educate themselves—to induce them to develope their mental and moral powers by the exercise of their own free energies, and thus acquire that habit of self-thinking and self-reliance which is the spring of all true manly action. In a word, he sought to bring out and invigorate the *character* of his pupils. He felt that he himself had been made stronger and better through his encounters with difficulty; and he would not have the road of knowledge made too smooth and easy for them. "Learn for yourselves,—think for yourselves," he would say :—"make yourselves masters of principles,—persevere,—be industrious,—and there is then no fear of you." And not the least emphatic proof of the soundness of this system of education, as conducted by Mr. Stephenson, was afforded by the after history of these pupils themselves. There was not one of those trained under his eye who did not rise to eminent usefulness and

distinction as an engineer.   He sent them forth into the
world braced with the spirit of manly self-help—inspired
by his own noble example ; and they repeated in their
after career the lessons of earnest effort and persistent
industry which his daily life had taught them.

Mr. Stephenson's evenings at home were not, however, exclusively devoted either to business or to the
graver exercises above referred to.   He would often
indulge in cheerful conversation and anecdote, falling
back from time to time upon the struggles and difficulties of his early life.   The not unfrequent winding up
of his story, addressed to the pupils about him, was—
" Ah ! ye young fellows don't know what *wark* is in
these days ! "   Mr. Swanwick delights recalling to mind
how seldom, if ever, a cross or captious word, or an
angry look, marred the enjoyment of those evenings.
The presence of Mrs. Stephenson gave them an additional charm : amiable, kind-hearted, and intelligent, she
shared quietly in the pleasure of the party ; and the
atmosphere of comfort which always pervaded her home
contributed in no small degree to render it a centre of
cheerful, hopeful intercourse, and of earnest, honest
industry.   She was a wife who well deserved, what
she through life retained, the strong and unremitting
affection of her husband.

When Mr. Stephenson retired for the night, it was
not always that he permitted himself to sink into slumber.   Like Brindley, he worked out many a difficult
problem in bed ; and for hours he would turn over in
his mind and study how to overcome some obstacle, or
to mature some project, on which his thoughts were
bent.   Some remark inadvertently dropped by him at the
breakfast-table in the morning, served to show that he
had been stealing some hours from the past night in
reflection and study.   Yet he would rise at his accustomed early hour, and there was no abatement of his
usual energy in carrying on the business of the day.

CHAPTER XIII.

Robert Stephenson's Residence in Colombia, and Return —
The Battle of the Locomotive — "The Rocket."

We return to the career of Robert Stephenson, who
had been absent from England during the construction
of the Liverpool railway, but was shortly about to join
his father and take part in "the battle of the locomotive," which was now impending.

We have seen that on his return from Edinburgh
College at the end of 1821, he had assisted in superintending the works of the Hetton railway until its opening in 1822, after which he proceeded to Liverpool to
take part with Mr. James in surveying the proposed railway there. In the following year we found him assisting
his father in the working survey of the Stockton and
Darlington Railway; and when the Locomotive Engine
Works were started in Forth-Street, Newcastle, he took
an active part in that concern. "The factory," he says,
"was in active operation early in 1824; I left England
for Colombia in June of that year, having finished
drawing the designs of the Brusselton stationary engines
for the Stockton and Darlington Railway before I left." [1]

Speculation was very rife at the time; and amongst
the most promising adventures were the companies
organized for the purpose of working the gold and silver mines of South America. Great difficulty was experienced in finding mining engineers capable of carrying
out those projects, and young men of even the most
moderate experience were eagerly sought after. The
Colombian Mining Association of London offered an

---

[1] Letter to the author.

engagement to young Stephenson, to go out to Mariquita and take charge of the engineering operations of that company. Robert was himself desirous of accepting it, but his father said it would first be necessary to ascertain whether the proposed change would be for his good. His health had been very delicate for some time, partly occasioned by his rapid growth, but principally because of his close application to work and study. Father and son proceeded together to call upon Dr. Headlam, the eminent physician of Newcastle, to consult him on the subject. During the examination which ensued, Robert afterwards used to say that he felt as if he were upon trial for life or death. To his great relief, the doctor pronounced that a temporary residence in a warm climate was the very thing likely to be most beneficial to him. The appointment was accordingly accepted, and, before many weeks had passed, Robert Stephenson set sail for South America.

After a tolerably prosperous voyage he landed at La Guayra, on the north coast of Venezuela, on the 23rd of July, from thence proceeding to Caraccas, the capital of the district, about fifteen miles inland. There he remained for two months, unable to proceed in consequence of the wretched state of the roads in the interior. He contrived, however, to make occasional excursions in the neighbourhood, with an eye to the mining business on which he had come. About the beginning of October he set out for Bogota, the capital of Colombia or New Granada. The distance was about twelve hundred miles, through a very difficult region, and it was performed entirely upon mule-back after the fashion of the country.

In the course of the journey Robert visited many of the districts reported to be rich in minerals, but he met with few traces except of copper, iron, and coal, with occasional indications of gold and silver. He found the people ready to furnish information, which, however, when tested, usually proved worthless. A guide, whom he employed for weeks, kept him buoyed up with the hope of richer mining quarters than he had yet seen; but when he professed to be able to show him mines of " brass, steel, alcohol, and pinchbeck," Stephenson discovered him to be an incorrigible rogue, and immediately dismissed him. At length our traveller reached Bogota, and after an interview with Mr. Illingworth, the commercial manager of the mining company, he proceeded to Honda, crossed the Magdalena, and shortly after reached the site of his intended operations on the eastern slope of the Andes.

Mr. Stephenson used afterwards to speak in glowing terms of this his first mule-journey in South America. Everything was entirely new to him. The variety and beauty of the indigenous plants, the luxurious tropical vegetation, the appearance, manners, and dress of the people, and the mode of travelling, were altogether different from everything he had before seen. His own travelling garb also must have been strange even to himself. " My hat," he says, " was of plaited grass, with a crown nine inches in height, surrounded by a brim of six inches; a white cotton suit; and a *ruana* of blue and crimson plaid, with a hole in the centre for the head to pass through. This cloak is admirably adapted for the purpose, amply covering the rider and mule, and at night answering the purpose of a blanket in the net-hammock, which is made from the fibres of the aloe, and which every traveller carries before him on his mule, and suspends to the trees or in houses, as occasion may require."[1] The part of the journey which seems to have made the most lasting impression on his mind

---

[1] Mr. Stephenson afterwards published an account of his journey from Caraccas to Sta. Bogota da Fé, in the 'National Magazine and Monthly Critic' (Mitchell, Red Lion Court, 1837), under the title of " Scraps from My Note-Book in Colombia." The articles indicate close and accurate observation of the scenery, climate, inhabitants, and productions of the country passed through, but do not possess sufficient interest to justify their republication.

was that between Bogota and the mining district in the neighbourhood of Mariquita. As he ascended the slopes of the mountain-range, and reached the first step of the table-land, he was struck beyond expression with the noble view of the valley of the Magdalena behind him, so vast that he failed in attempting to define the point at which the course of the river blended with the horizon. Like all travellers in the district, he noted the remarkable changes of climate and vegetation, as he rose from the burning plains towards the fresh breath of the mountains. From an atmosphere as hot as that of an oven he passed into delicious cool air; until, in his onward and upward journey, a still more temperate region was reached, the very perfection of climate. Before him rose the majestic Cordilleras, forming a rampart against the western skies, and at certain times of the day looking positively black, sharp, and even at their summit, like a wall.

Our engineer took up his abode for a time at Mariquita, a fine old city, though then greatly fallen into decay. During the period of the Spanish dominion, it was an important place, most of the gold and silver convoys passing through it on their way to Cartagena, there to be shipped in galleons for Europe. The mountainous country to the west was rich in silver, gold, and other metals, and it was Mr. Stephenson's object to select the best site for commencing operations for the Company. With this object he "prospected" about in all directions, visiting long-abandoned mines, and. analyzing specimens obtained from many quarters. The mines eventually fixed upon as the scene of his operations were those of La Manta and Santa Anna, long before worked by the Spaniards, though, in consequence of the luxuriance and rapidity of the vegetation, all traces of the old workings had become completely overgrown and lost. Everything had to be begun anew. Roads had to be cut to open a way to the mines, machinery had to be erected, and the ground opened up, when some of the old adits were eventually hit upon. The native peons or labourers were not accustomed to work, and at first they usually contrived to desert when they were not watched, so that very little progress could be made until the arrival of the expected band of miners from England. The authorities were by no means helpful, and the engineer was driven to an old expedient with the object of overcoming this difficulty. " We endeavour all we can," he says, in one of his letters, " to make ourselves popular, and this we find most effectually accomplished by ' regaling the venal beasts.' " He also gave a ball at Mariquita, which passed off with éclat, the governor from Honda, with a host of friends, honouring it with their presence.[1] It was, indeed, necessary to " make a party" in this way, as other schemers were already trying to undermine the Colombian Company in influential directions. The engineer did not exaggerate when he said, " The uncertainty of transacting business in this country is perplexing beyond description." In the mean time labourers had been attracted to Santa Anna, which became, the engineer wrote, " like an English fair on Sundays : people flock to it from all quarters, to buy beef and chat with their friends. Sometimes three or four torros are slaughtered in a day. The people now eat more beef in a week than they did in two months before, and they are consequently getting fat." [2]

---

[1] During his short residence on the Colombian table-land, Mr. Stephenson made the acquaintance of several native families of distinction. Nor did his connexion with them altogether cease upon his return to England ; for when he went over the scenes of his youth at Killingworth with the author, in 1854, he was accompanied by a young gentleman, then learning engineering in the Newcastle factory, the son of one of the gentlemen whose friendship he had formed during his American residence.

[2] Letter to Mr. Illingworth, September 25th, 1825. The reports made to the directors and officers of the company, which we have seen, contain the details of the operations carried on at the mines ; but they are as dry and uninteresting as such reports usually are, and furnish no materials calculated to illustrate the subject of the text.

At last, his party of miners arrived from England, but they gave him even more trouble than the peons had done. They were rough, drunken, and sometimes altogether ungovernable. He set them to work at the Santa Anna mine without delay, and at the same time took up his abode amongst them, " to keep them," he said, " if possible, from indulging in the detestable vice of drunkenness, which, if not put a stop to, will eventually destroy themselves, and involve the mining association in ruin." To add to his troubles, the captain of the miners displayed a very hostile and insubordinate spirit, quarrelled and fought with the men, and was insolent to the engineer himself. The captain and his gang, being Cornish men, told Robert to his face that because he was a North-country man, and not born in Cornwall, it was impossible he should know anything of mining. Disease also fell upon him,—first fever, and then visceral derangement, followed by a return of his " old complaint, a feeling of oppression in the breast." No wonder that in the midst of these troubles he should longingly speak of returning to his native land. But he stuck to his post and his duty, kept up his courage, and by a mixture of mildness and firmness, and the display of great coolness of judgment, he contrived to keep the men to their work, and gradually to carry forward the enterprise which he had undertaken. By the beginning of July, 1826, we find that quietness and order had been restored, and the works were proceeding more satisfactorily, though the yield of silver was not as yet very promising. Mr. Stephenson calculated that at least three years' diligent and costly operations would be needed to render the mines productive.

In the mean time he removed to the dwelling which had been erected for his accommodation at Santa Anna. It was a structure speedily raised after the fashion of

the country. The walls were of split and flattened
bamboo, tied together with the long fibres of a dried
climbing plant ; the roof was of palm-leaves, and the
ceiling of reeds. When an earthquake shook the
district—for earthquakes were frequent—the inmates of
such a fabric merely felt as if shaken in a basket, without sustaining any harm. In front of the cottage lay
a woody ravine, extending almost to the base of the
Andes, gorgeously clothed in primeval vegetation—

ROBERT STEPHENSON'S COTTAGE AT SANTA ANNA.

magnolias, palms, bamboos, tree-ferns, acacias, cedars ;
and, towering over all, the great almendrons, with their
smooth, silvery stems, bearing aloft noble clusters of
pure white blossom. The forest was haunted by my-
riads of gay insects, butterflies with wings of dazzling
lustre, birds of brilliant plumage, humming-birds,
golden orioles, toucans, and a host of solitary warblers.
But the glorious sunsets seen from his cottage-porch
more than all astonished and delighted the young en-

gineer ; and he was accustomed to say that, after having
witnessed them, he was reluctant to accuse the ancient
Peruvians of idolatry.

But all these natural beauties failed to reconcile him
to the harassing difficulties of his situation, which
continued to increase rather than diminish. He was
hampered by the action of the Board at home, who
gave ear to hostile criticisms on his reports ; and although
they afterwards made handsome acknowledgment of his
services, he felt his position to be altogether unsatisfactory. He therefore determined to leave at the expiry
of his three years' engagement, and communicated his
decision to the directors accordingly.[1] On receiving his

[1] In a letter to Mr. Illingworth, then resident at Bogota, dated the 24th March, 1826, Robert wrote as follows :—" Nothing but the fullest consent of my partners in England could induce me to stay in this country, and the assurance that no absolute necessity existed to call me home. I must also have the consent of my father. I know that he must have suffered severely from my absence, but that having been extended so far beyond the period he was led to expect, may have induced him to curtail his plans; which, had they been accomplished, as they would have been by my assistance, would have placed us both in a situation far superior to anything that I can hope for as the servant of an Association however wealthy and liberal. What I might do in England is, perhaps, known to myself only; it is difficult, therefore, for the Association to calculate upon rewarding me to the full extent of my prospects at home. My prosperity is involved in that of my father, whose property was sacrificed in laying the foundations of an establishment for me; his capital being invested in a concern which requires the greatest attention, and which, with our personal superintendence, could not fail to secure that independence which forms so principally the object of all our toil. Ignorant as I am of the present state of affairs in England, it would be inconsiderate on my part to enter upon any further engagement; but I have the prosperity of my present task so much at heart, that my duty only would induce me to abandon it. My residence in this country, and the work I have had to perform, would have been irksome in the extreme, had I not been fortunate in meeting you, whose acquaintance and generous kindness to me has comparatively lightened my task, and of which believe me to be gratefully sensible. My experience in Colombia has, of course, led me to a knowledge of all that can be alleged against my prolonging my stay, even supposing that my duties should not call me to England. I should be shut up in Sta. Anna, where no desirable society exists, excepting that of my friend Empson.[a] I should be completely debarred from following out my studies; in short, the faculties of my mind must become dormant, ex-

[a] Charles Empson accompanied Robert Stephenson to Colombia, as his secretary and book-keeper. He afterwards published a book, entitled, ' Narratives of South America,. illustrating Manners, Customs, and Scenery.' London: 1836. He died at Bath, in the autumn of 1861.

letter, the Board, through Mr. Richardson, of Lombard-street, one of the directors, communicated with his father at Newcastle, representing that if he would allow his son to remain in Colombia the Company would make it "worth his while." To this the father gave a decided negative, and intimated that he himself urgently needed his son's assistance, and that he must return at the expiry of his three years' term,—a decision, writes Robert, " at which I feel much gratified, as it is clear that he is as anxious to have me back in England as I am to get there." [1] At the same time, Edward Pease, a principal partner in the Newcastle firm, privately wrote Robert to the following effect, urging his return home :—" I can assure thee that your business at New-castle, as well as thy father's engineering, have suffered very much from thy absence, and, unless thou soon return, the former will be given up, as Mr. Longridge is not able to give it that attention it requires ; and what *is* done is not done with credit to the house." The idea of the manufactory being given up, which Robert had laboured so hard to establish before leaving England, was painful to him in the extreme, and he wrote Mr. Illingworth, strongly urging that arrangements should be made for enabling him to leave without delay. In the mean time he was again laid prostrate by another violent attack of aguish fever ; and when able to write,

---

cepting what were called into exercise in the monotonous routine of mining, in which variety is scarcely known. I mean not to imply that the art of mining is devoid of interest; on the contrary, its pursuit always afforded me pleasure, but I should wish to blend other studies with it, and I know it could be done with advantage, and without detracting from the attention due to operative mining. To be open, should I remain here I would erect a complete laboratory for performing all the necessary kinds of metallurgical operations. I would have a liberal supply of scientific

journals, as well as standard works on chemistry and mineralogy. These, the superintendence of the mines, and the engineering department, would form ample resources for the mind, and render a four years' residence bearable, otherwise it would be intolerable. With these privileges, an adequate remuneration, and the consent of my friends, perhaps I might remain; but my feelings and ideas will be entirely guided by future communications from England."

[1] Letter to Mr. Illingworth, April 9th, 1827.

in June, 1827, he expressed himself as "completely wearied and worn down with vexation."

At length, when he was sufficiently recovered from his attack and able to travel, he set out on his voyage homeward in the beginning of August. At Mompox, on his way down the river Magdalena, he met Mr. Bodmer, his successor, with a fresh party of miners from England, on their way up the country to the quarters which he had just quitted. Next day, six hours after leaving Mompox, a steamboat was met ascending the river, with Bolivar the Liberator on board, on his way to St. Bogota; and it was a mortification to our engineer that he had only a passing sight of that distinguished person. It was his intention, on leaving Mariquita, to visit the Isthmus of Panama on his way home, for the purpose of inquiring into the practicability of cutting a canal to unite the Atlantic and Pacific—a project which then formed the subject of considerable public discussion; but Mr. Bodmer having informed him, at Mompox, that such a visit would be inconsistent with the statements made to the London Board that his presence was so anxiously desired at home, he determined to embrace the first opportunity of proceeding to New York.

Arrived at the port of Cartagena, he found himself under the necessity of waiting some time for a ship. The delay was very irksome to him, the more so as the city was then desolated by the ravages of the yellow fever. While sitting one day in the large, bare, comfortless public room of the miserable hotel at which he put up, he observed two strangers, whom he at once perceived to be English. One of the strangers was a tall, gaunt man, shrunken and hollow-looking, shabbily dressed, and apparently poverty-stricken. On making inquiry, he found it was Trevithick, the builder of the first railroad locomotive! He was returning home from the gold mines of Peru penniless. He had left England in 1816, with powerful steam-engines, intended for the drainage and working of the Peruvian mines. He met with almost a royal reception on his landing at Lima. A guard of honour was appointed to attend him, and it was even proposed to erect a statue of Don Ricardo Trevithick in solid silver. It was given forth in Cornwall that his emoluments amounted to 100,000*l.* a year,[1] and that he was making a gigantic fortune. Great, therefore, was Robert Stephenson's surprise to find this potent Don Ricardo in the inn at Cartagena, reduced almost to his last shilling, and unable to proceed further. He had indeed realized the truth of the Spanish proverb, that " a silver mine brings misery, a gold mine ruin." He and his friend had lost everything in their journey across the country from Peru. They had forded rivers and wandered through forests, leaving all their baggage behind them, and had reached thus far only with the clothes upon their backs. Almost the only remnant of precious metal saved by Trevithick was a pair of silver spurs, which he took back with him to Cornwall. Robert Stephenson lent him 50*l.* to enable him to reach England ; and though he was afterwards heard of as an inventor there, he had no further part in the ultimate triumph of the locomotive.

But Trevithick's misadventures on this occasion had not yet ended, for before he reached New York he was wrecked, and Robert Stephenson with him. The following is the account of the voyage, " big with adventures," as given by the latter in a letter to his friend Illingworth :—" At first we had very little foul weather, and indeed were for several days becalmed amongst the islands, which was so far fortunate, for a few degrees farther north the most tremendous gales were blowing, and they appear (from our future information) to have wrecked every vessel exposed to their violence. We

---

[1] ' Geological Transactions of Cornwall,' i., 222.

had two examples of the effects of the hurricane; for,
as we sailed north we took on board the remains of two
crews found floating about on dismantled hulls. The
one had been nine days without food of any kind, except
the carcasses of two of their companions who had died a
day or two previously from fatigue and hunger. The
other crew had been driven about for six days, and were
not so dejected, but reduced to such a weak state that
they were obliged to be drawn on board our vessel by
ropes. A brig bound for Havannah took part of the
men, and we took the remainder. To attempt any
description of my feelings on witnessing such scenes
would be in vain. You will not be surprised to learn
that I felt somewhat uneasy at the thought that we
were so far from England, and that I also might possibly
suffer similar shipwreck; but I consoled myself with
the hope that fate would be more kind to us. It was
not so much so, however, as I had flattered myself; for
on voyaging towards New York, after we had made
the land, we ran aground about midnight. The vessel
soon filled with water, and, being surrounded by the
breaking surf, the ship was soon split up, and before
morning our situation became perilous. Masts and all
were cut away to prevent the hull rocking; but all we
could do was of no avail. About 8 o'clock on the following morning, after a most miserable night, we were
taken off the wreck, and were so fortunate as to reach
the shore. I saved my minerals, but Empson lost part
of his botanical collection. Upon the whole, we got off
well; and, had I not been on the American side of the
Atlantic, I 'guess' I would not have gone to sea
again."

After a short tour in the United States and Canada,
Robert Stephenson and his friend took ship for Liverpool, where they arrived at the end of November, and
at once proceeded to Newcastle. The factory, we have
seen, was by no means in a prosperous state. During

the time Robert had been in America it had been carried
on at a considerable loss; and Edward Pease, very
much disheartened, wished to retire from it, but George
Stephenson being unable to raise the requisite money
to buy him out, the establishment was of necessity carried
on by its then partners until the locomotive could be
established in public estimation as a practicable and
economical working power. Robert Stephenson immediately instituted a rigid inquiry into the working of
the concern, unravelled the accounts, which had been
allowed to fall into confusion during his father's absence
at Liverpool, and very shortly succeeded in placing the
affairs of the factory in a more healthy condition. In
all this he had the hearty support of his father, as well
as of the other partners.

The works of the Liverpool and Manchester Railway
were now approaching completion. But, singular to
say, the directors had not yet decided as to the tractive
power to be employed in working the line when opened
for traffic. The differences of opinion among them were
so great as apparently to be irreconcilable. It was
necessary, however, that they should come to some
decision without further loss of time; and many board
meetings were accordingly held to discuss the subject.
The old-fashioned and well-tried system of horse haulage
was not without its advocates; but, looking at the large
amount of traffic which there was to be conveyed, and
at the probable delay in the transit from station to
station if this method were adopted, the directors, after
a visit made by them to the Northumberland and
Durham railways in 1828, came to the conclusion that
the employment of horse power was inadmissible.

Fixed engines had many advocates; the locomotive
very few: it stood as yet almost in a minority of one—
George Stephenson. The prejudice against the employment of the latter power had even increased since the
Liverpool and Manchester Bill underwent its first ordeal

in the House of Commons.   In proof of this, we may
mention that the Newcastle and Carlisle Railway Act
was conceded in 1829, on the express condition that it
should *not* be worked by locomotives, but by horses
only.

Grave doubts existed as to the practicability of working a large traffic by means of travelling engines.   The
most celebrated engineers offered no opinion on the
subject.   They did not believe in the locomotive, and
would scarcely take the trouble to examine it.   The
ridicule with which George Stephenson had been assailed
by the barristers before the Parliamentary Committee
had not been altogether distasteful to them.   Perhaps
they did not relish the idea of a man who had picked
up his experience in Newcastle coal-pits appearing in
the capacity of a leading engineer before Parliament,
and attempting to establish a new system of internal
communication in the country.   Mr. Telford, the Government engineer, was consulted by his employers on
the occasion of the Company applying to the Exchequer
Loan Commissioners to forego their security of 30 per
cent. of the calls, which the Directors wished to raise to
enable them to proceed more expeditiously with the
works.   But his Report was considered so unsatisfactory
that the Commissioners would not release any part of
the calls.   All that Mr. Telford would say on the subject
of the power to be employed was, that the use of horses
had been done away with by introducing two sets of
inclined planes, and he considered this an evil, inasmuch
as the planes must be worked either by locomotive or
fixed engines ; " but," he said, " which of the two latter
modes shall be adopted, I understand has not yet been
finally determined ; and both being recent projects, in
which I have had no experience, I cannot take upon me
to say whether either will fully answer in practice."
The directors could not disregard the adverse and con-
flicting views of the professional men whom they

consulted. But Mr. Stephenson had so repeatedly and earnestly urged upon them the propriety of making a trial of the locomotive before coming to any decision against it, that they at length authorised him to proceed with the construction of one of his engines by way of experiment. In their report to the proprietors at their annual meeting on the 27th March, 1828, they state that they had, after due consideration, authorised the engineer " to prepare a locomotive engine, which, from the nature of its construction and from the experiments already made, he is of opinion will be effective for the purposes of the company, without proving an annoyance to the public." The locomotive thus ordered was placed upon the line in 1829, and was found of great service in drawing the waggons full of marl from the two great cuttings.

In the mean time the discussion proceeded as to the kind of power to be permanently employed for the working of the railway. The directors were inundated with schemes of all sorts for facilitating locomotion. The projectors of England, France, and America, seemed to be let loose upon them. There were plans for working the waggons along the line by water power. Some proposed hydrogen, and others carbonic acid gas. Atmospheric pressure had its eager advocates. And various kinds of fixed and locomotive steam power were suggested. Thomas Gray urged his plan of a greased road with cog rails; and Messrs. Vignolles and Ericsson recommended the adoption of a central friction rail, against which two horizontal rollers under the locomotive, pressing upon the sides of this rail, were to afford the means of ascending the inclined planes. The directors felt themselves quite unable to choose from amidst this multitude of projects. Their engineer expressed himself as decidedly as heretofore in favour of smooth rails and locomotive engines, which, he was confident, would be found the most economical and by far the most convenient moving power that could be employed. The Stockton and Darlington Railway being now at work, another deputation went down personally to inspect the fixed and locomotive engines on that line, as well as at Hetton and Killingworth. They returned to Liverpool with much information ; but their testimony as to the relative merits of the two kinds of engines was so contradictory, that the directors were as far from a decision as ever.

They then resolved to call to their aid two professional engineers of high standing, who should visit the Darlington and Newcastle railways, carefully examine both modes of working—the fixed and the locomotive,—and report to them fully on the subject. The gentlemen selected were Mr. Walker of Limehouse, and Mr. Rastrick of Stourbridge. After carefully examining the modes of working the northern railways, they made their report to the directors in the spring of 1829. They concurred in the opinion that the cost of an establishment of fixed engines would be somewhat greater than that of locomotives to do the same work ; but thought the annual charge would be less if the former were adopted. They calculated that the cost of moving a ton of goods thirty miles by fixed engines would be 6·40d., and by locomotives, 8·36d.,—assuming a profitable traffic to be obtained both ways. At the same time it was admitted that there appeared more ground for expecting improvements in the construction and working of locomotives than of stationary engines. " On the whole, however, and looking especially at the computed annual charge of working the road on the two systems on a large scale, Messrs. Walker and Rastrick were of opinion that fixed engines were preferable, and accordingly recommended their adoption to the directors." [1]   And in order to

---

[1] Mr. Booth's Account, pp. 70-1. While concurring with Mr. Rastrick in recommending " the stationary reci-   procating system as the best," if it was the directors' intention to make the line complete at once, so as to accom-

carry the system recommended by them into effect, they proposed to divide the railroad between Liverpool and Manchester into nineteen stages of about a mile and a half each, with twenty-one engines fixed at the different points to work the trains forward.

Such was the result, so far, of George Stephenson's labours. Two of the best practical engineers of the day concurred in reporting substantially in favour of the employment of fixed engines. Not a single professional man of eminence could be found to coincide with the engineer of the railway in his preference for locomotive over fixed engine power. He had scarcely a supporter, and the locomotive system seemed on the eve of being abandoned. Still he did not despair. With the profession against him, and public opinion against him—for the most frightful stories were abroad respecting the dangers, the unsightliness, and the nuisance which the locomotive would create—Stephenson held to his purpose. Even in this, apparently the darkest hour of the locomotive, he did not hesitate to declare that locomotive railroads would, before many years had passed, be " the great highways of the world."

He urged his views upon the directors in all ways, and, as some of them thought, at all seasons. He pointed out the greater convenience of locomotive power for the purposes of a public highway, likening it to a series of short unconnected chains, any one of which could be

modate the traffic expected by them, or a quantity approaching to it (*i. e.*, 3750 tons of goods and passengers from Liverpool towards Manchester, and 3950 tons from Manchester towards Liverpool), Mr. Walker added, —" but if any circumstances should induce the directors to proceed by degrees, and to proportion the power of conveyance to the demand, then we recommend locomotive-engines upon the line generally ; and two fixed engines upon Rainhill and Sutton planes, to draw up the locomotive-engines as well as the goods and carriages." And " if on any occasion the trade should get beyond the supply of locomotives, the horse might form a temporary substitute." As, however, it was the directors' determination, with a view to the success of their experiment, to open the line complete for working, they felt that it would be unadvisable to adopt this partial experiment ; and it was still left for them to decide whether they would adopt or not the substantial recommendation of the reporting engineers in favour of the stationary engine system for the complete accommodation of the expected traffic.

removed and another substituted without interruption to the traffic; whereas the fixed engine system might be regarded in the light of a continuous chain extending between the two termini, the failure of any link of which would derange the whole.[1] But the fixed engine party were very strong at the board, and, led by Mr. Cropper, they urged the propriety of forthwith adopting the report of Messrs. Walker and Rastrick. Mr. Sandars and Mr. William Rathbone, on the other hand, desired that a fair trial should be given to the locomotive; and they with reason objected to the expenditure of the large capital necessary to construct the proposed engine-houses, with their fixed engines, ropes, and machinery, until they had tested the powers of the locomotive as recommended by their own engineer. Mr. Stephenson continued to urge upon them that the locomotive was yet capable of great improvements, if proper induce-ments were held out to inventors and machinists to make them; and he pledged himself that, if time were given him, he would construct an engine that should satisfy their requirements, and prove itself capable of working heavy loads along the railway with speed, regularity, and safety. At length, influenced by his persistent earnestness not less than by his arguments, the directors, at the suggestion of Mr. Harrison, determined to offer a prize of 500l. for the best locomotive engine, which, on a certain day, should be produced on the rail-way, and perform certain specified conditions in the most satisfactory manner.[2]

---

[1] The arguments used by Mr. Ste-phenson with the directors, in favour of the locomotive engine, were after-wards collected and published in 1830 by Robert Stephenson and Joseph Locke, as " compiled from the Reports of Mr. George Stephenson." The pamphlet was entitled, ' Observations on the Comparative Merits of Locomo-tive and Fixed Engines.' Robert Ste-phenson, speaking of the authorship many years after, said, 'I believe I furnished the facts and the arguments, and Locke put them into shape. Locke was a very flowery writer, whereas my style was rather bald and unattractive; so he was the editor of the pamphlet, which excited a good deal of attention amongst engineers at the time."

[2] The conditions were these:—

1. The engine must effectually con-sume its own smoke.

2. The engine, if of six tons weight,

The requirements of the directors as to speed were not excessive. All that they asked for was, that ten miles an hour should be maintained. Perhaps they had in mind the animadversions of the ' Quarterly Reviewer ' on the absurdity of travelling at a greater velocity, as well as the remarks published by Mr. Nicholas Wood, whom they selected to be one of the judges of the competition, in conjunction with Mr. Rastrick of Stourbridge and Mr. Kennedy of Manchester.

It was now felt that the fate of railways in a great measure depended upon the issue of this appeal to the mechanical genius of England. When the advertisement of the prize for the best locomotive was published, scientific men began more particularly to direct their attention to the new power which was thus struggling into existence. In the mean time public opinion on the subject of railway working remained suspended, and the progress of the undertaking was watched with the most intense interest.

---

must be able to draw after it, day by day, twenty tons weight (including the tender and water-tank) at *ten miles* an hour, with a pressure of steam on the boiler not exceeding fifty pounds to the square inch.

3. The boiler must have two safety valves, neither of which must be fastened down, and one of them be completely out of the control of the engineman.

4. The engine and boiler must be supported on springs, and rest on six wheels, the height of the whole not exceeding fifteen feet to the top of the chimney.

5. The engine, with water, must not weigh more than six tons; but an engine of less weight would be preferred on its drawing a proportionate load behind it; if of only four and a half tons, then it might be put on only four wheels. The Company to be at liberty to test the boiler, &c., by a pressure of one hundred and fifty pounds to the square inch.

6. A mercurial gauge must be affixed to the machine, showing the steam pressure above forty-five pounds per square inch.

7. The engine must be delivered, complete and ready for trial, at the Liverpool end of the railway, not later than the 1st of October, 1829.

8. The price of the engine must not exceed 550*l.*

Many persons of influence declared the conditions published by the directors of the railway chimerical in the extreme. One gentleman of some eminence in Liverpool, Mr. P. Ewart, who afterwards filled the office of Government Inspector of Post Office Steam-packets, declared that only a parcel of charlatans would ever have issued such a set of conditions; that it had been *proved* to be impossible to make a locomotive engine go at ten miles an hour; but if it ever was done, he would eat a stewed engine-wheel to his breakfast.

During the progress of the above important discussion with reference to the kind of power to be employed in working the railway, Mr. Stephenson was in constant communication with his son Robert, who made frequent visits to Liverpool for the purpose of assisting his father in the preparation of his reports to the board on the subject. Mr. Swanwick remembers the vivid interest of the evening conversations which took place between father and son as to the best mode of increasing the powers and perfecting the mechanism of the locomotive. He wondered at their quick perception and rapid judgment on each other's suggestions, at the mechanical difficulties which they anticipated and provided for in the practical arrangement of the machine ; and he speaks of these evenings as most interesting displays of two actively ingenious and able minds, stimulating each other to feats of mechanical invention, by which it was ordained that the locomotive engine should become what it now is. These discussions became more frequent, and still more interesting, after the public prize had been offered for the best locomotive by the directors of the railway, and the working plans of the engine which they proposed to construct had to be settled.

One of the most important considerations in the new engine was the arrangement of the boiler and the extension of its heating surface to enable steam enough to be raised rapidly and continuously, for the purpose of maintaining high rates of speed,—the effect of high-pressure engines being ascertained to depend mainly upon the quantity of steam which the boiler can generate, and upon its degree of elasticity when produced. The quantity of steam so generated, it will be obvious, must chiefly depend upon the quantity of fuel consumed in the furnace, and, by necessary consequence, upon the high rate of temperature maintained there.

It will be remembered that in Stephenson's first Killingworth engines he invented and applied the ingenious method of stimulating combustion in the furnace, by throwing the waste steam into the chimney after performing its office in the cylinders, thus accelerating the ascent of the current of air, greatly increasing the draught, and consequently the temperature of the fire. This plan was adopted by him, as we have already seen, as early as 1815 ; and it was so successful that he himself attributed to it the greater economy of the locomotive as compared with horse power. Hence, the continuance of its use upon the Killingworth Railway.

Though the adoption of the steam blast greatly quickened combustion and contributed to the rapid production of high-pressure steam, the limited amount of heating surface presented to the fire was still felt to be an obstacle to the complete success of the locomotive engine. Mr. Stephenson endeavoured to overcome this by lengthening the boilers and increasing the surface presented by the flue tubes. The "Lancashire Witch," which he built for the Bolton and Leigh Railway, and used in forming the Liverpool and Manchester Railway embankments, was constructed with a double tube, each of which contained a fire and passed longitudinally through the boiler. But this arrangement necessarily led to a considerable increase in the weight of the engine, which amounted to about twelve tons each ; and as six tons was the limit allowed for engines admitted to the Liverpool competition, it was clear that the time was come when the Killingworth loco- motive must undergo a further important modification.

For many years previous to this period, ingenious mechanics had been engaged in attempting to solve the problem of the best and most economical boiler for the production of high-pressure steam. As early as 1803, Mr. Woolf patented a tubular boiler, which was extensively employed at the Cornish mines, and was found greatly to facilitate the production of steam, by the extension of the heating surface. The ingenious

Trevithick, in his patent of 1815, seems also to have entertained the idea of employing a boiler constructed of " small perpendicular tubes," with the same object of increasing the heating surface.   These tubes were to be closed at the bottom, and open into a common reservoir, from which they were to receive their water, and where the steam of all the tubes was to be united.   It does not, however, appear that any locomotive was ever constructed according to this patent.   Mr. Goldsworthy Gurney, the persevering adaptor of steam-carriages to travelling on common roads, applied the tubular principle in the boiler of his engine, in which the steam was generated *within* the tubes; whilst the boiler invented by Messrs. Summers and Ogle for their turnpike-road steam-carriage, consisted of a series of tubes placed vertically over the furnace, through which the heated air passed before reaching the chimney.

About the same time George Stephenson was trying the effect of introducing small tubes in the boilers of his locomotives, with the object of increasing their evaporative power.   Thus, in 1829, he sent to France two engines constructed at the Newcastle works for the Lyons and St. Etienne Railway, in the boilers of which tubes were placed containing water.   The heating surface was thus found to be materially increased; but the expedient was not successful, for the tubes, becoming furred with deposit, shortly burned out and were removed.   It was then that M. Seguin, the engineer of the railway, pursuing the same idea, is said to have adopted his plan of employing horizontal tubes through which the heated air passed in streamlets.   Mr. Henry Booth, the secretary of the Liverpool and Manchester Railway, without any knowledge of M. Seguin's proceedings, next devised his plan of a tubular boiler, which he brought under the notice of Mr. Stephenson, who at once adopted it, and settled the mode in which the firebox and tubes were to be mutually arranged and con-

nected. This plan was adopted in the construction of the
celebrated "Rocket" engine, the building of which was
immediately proceeded with at the Newcastle works.

The principal circumstances connected with the construction of the "Rocket," as described by Robert Stephenson to the author, may be thus briefly stated.   The
tubular principle was adopted in a more complete manner than had yet been attempted.   Twenty-five copper
tubes, each three inches in diameter, extended from one
end of the boiler to the other, the heated air passing
through them on its way to the chimney; and the tubes
being surrounded by the water of the boiler, it will be
obvious that a large extension of the *heating surface* was
thus effectually secured.   The principal difficulty was
in fitting the copper tubes within the boiler so as to prevent leakage.   They were manufactured by a Newcastle
coppersmith, and soldered to brass screws which were
screwed into the boiler ends, standing out in great
knobs.   When the tubes were thus fitted, and the
boiler was filled with water, hydraulic pressure was
applied; but the water squirted out at every joint, and
the factory floor was soon flooded.   Robert went home
in despair; and in the first moment of grief, he wrote to
his father that the whole thing was a failure.   By
return of post came a letter from his father, telling
him that despair was not to be thought of—that he
must "try again;" and he suggested a mode of overcoming the difficulty, which his son had already anticipated and proceeded to adopt.   It was, to bore clean
holes in the boiler ends, fit in the smooth copper tubes
as tightly as possible, solder up, and then raise the
steam.   This plan succeeded perfectly, the expansion
of the copper tubes completely filling up all interstices,
and producing a perfectly watertight boiler, capable of
withstanding extreme internal pressure.

The mode of employing the steam-blast for the purpose of increasing the draught in the chimney, was also

the subject of numerous experiments.[1]   When the
engine was first tried, it was thought that the blast
in the chimney was not sufficiently strong for the purpose of keeping up the intensity of the fire in the furnace,
so as to produce high-pressure steam with the required
velocity.   The expedient was therefore adopted of
hammering the copper tubes at the point at which
they entered the chimney, whereby the blast was considerably sharpened; and on a further trial it was found
that the draught was increased to such an extent as to
enable abundance of steam to be raised.   The rationale
of the blast may be simply explained by referring to the
effect of contracting the pipe of a water-hose, by which
the force of the jet of water is proportionately increased.
Widen the nozzle of the pipe, and the force is in like
manner diminished.   So is it with the steam-blast in
the chimney of the locomotive.

Doubts were, however, expressed whether the greater
draught secured by the contraction of the blast-pipe was
not counterbalanced in some degree by the negative
pressure upon the piston.   Hence a series of experiments
was made with pipes of different diameters; and their
efficiency was tested by the amount of vacuum that
was produced in the smoke-box.   The degree of rarefaction was determined by a glass tube fixed to the
bottom of the smoke-box, and descending into a bucket
of water, the tube being open at both ends.   As the
rarefaction took place, the water would of course rise in
the tube; and the height to which it rose above the surface of the water in the bucket was made the measure of
the amount of rarefaction.   These experiments proved
that a considerable increase of draught was obtained by
the contraction of the orifice; accordingly, the two blast-
pipes opening from the cylinders into either side of the

---

[1] For further details as to the steam-blast, see Robert Stephenson's Account,
given in the Appendix to this volume.

"Rocket" chimney, and turned up within it, were contracted slightly below the area of the steam-ports; and before the engine left the factory, the water rose in the glass tube three inches above the water in the bucket.

The other arrangements of the "Rocket" were briefly these :—the boiler was cylindrical with flat ends, six feet in length, and three feet four inches in diameter. The upper half of the boiler was used as a reservoir for the steam, the lower half being filled with water. Through the lower part, twenty-five copper tubes of three inches

THE "ROCKET."

diameter extended, which were open to the fire-box at one end, and to the chimney at the other. The fire-box, or furnace, two feet wide and three feet high, was attached immediately behind the boiler, and was also surrounded with water. The cylinders of the engine were placed on each side of the boiler, in an oblique position, one end being nearly level with the top of the boiler at its after end, and the other pointing towards the centre of the foremost or driving pair of wheels, with which the connection was directly made from the piston-rod, to a pin on the outside of the wheel. The engine, together with its load of water, weighed only four tons and a quarter; and it was supported on four wheels, not coupled. The tender was four-wheeled, and similar in shape to a waggon,—the foremost part holding the fuel, and the hind part a water-cask.

When the "Rocket" was finished, it was placed upon the Killingworth railway for the purpose of experiment. The new boiler arrangement was found perfectly successful. The steam was raised rapidly and continuously, and in a quantity which then appeared marvellous. The same evening Robert dispatched a letter to his father at Liverpool, informing him, to his great joy, that the "Rocket" was "all right," and would be in complete working trim by the day of trial. The engine was shortly after sent by waggon to Carlisle, and thence shipped for Liverpool.

The time so much longed for by George Stephenson had now arrived, when the merits of the passenger locomotive were to be put to a public test. He had fought the battle for it until now almost single-handed. Engrossed by his daily labours and anxieties, and harassed by difficulties and discouragements which would have crushed the spirit of a less resolute man, he had held firmly to his purpose through good and through evil report. The hostility which he experienced from some of the directors opposed to the adoption of the locomotive, was the circumstance that caused him the greatest grief of all; for where he had looked for encouragement, he found only carping and opposition. But his pluck never failed him; and now the "Rocket" was upon the ground,—to prove, to use his own words, "whether he was a man of his word or not."

Great interest was felt at Liverpool, as well as throughout the country, in the approaching compe-

tition.   Engineers, scientific men, and mechanics,
arrived from all quarters to witness the novel display
of mechanical ingenuity on which such great results
depended.   The public generally were no indifferent
spectators either.   The inhabitants of Liverpool, Manchester, and the adjacent towns felt that the successful
issue of the experiment would confer upon them individual benefits and local advantages almost incalculable,
whilst populations at a distance waited for the result
with almost equal interest.

On the day appointed for the great competition of
locomotives at Rainhill, the following engines were
entered for the prize :—

1. Messrs. Braithwaite and Ericsson's[1] " Novelty."
2. Mr. Timothy Hackworth's " Sanspareil."
3. Messrs. R. Stephenson and Co.'s " Rocket."
4. Mr. Burstall's " Perseverance."

Another engine was entered by Mr. Brandreth of
Liverpool — the " Cycloped," weighing three tons,
worked by a horse in a frame, but it could not be
admitted to the competition.   The above were the
only four exhibited, out of a considerable number of
engines constructed in different parts of the country
in anticipation of this contest, many of which could not
be satisfactorily completed by the day of trial.

The ground on which the engines were to be tried
was a level piece of railroad, about two miles in length.
Each was required to make twenty trips, or equal to a
journey of seventy miles, in the course of the day ; and
the average rate of travelling was to be not under ten
miles an hour.   It was determined that, to avoid confusion, each engine should be tried separately, and on
different days.

---

[1] The inventor of this engine was a
Swede, who afterwards proceeded to
the United States, and there achieved
considerable distinction as an engineer.
His Caloric Engine has so far proved a
failure, but his iron cupola vessel, the
" Monitor," must be admitted to have
been a remarkable success in its way.

The day fixed for the competition was the 1st of October, but to allow sufficient time to get the locomotives into good working order, the directors extended it to the 6th. On the morning of the 6th, the ground at Rainhill presented a lively appearance, and there was as much excitement as if the St. Leger were about to be run. Many thousand spectators looked on, amongst whom were some of the first engineers and mechanicians of the day. A stand was provided for the ladies; the "beauty and fashion" of the neighbourhood were present, and the side of the railroad was lined with carriages of all descriptions.

LOCOMOTIVE COMPETITION AT RAINHILL.

It was quite characteristic of the Stephensons, that, although their engine did not stand first on the list for trial, it was the first that was ready; and it was accordingly ordered out by the judges for an experimental trip. Yet the "Rocket" was by no means "the favourite" with either the judges or the spectators. Nicholas Wood has since stated that a majority of the judges were strongly predisposed in favour of the "Novelty," and that "nine-tenths, if not ten-tenths, of the persons present, were against the "Rocket" because of its appearance."[1]   Nearly every person favoured some other

---

[1] Mr. Wood's speech at Newcastle, 26th October, 1858.

engine, so that there was nothing for the "Rocket" but the practical test. The first trip which it made was quite successful. It ran about twelve miles, without interruption, in about fifty-three minutes.

The "Novelty" was next called out. It was a light engine, very compact in appearance, carrying the water and fuel upon the same wheels as the engine. The weight of the whole was only three tons and one hundredweight. A peculiarity of this engine was that the air was driven or *forced* through the fire by means of bellows. The day being now far advanced, and some dispute having arisen as to the method of assigning the proper load for the "Novelty," no particular experiment was made, further than that the engine traversed the line by way of exhibition, occasionally moving at the rate of twenty-four miles an hour.

The "Sanspareil," constructed by Mr. Timothy Hack-worth, was next exhibited; but no particular experiment was made with it on this day. This engine differed but little in its construction from the locomotive last supplied by George Stephenson to the Stockton and Darlington Railway, of which Hackworth was the locomotive foreman. It had the double tube containing the fire passing along the inside of the boiler, and returning back to the same end at which it entered. It had also the steam blast in the chimney; but the contraction of the orifice by which the steam was thrown into the chimney, for the purpose of intensifying the draught, being a favourite idea of Mr. Hackworth (though of this Mr. Goldsworthy Gurney claims the credit), he had sharpened the blast of his engine in a remarkable degree; and this was perhaps the only noticeable feature in the "Sanspareil."

The contest was postponed until the following day; but before the judges arrived on the ground, the bellows for creating the blast in the "Novelty" gave way, and it was found incapable of going through its performance.

A defect was also detected in the boiler of the "Sans-pareil;" and Mr. Hackworth was allowed some further time to get it repaired. The large number of spectators who had assembled to witness the contest were greatly disappointed at this postponement; but, to lessen it, Stephenson again brought out the "Rocket," and, attaching to it a coach containing thirty persons, he ran them along the line at the rate of from twenty-four to thirty miles an hour, much to their gratification and amazement. Before separating, the judges ordered the engine to be in readiness by eight o'clock on the follow-ing morning, to go through its definitive trial according to the prescribed conditions.

On the morning of the 8th of October, the "Rocket" was again ready for the contest. The engine was taken to the extremity of the stage, the fire-box was filled with coke, the fire lighted, and the steam raised until it lifted the safety-valve loaded to a pressure of fifty pounds to the square inch. This proceeding occupied fifty-seven minutes. The engine then started on its journey, dragging after it about thirteen tons weight in waggons, and made the first ten trips backwards and forwards along the two miles of road, running the thirty-five miles, including stoppages, in an hour and forty-eight minutes. The second ten trips were in like manner performed in two hours and three minutes. The maxi-mum velocity attained during the trial trip was twenty-nine miles an hour, or about three times the speed that one of the judges of the competition had declared to be the limit of possibility. The average speed at which the whole of the journeys were performed was fifteen miles an hour, or five miles beyond the rate specified in the conditions published by the company. The entire performance excited the greatest astonishment amongst the assembled spectators; the directors felt confident that their enterprise was now on the eve of success; and George Stephenson rejoiced to think that in spite of all false prophets and fickle counsellors, the locomotive system was now safe. When the "Rocket," having performed all the conditions of the contest, arrived at the "grand stand" at the close of its day's successful run, Mr. Cropper—one of the directors favourable to the fixed-engine system—lifted up his hands, and exclaimed, "Now has George Stephenson at last delivered himself."

Neither the "Novelty" nor the "Sanspareil" was ready for trial until the 10th, on the morning of which day an advertisement appeared, stating that the former engine was to be tried on that day, when it would perform more work than any engine upon the ground. The weight of the carriages attached to it was only about seven tons. The engine passed the first post in good style; but in returning, the pipe from the forcing-pump burst and put an end to the trial. The pipe was afterwards repaired, and the engine made several trips by itself, in which it was said to have gone at the rate of from twenty-four to twenty-eight miles an hour.

The "Sanspareil" was not ready until the 13th; and when its boiler and tender were filled with water, it was found to weigh four hundredweight beyond the weight specified in the published conditions as the limit of four-wheeled engines; nevertheless the judges allowed it to run on the same footing as the other engines, to enable them to ascertain whether its merits entitled it to favourable consideration. It travelled at the average speed of about fourteen miles an hour, with its load attached; but at the eighth trip the cold-water pump got wrong, and the engine could proceed no further.

It was determined to award the premium to the successful engine on the following day, the 14th, on which occasion there was an unusual assemblage of spectators. The owners of the "Novelty" pleaded for another trial; and it was conceded. But again it broke down. Then Mr. Hackworth requested the opportunity for making another trial of his " Sanspareil." But the judges had now had enough of failures; and they declined, on the ground that not only was the engine above the stipulated weight, but that it was constructed on a plan which they could not recommend for adoption by the directors of the Company. One of the principal practical objections to this locomotive was the enormous quantity of coke consumed or wasted by it—about 692 lbs. per hour when travelling—caused by the sharpness of the steam blast in the chimney, which blew a large proportion of the burning coke into the air.

The " Perseverance " of Mr. Burstall was found unable to move at more than five or six miles an hour; and it was withdrawn from the contest at an early period. The " Rocket " was thus the only engine that had performed, and more than performed, all the stipulated conditions; and it was declared to be fully entitled to the prize of 500l., which was awarded to the Messrs. Stephenson and Booth accordingly. And further to show that the engine had been working quite within its powers, Mr. Stephenson ordered it to be brought upon the ground and detached from all incumbrances, when, in making two trips, it was found to travel at the astonishing rate of thirty-five miles an hour.

The " Rocket " had thus eclipsed the performances of all locomotive engines that had yet been constructed, and outstripped even the sanguine anticipations of its constructors. It satisfactorily answered the report of Messrs. Walker and Rastrick; and established the efficiency of the locomotive for working the Liverpool and Manchester Railway, and indeed all future railways. The " Rocket " showed that a new power had been born into the world, full of activity and strength, with boundless capability of work. It was the simple but admirable contrivance of the steam-blast, and its combination with the multitubular boiler, that at once gave locomotion a vigorous life, and secured the triumph of the railway system.[1] As has been well observed, this wonderful ability to increase and multiply its powers of performance with the emergency that demands them, has made this giant engine the noblest creation of human wit, the very lion among machines. The success of the Rainhill experiment as judged by the public, may be inferred from the fact that the shares of the Company immediately rose ten per cent., and nothing further was heard of the proposed twenty-one fixed engines, engine-houses, ropes, &c. All this cumbersome apparatus was thenceforward effectually disposed of.

Very different now was the tone of those directors who had distinguished themselves by the persistency of their opposition to Mr. Stephenson's plans. Coolness gave way to eulogy, and hostility to unbounded offers of friendship; after the manner of many men who run to the help of the strong. Deeply though the engineer had felt aggrieved by the conduct pur-

---

[1] The immense consequences involved in the success of the "Rocket," and the important influence the above contest, in which it came off the victor, exercised upon the future development of the railway system, might have led one to suppose that the directors of the Liverpool and Manchester Railway would have regarded the engine with pride, and cherished it with care, as warriors prize a trusty weapon which has borne them victoriously through some grand historical battle. The French preserve with the greatest care the locomotive constructed by Cugnot, which is to this day to be seen in the Conservatoire des Arts et Métiers at Paris. But the "Rocket" was an engine of much greater historical interest. And what became of the "Rocket"? When heavier and more powerful engines were brought upon the road, the old "Rocket," becoming regarded as a thing of no value, was sold in 1837. It was purchased by Mr. Thompson, of Kirkhouse, the lessee of the Earl of Carlisle's coal and lime works near Carlisle. He worked the engine on the Midgeholme Railway for five or six years, during which it hauled coals from the pits to the town. There was wonderful vitality in the old engine, as the following circumstance proves. When the great contest for the representation of East Cumberland took place, and Sir James Graham was superseded by Major Aglionby, the "Rocket" was employed to convey the Alston express with the state of the poll from Midgeholme to Kirkhouse. On that occasion the engine was driven by Mr. Mark Thompson, and it ran the distance of upwards of four miles in four and a-half minutes, thus reaching a speed of nearly sixty miles an hour—proving its still admirable qualities as an engine. But again it was superseded by heavier engines; for it only weighed about four tons, whereas the new engines were at least three times that weight. The "Rocket" was consequently laid up in ordinary in the yard at Kirkhouse, where, we believe, it still remains.

sued towards him during this eventful struggle, by
some from whom forbearance was to have been expected,
he never entertained towards them in after life any
angry feelings ; on the contrary, he forgave all.   But
though the directors afterwards passed unanimous resolutions eulogising " the great skill and unwearied energy"
of their engineer, he himself, when speaking confiden-
tially to those with whom he was most intimate, could
not help pointing out the difference between his " foul-
weather and fair-weather friends."   Mr. Gooch says,
that though naturally most cheerful and kind-hearted in
his disposition, the anxiety and pressure which weighed
upon his mind during the construction of the railway,
had the effect of making him occasionally impatient and
irritable, like a spirited horse touched by the spur ;
though his original good nature from time to time shone
through it all.   When the line had been brought to a
successful completion, a very marked change in him
became visible.   The irritability passed away, and when
difficulties and vexations arose they were treated by him
as matters of course, and with perfect composure and
cheerfulness.

RAILWAY versus ROAD

CHAPTER XIV.

OPENING OF THE LIVERPOOL AND MANCHESTER RAILWAY, AND
EXTENSION OF THE RAILWAY SYSTEM.

THE directors of the Railway now began to see daylight;
and they derived encouragement from the skilful manner
in which their engineer had overcome the principal
difficulties of the undertaking. He had formed a solid
road over Chat Moss, and thus achieved one "impossibility;" and he had constructed a locomotive that
could run at a speed of thirty miles an hour, thus
vanquishing a still more formidable difficulty.

About the middle of 1829 the tunnel at Liverpool was
finished; and being lit up with gas, it was publicly
exhibited one day in each week. Many thousand persons visited it at the charge of a shilling a head,—the
fund thus raised being appropriated partly to the support
of the families of labourers who had been injured upon
the line, and partly in contributions to the Manchester
and Liverpool infirmaries. As promised by the engineer, a single line of way was completed over Chat
Moss by the 1st of January, 1830; and on that day,
the "Rocket" with a carriage full of directors, engineers, and their friends, passed along the greater part
of the road between Liverpool and Manchester. Mr.
Stephenson continued to direct his close attention to
the improvement of the details of the locomotive, every
successive trial of which proved more satisfactory. In
this department, he had the benefit of the able and
unremitting assistance of his son, who, in the workshops
at Newcastle, directly superintended the construction of
the new engines required for the public working of the

railway. He did not by any means rest satisfied with
the success, decided though it was, which had been
achieved by the "Rocket." He regarded it but in the
light of a successful experiment; and every succeeding
engine placed upon the railway exhibited some improvement on its predecessors. The arrangement of the parts,
and the weight and proportions of the engines, were
altered, as the experience of each successive day, or
week, or month, suggested; and it was soon found that
the performances of the "Rocket" on the day of trial
had been greatly within the powers of the locomotive.

The first entire trip between Liverpool and Manchester
was performed on the 14th of June, 1830, on the occasion of a board meeting being held at the latter town.
The train was on this occasion drawn by the "Arrow,"
one of the new locomotives, in which the most recent
improvements had been adopted. Mr. Stephenson himself drove the engine, and Captain Scoresby, the circum-
polar navigator, stood beside him on the foot-plate, and
minuted the speed of the train. A great concourse of
people assembled at both termini, as well as along the
line, to witness the novel spectacle of a train of carriages
dragged by an engine at a speed of seventeen miles an
hour. On the return journey to Liverpool in the
evening, the "Arrow" crossed Chat Moss at a speed of
nearly twenty-seven miles an hour, reaching its destina-
tion in about an hour and a half.

In the mean time Mr. Stephenson and his assistant,
Mr. Gooch, were diligently occupied in making the
necessary preliminary arrangements for the conduct of
the traffic against the time when the line should be ready
for opening. The experiments made with the object of
carrying on the passenger traffic at quick velocities
were of an especially harassing and anxious character.
Every week, for nearly three months before the opening,
trial trips were made to Newton and back, generally
with two or three trains following each other, and carry-

ing altogether from two to three hundred persons. These trips were usually made on Saturday afternoons, when the works could be more conveniently stopped and the line cleared for the occasion. In these experiments Mr. Stephenson had the able assistance of Mr. Henry Booth, the secretary of the Company, who contrived many of the arrangements in the passenger carriages, not the least valuable of which was his invention of the coupling screw, still in use on all passenger railways.

At length the line was finished, and ready for the public ceremony of the opening, which took place on the 15th of September, 1830, and attracted a vast number of spectators from all parts of the country. The completion of the railway was justly regarded as an important national event, and the ceremony of the opening was celebrated accordingly. The Duke of Wellington, then Prime Minister, Sir Robert Peel, Secretary of State, Mr. Huskisson, one of the members for Liverpool and an earnest supporter of the project from its commencement, were amongst the number of distinguished public personages present.

Eight locomotive engines, constructed at the Stephenson works, had been delivered and placed upon the line, the whole of which had been tried and tested, weeks before, with perfect success. The several trains of carriages accommodated in all about six hundred persons. The " Northumbrian " engine, driven by George Stephenson himself, headed the line of trains; then followed the " Phœnix," driven by Robert Stephenson; the " North Star," by Robert Stephenson, senior (brother of George); the " Rocket," by Joseph Locke; the " Dart," by Thomas L. Gooch; the " Comet " by William Allcard; the " Arrow," by Frederick Swanwick; and the " Meteor," by Anthony Harding. The procession was cheered in its progress by thousands of spectators—through the deep ravine of Olive Mount; up the Sutton incline; over the great Sankey viaduct, beneath which a multitude of persons had assembled,—carriages filling the narrow lanes, and barges crowding the river; the people below gazing with wonder and admiration at the trains which sped along the line, far above their heads, at the rate of some twenty-four miles an hour.

At Parkside, about seventeen miles from Liverpool, the engines stopped to take in water. Here a deplorable accident occurred to one of the illustrious visitors, which threw a deep shadow over the subsequent proceedings of the day. The "Northumbrian" engine, with the carriage containing the Duke of Wellington, was drawn up on one line, in order that the whole of the trains on the other line might pass in review before him and his party. Mr. Huskisson had alighted from the carriage, and was standing on the opposite road, along which the "Rocket" was observed rapidly coming up. At this moment the Duke of Wellington, between whom and Mr. Huskisson some coolness had existed, made a sign of recognition, and held out his hand. A hurried but friendly grasp was given; and before it was loosened there was a general cry from the bystanders of "Get in, get in!" Flurried and confused, Mr. Huskisson endeavoured to get round the open door of the carriage, which projected over the opposite rail; but in so doing he was struck down by the "Rocket," and falling with his leg doubled across the rail, the limb was instantly crushed. His first words, on being raised, were, "I have met my death," which unhappily proved true, for he expired that same evening in the parsonage of Eccles. It was cited at the time as a remarkable fact, that the "Northumbrian" engine, driven by George Stephenson himself, conveyed the wounded body of the unfortunate gentleman a distance of about fifteen miles in twenty-five minutes, or at the rate of thirty-six miles an hour. This incredible speed burst upon the world with the effect of a new and unlooked-for phenomenon.

The accident threw a gloom over the rest of the day's proceedings.   The Duke of Wellington and Sir Robert Peel expressed a wish that the procession should return to Liverpool.   It was, however, represented to them that a vast concourse of people had assembled at Manchester to witness the arrival of the trains; that report would exaggerate the mischief, if they did not complete the journey; and that a false panic on that day might seriously affect future railway travelling and the value of the Company's property.   The party consented accordingly to proceed to Manchester, but on the understanding that they should return as soon as possible, and refrain from further festivity.

As the trains approached Manchester, crowds of people were found covering the banks, the slopes of the cuttings, and even the railway itself.   The multitude, become impatient and excited by the rumours which reached them, had outflanked the military, and all order was at an end.   The people clambered about the carriages, holding on by the door handles, and many were tumbled over; but, happily, no fatal accident occurred.   At the Manchester station, the political element began to display itself; placards about " Peterloo," &c., were exhibited, and brickbats were thrown at the carriage containing the Duke.   On the carriages coming to a stand in the Manchester station the Duke did not descend, but remained seated, shaking hands with the women and children who were pushed forward by the crowd.   Shortly after, the trains returned to Liverpool, which they reached, after considerable interruptions, in the dark, at a late hour.

On the following morning the railway was opened for public traffic.   The first train of 140 passengers was booked and sent on to Manchester, reaching it in the allotted time of two hours; and from that time the traffic has regularly proceeded from day to day until now.

It is scarcely necessary that we should speak at any length of the commercial results of the Liverpool and Manchester Railway. Suffice it to say that its success was complete and decisive. The anticipations of its projectors were, however, in many respects at fault. They had based their calculations almost entirely on the heavy merchandise traffic—such as coal, cotton, and timber,—relying little upon passengers; whereas the receipts derived from the conveyance of passengers far exceeded those derived from merchandise of all kinds, which, for a time, continued a subordinate branch of the traffic. In the evidence given before the committee of the House of Commons, the promoters stated their expectation of obtaining about one-half of the whole number of passengers which the coaches then running could carry, or about 400 a day. But the railway was scarcely opened before it carried on an average about 1200 passengers daily; and five years after the opening, it carried nearly half a million of persons yearly. So successful, indeed, was the passenger traffic, that it engrossed the whole of the Company's small stock of engines.

For some time after the public opening of the line, Mr. Stephenson's ingenuity continued to be employed in devising improved methods for securing the safety and comfort of the travelling public. Few are aware of the thousand minute details which have to be arranged— the forethought and contrivance that have to be exercised—to enable the traveller by railway to accomplish his journey in safety. After the difficulties of constructing a level road over bogs, across valleys, and through deep cuttings, have been overcome, the maintenance of the way has to be provided for with continuous care. Every rail with its fastenings must be complete to prevent risk of accident, and the road must be kept regularly ballasted up to the level to diminish the jolting of vehicles passing over it at high speeds.

Then the stations must be protected by signals observable from such a distance as to enable the train to be stopped in event of an obstacle, such as a stopping or shunting train being in the way. For some years the signals employed on the Liverpool railway were entirely given by men with flags of different colours stationed along the line; there were no fixed signals, nor electric telegraphs; but the traffic was nevertheless worked quite as safely as under the more elaborate and complicated system of telegraphing which has since been established.

From an early period it became obvious that the iron road as originally laid down was quite insufficient for the heavy traffic which it had to carry. The line was in the first place laid with fish-bellied rails of thirty-five pounds to the yard, calculated only for horse-traffic, or, at most, for engines like the " Rocket," of very light weight. But as the power and the weight of the locomotives were increased, it was found that such rails were quite insufficient for the safe conduct of the traffic, and it therefore became necessary to re-lay the road with heavier and stronger rails at considerable expense.

The details of the carrying stock had in like manner to be settled by experience. Everything had, as it were, to be begun from the beginning. The coal-waggon, it is true, served in some degree as a model for the railway-truck; but the railway passenger-carriage was an entirely novel structure. It had to be mounted upon strong framing, of a peculiar kind, supported on springs to prevent jolting. Then there was the necessity for contriving some method of preventing hard bumping of the carriage-ends when the train was pulled up; and hence the contrivance of buffer-springs and spring frames. For the purpose of stopping the train, brakes on an improved plan were also contrived, with new modes of lubricating the carriage-axles, on which the wheels revolved at an unusually high velocity. In all these contrivances, Mr. Stephenson's inventiveness was kept constantly on the stretch; and though many improvements in detail have been effected since his time, the foundations were then laid by him of the present system of conducting railway traffic. As a curious illustration of the inventive ingenuity which he displayed in contriving the working of the Liverpool line, we may mention his invention of the Self-acting Brake. He early entertained the idea that the momentum of the running train might itself be made available for the purpose of checking its speed. He proposed to fit each carriage with a brake which should be called into action immediately on the locomotive at the head of the train being pulled up. The impetus of the carriages carrying them forward, the buffer-springs would be driven home, and, at the same time, by a simple arrangement of the mechanism, the brakes would be called into simultaneous action; thus the wheels would be brought into a state of sledge, and the train speedily stopped. This plan was adopted by Mr. Stephenson before he left the Liverpool and Manchester Railway, though it was afterwards discontinued; and it is a remarkable fact, that this identical plan, with the addition of a centrifugal apparatus, has quite recently been revived by M. Guérin, a French engineer, and extensively employed on foreign railways, as the best method of stopping railway trains in the most efficient manner and in the shortest time.

Finally, Mr. Stephenson had to attend to the improvement of the power and speed of the locomotive—always the grand object of his study,—with a view to economy as well as regularity in the working of the railway. In the " Planet " engine, delivered upon the line immediately subsequent to the public opening, all the improvements which had up to this time been contrived by him and his son were introduced in combination—the blast-pipe, the tubular boiler, horizontal cylinders inside the smoke-box, the cranked axle, and the fire-box firmly fixed to the boiler. The first load of goods conveyed from Liverpool to Manchester by the "Planet" was eighty tons in weight, and the engine performed the journey against a strong head wind in two hours and a half. On another occasion, the same engine brought up a cargo of voters from Manchester to Liverpool, during a contested election, within a space of sixty minutes. The "Samson," delivered in the following year, exhibited still further improvements, the most important of which was that of *coupling* the fore and hind wheels of the engine. By this means, the adhesion of the wheels on the rails was more effectually secured, and thus the full hauling power of the locomotive was made available. The "Samson," shortly after it was placed upon the line, dragged after it a train of waggons weighing one hundred and fifty tons, at a speed of about twenty miles an hour; the consumption of coke being reduced to only about a third of a pound per ton per mile.

The rapid progress thus made will show that the inventive faculties of Mr. Stephenson and his son were kept fully on the stretch; but their labours were amply repaid by the result. They were, doubtless, to some extent stimulated by the number of competitors who about the same time appeared as improvers of the locomotive engine. Of these, the most prominent were the Messrs. Braithwaite and Ericsson, whose engine, the "Novelty," had excited such high expectations at the Rainhill competition. The directors of the railway, desirous of giving all parties a fair chance, ordered from those makers two engines on the same model; but their performances not proving satisfactory, they were finally withdrawn. One of them slipped off the rails near the Sankey viaduct, and was nearly thrown over the embankment. The superiority of Mr. Stephenson's locomotives over all others that had yet been tried, induced the directors of the railway to require that the engines

supplied to them by other builders should be constructed
after the same model.   Mr. Stephenson himself always
had the greatest faith in the superiority of his own
engines over all others, and did not hesitate strongly to
declare it.   When it was once proposed to introduce the
engines of another maker on the Manchester and Leeds
line, he said, " Very well; I have no objection: but put
them to this fair test.   Hang one of ——'s engines on
to one of mine, back to back.   Then let them go at it;
and whichever walks away with the other, *that's the
engine.*"

The engineer had also to seek out the proper men to
maintain and watch the road, and more especially to
work the locomotive engines.   Steadiness, sobriety, common sense, and practical experience, were the qualities
which he especially valued in those selected by him for
that purpose.   But where were the men of experience
to be found ?   Very few railways were yet at work, and
these were almost exclusively confined to the northern
coal counties; hence a considerable proportion of the
drivers and firemen employed on the Liverpool line
were brought from the neighbourhood of Newcastle.
Mr. Stephenson was, however, severely censured in the
' Edinburgh Review' for the alleged preference shown
by him in selecting workmen from his own county.   It
was there insisted that the local population had the first
claim to be employed, and he was blamed for " introducing into the country a numerous body of workmen,
in various capacities, strangers to the soil and to the
surrounding population; thus wresting from the hands
of those to whom they had naturally belonged, all the
benefits which the enterprise and capital of the district
had conferred."   In the case of the drivers of stage-
coaches, it was never regarded as a qualification for the
performance of their duties that they should be natives
of the parishes through which the coaches ran, but
mainly that they should know something of the business

of coach-driving. Mr. Stephenson merely adopted the same course in selecting his drivers and firemen; and though Durham and Northumberland supplied a considerable proportion of them in the first instance, he could not always find skilled workmen enough for the important and responsible duties to be performed. It was a saying of his, that "he could engineer matter very well, and make it bend to his purpose, but his greatest difficulty was in engineering *men*."

Mr. Stephenson did not think it necessary to vindicate himself from the above charge, but Mr. Hardman Earle, one of the directors of the Company, did so in an effectual manner, showing that of the six hundred persons employed in the working of the Liverpool line, not more than sixty had been recommended by their engineer, and of these a considerable number were personally unknown to him. Some of them, indeed, had been brought up under his own eye, and were men whose character and qualifications he could vouch for. But these were not nearly enough for his purpose; and he often wished that he could contrive heads and hands on which he might rely, as easily as he could construct railways and manufacture locomotives. As it was, Stephenson's mechanics were in request all over England; the Newcastle workshops continuing for many years to perform the part of a training school for engineers, and to supply locomotive superintendents and drivers, not only for England but for nearly every country in Europe; preference being given to them by the directors of railways, in consequence of their previous training and experience, as well as because of their generally excellent qualities as steady and industrious workmen.

The success of the Liverpool and Manchester experiment naturally excited great interest. People flocked to Lancashire from all quarters to see the steam-coach running upon a railway at three times the speed of a mail-coach, and to enjoy the excitement of actually travelling in the wake of an engine at that incredible' velocity. The travellers returned to their respective districts full of the wonders of the locomotive, considering it to be the greatest marvel of the age. Railways are familiar enough objects now, and our children who grow up in their midst may think little of them; but thirty years since it was an event in one's life to see a locomotive, and to travel for the first time upon a public railroad.

In remote districts, however, the stories told about the benefits conferred by the Liverpool railway were received with considerable incredulity, and the proposal to extend such roads in all directions throughout the country caused great alarm. In the districts through which stage-coaches ran, giving employment to large numbers of persons, it was apprehended that, if railways were established, the turnpike-roads would become deserted and grown over with grass, country inns and their buxom landladies would be ruined, the race of coach-drivers and hostlers would become extinct, and the breed of horses be entirely destroyed. But there was hope for the coaching interest, in the fact that the Government were employing their engineers to improve the public high roads so as to render railways unnecessary. It was announced in the papers that a saving of thirty miles would be effected by the new road between London and Holyhead, and an equal saving between London and Edinburgh. And to show what the speed of horses could accomplish, we find it set forth as an extraordinary fact, that the " Patent Tallyho Coach," in the year 1830 (when the Birmingham line had been projected), performed the entire journey of 109 miles between London and Birmingham —breakfast included—in seven hours and fifty minutes! Great speed was also recorded on the Brighton road, the " Red Rover " doing the distance between London and Brighton in four hours and a half. These speeds were

not, however, secured without accidents, for there was
scarcely a newspaper of the period that did not contain
one or more paragraphs headed, " Another dreadful coach
accident."

The practicability of railway locomotion being now
proved, and its great social and commercial advantages
ascertained, the extension of the system was merely a
question of time, money, and labour. A fine opportunity presented itself for the wise and judicious action
of the Government in the matter,—the improvement of
the internal communications of a country being really
one of its most important functions. But the Government of the day, though ready enough to spend money
in improvements of the old turnpike roads, regarded the
railroads with hostility, and met them with obstructions
of all kinds. They seemed to think it their duty to
protect the turnpike trusts, disregarding the paramount
interest of the public. This may possibly account for
the singular circumstance that, at the very time they
were manifesting indifference or aversion to the locomotive on the railroad, they were giving every encouragement to the locomotive on turnpike roads. In 1831, we
find a Committee of the House of Commons appointed
to inquire into and report upon—not the railway system
—but the applicability of the steam carriage to common
roads ; and, after investigation, the committee were so
satisfied with the evidence taken, that they reported
decidedly in favour of the road locomotive system.
Though they ignored the railway, they recognised the
steam carriage.

But even a Report of the House of Commons—powerful though it be—cannot alter the laws of gravity and
friction ; and the road locomotive remained, what it ever
will be, an impracticable machine. Not that it is impossible to work a locomotive upon a common road ;
but to work it to any profit at all as compared with the
locomotive upon a railway. Numerous trials of steam

carriages were made at the time by Sir Charles Dance, Mr. Hancock, Mr. Gurney, Sir James Anderson, and other distinguished gentlemen of influence. Journalists extolled their utility, compared with "the much-boasted application on railroads."[1] But notwithstanding all this, and the House of Commons' Report in its favour, Mr. Stephenson's first verdict, pronounced upon the road locomotive many years before, when he was only an engine-wright at Killingworth, was fully borne out by the result; and it became day by day clearer that the attempt to introduce the engine into general use upon turnpike roads could only prove a delusion and a snare.

Although the legislature took no initiative step in the direction of railway extension, the public spirit and enterprise of the country did not fail it at this juncture. The English people, though they may be defective in their capacity for organization, are strong in individualism; and not improbably their admirable qualities in the latter respect detract from their efficiency in the former. Thus, in all times, their greatest national enterprises have not been planned by officialism and carried out upon any regular system, but have sprung, like their constitution, their laws, and their entire industrial arrangements, from the force of circumstances and the individual energies of the people. Hence railway extension, like so many other great English enterprises, was now left to be carried out by the genius of English engineers, backed by the energy of the English public.

The mode of action was characteristic and national. The execution of the new lines was undertaken entirely by joint-stock associations of proprietors, after the manner of the Stockton and Darlington, and Liverpool and Manchester companies. These associations are con-

---

[1] Letter of Mr. John Herapath in 'Mechanics' Magazine,' vol. xv. p. 123. For full information as to the various trials made with steam-carriages, see 'The Economy of Steam-power on Common Roads,' by C. F. T. Young, C.E. London, 1861.

formable to our national habits, and fit well into our system of laws. They combine the power of vast resources with individual watchfulness and motives of self-interest; and by their means gigantic undertakings, which elsewhere would be impossible to any but kings and emperors with great national resources at command, were carried out by the co-operation of private persons. And the results of this combination of means and of enterprise have been truly marvellous. Within the life of the present generation, the private citizens of England engaged in railway extension have, in the face of Government obstructions, and without taking a penny from the public purse, executed a system of communications involving works of the most gigantic kind, which, in their total mass, their cost, and their eminent public utility, far exceed the most famous national undertakings of any age or country.

Mr. Stephenson was, of course, actively engaged in the construction of the numerous railways now projected by the joint-stock companies. During the formation of the Manchester and Liverpool line, he had been consulted respecting many projects of a similar kind. One of these was a short railway between Canterbury and Whitstable, about six miles in length. He was too much occupied with the works at Liverpool to give this scheme much of his personal attention. But he sent his assistant, Mr. John Dixon, to survey the line; and afterwards Mr. Locke to superintend the execution of the works. The act was obtained in 1826, and the line was opened for traffic in 1830. It was partly worked by fixed engine-power, and partly by Stephenson's locomotives, similar to the engines used upon the Stockton and Darlington Railway.

But the desire for railway extension principally pervaded the manufacturing districts, especially after the successful opening of the Liverpool and Manchester line. The commercial classes of the larger towns soon became eager for a participation in the good which they had so recently derided. Railway projects were set on foot in great numbers, and Manchester became a centre from which main lines and branches were started in all directions. The interest, however, which attaches to these later schemes is of a much less absorbing kind than that which belongs to the early history of the railway and the steps by which it was mainly established. We naturally sympathise more keenly with the early struggles of a great principle, its trials and its difficulties, than with its after stages of success; and, however gratified and astonished we may be at its consequences, the interest is in a great measure gone when its triumph has become a matter of certainty.

The commercial results of the Liverpool and Manchester line were so satisfactory, and indeed so greatly exceeded the expectations of its projectors, that many of the abandoned projects of the speculative year 1825 were forthwith revived. An abundant crop of engineers sprang up, ready to execute railways of any extent. Now that the Liverpool and Manchester line had been made, and the practicability of working it by locomotive power had been proved, it was as easy for engineers to make railways and to work them, as it was for navigators to find America after Columbus had made the first voyage. George Stephenson had shown the way, and engineers forthwith crowded after him full of great projects. Mr. Francis Giles himself took the field as a locomotive railway engineer, attaching himself to the Newcastle and Carlisle and London and Southampton projects. Mr. Brunel appeared, in like manner, as the engineer of the line projected between London and Bristol; and Mr. Braithwaite, the builder of the " Novelty" engine, as the engineer of a line from London to Colchester.

The first lines, however, which were actually constructed subsequent to the opening of the Liverpool and

Manchester Railway, were in connexion with it, and principally in the county of Lancaster. Thus a branch was formed from Bolton to Leigh, and another from Leigh to Kenyon, where it formed a junction with the main line between Liverpool and Manchester. Branches to Wigan on the north, and to Runcorn Gap and Warrington on the south of the same line, were also formed. A continuation of the latter, as far south as Birmingham, was shortly after projected under the name of the Grand Junction Railway.

The Grand Junction line was projected as early as the year 1824, when the Liverpool and Manchester scheme was under discussion, and Mr. Stephenson then published a report on the subject. The plans were deposited, but the bill was thrown out on the opposition of the landowners and canal proprietors. When engaged in making the survey, Mr. Stephenson called upon some of the landowners in the neighbourhood of Nantwich to obtain their assent, and was somewhat disgusted to learn that the agents of the canal companies had been before him, and described the locomotive to the farmers as a most frightful machine, emitting a breath as poisonous as the fabled dragon of old; and telling them that if a bird flew over the district where one of these engines passed, it would inevitably drop down dead! The application for the bill was renewed in 1826, and again failed; and at length it was determined to wait the issue of the Liverpool and Manchester experiment. The act was eventually obtained in 1833, by which time the projectors of railways had learnt the art of "conciliating" the landlords,—and a very expensive process it proved. But it was the only mode of avoiding a still more expensive parliamentary opposition.

When it was proposed to extend the advantages of railways to the population of the midland and southern counties of England, an immense amount of alarm was created in the minds of the country gentlemen. They did not relish the idea of private individuals, principally resident in the manufacturing districts, invading their domains; and they everywhere rose up in arms against the "new-fangled roads." Colonel Sibthorpe openly declared his hatred of the "infernal railroads," and said that he "would rather meet a highwayman, or see a burglar on his premises, than an engineer!" Mr. Berkeley, the member for Cheltenham, at a public meeting in that town, re-echoed Colonel Sibthorpe's sentiments, and "wished that the concoctors of every such scheme, with their solicitors and engineers, were at rest in Paradise!" The impression prevailed amongst the rural classes, that fox-covers and game-preserves would be seriously prejudiced by the formation of railroads; that agricultural communications would be destroyed, land thrown out of cultivation, landowners and farmers reduced to beggary, the poor-rates increased through the number of persons thrown out of employment by the railways,—and all this in order that Liverpool, Manchester, and Birmingham shopkeepers and manufacturers might establish a monstrous monopoly in railway traffic.

The inhabitants of even some of the large towns were thrown into a state of consternation by the proposal to provide them with the accommodation of a railway. The line from London to Birmingham would naturally have passed close to the handsome town of Northampton, and was so projected. But the inhabitants of the shire, urged on by the local press, and excited by men of influence and education, opposed the project, and succeeded in forcing the promoters, in their survey of the line, to pass the town at a distance. The necessity was thus involved of distorting the line, by which the enormous expense of constructing the Kilsby Tunnel was incurred. Not many years elapsed before the inhabitants of Northampton became clamorous for rail-

way accommodation, and a special branch was constructed for them.   The additional cost involved by this
forced deviation of the line could not have amounted to
less than half a million sterling ; the loss falling, not upon
the shareholders only, but mainly upon the public.

Other towns in the south followed the example of
Northampton in howling down the railways.   Thus,
when it was proposed to carry a line through Kent, by
the populous county town of Maidstone, a public meeting
was held to oppose the project, and the railway had not
a single supporter amongst the townspeople.   When at
length formed through Kent, it passed Maidstone at a
distance ; but in a few years the Maidstone burgesses,
like those of Northampton, became clamorous for a
railway ; and a branch was formed for their accommodation.   Again, in a few years, they complained that
the route was circuitous, as they had compelled it to be ;
consequently another and shorter line was formed, to
bring Maidstone into more direct communication with
the metropolis ; and it is expected that even a *third* line
to the same place will shortly be under construction !
In like manner the London and Bristol (afterwards the
Great Western) Railway was vehemently opposed by
the people of the towns through which the line was
projected to pass ; and when the bill was thrown out by
the Lords—after 30,000*l.* had been expended by the
promoters—the inhabitants of Eton assembled, under
the presidency of the Marquis of Chandos, to rejoice
and congratulate themselves and the country on the
defeat of the measure.   Eton, however, has now the
convenience of two railways to the metropolis.

During the time that the works of the Liverpool
and Manchester line were in progress, our engineer was
consulted respecting a short railway proposed to be
formed between Leicester and Swannington, for the
purpose of opening up a communication between the
town of Leicester and the coal-fields in the western part

of the county.   Mr. Ellis, afterwards chairman of the
Midland Railway—like Edward Pease, a member of the
Society of Friends—was the projector of this undertaking.   He had some difficulty, however, in getting
the requisite capital subscribed for, the Leicester townspeople who had money being for the most part interested
in canals.   Mr. Ellis went over to Liverpool to invite
George Stephenson to come upon the ground and survey
the line.   He did so, and then the projector told him of
the difficulty he had in finding subscribers to the concern.   " Give me a sheet," said Stephenson, " and I will
raise the money for you in Liverpool."   The engineer
was as good as his word, and in a short time the sheet
was returned with the subscription complete.   Mr.
Stephenson was then asked to undertake the office of
engineer for the line, but his answer was that he had
thirty miles of railway in hand, which were enough for
any engineer to attend to properly.   Was there any person he could recommend ?
" Well," said he, " I think
my son Robert is com-
petent to undertake the
thing."   Would Mr. Ste-
phenson be answerable for
him ? " Oh, yes, certainly."
And Robert Stephenson, at
twenty-seven years of age,
was installed engineer of
the line accordingly.

The requisite
Parliamentary
powers having
been obtained,
Robert   Ste-
phenson   pro-
ceeded with the
construction of
the   railway,
about   sixteen

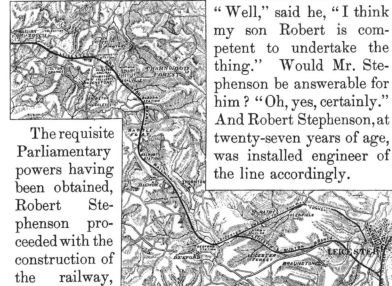

MAP OF LEICESTER AND SWANNINGTON RAILWAY.

miles in length, towards the end of 1830. The works were comparatively easy, excepting at the Leicester end, where the young engineer encountered his first stiff bit of tunnelling. The line passed underground for a mile and three-quarters, and 500 yards of its course lay through loose running sand. The presence of this material rendered it necessary for the engineer, in the first place, to construct a wooden tunnel to support the soil while the brickwork was being executed. This measure proved sufficient, and the whole was brought to a successful termination within a reasonable time. While the works were in progress, Robert kept up a regular correspondence with his father at Liverpool, consulting him on all points in which his greater experience was likely to be of service. Like his father, Robert was very observant, and always ready to seize opportunity by the forelock. It happened that the estate of Snibston, near Ashby-de-la-Zouch, was advertised for sale; and the young engineer's experience as a coal-viewer and practical geologist suggested to his mind that coal was most probably to be found underneath. He communicated his views to his father on the subject.[1] The estate lay in the immediate neighbourhood of the railway; and if the conjecture proved correct, the finding of the coal must necessarily prove a most fortunate circumstance for the purchasers

---

[1] George Stephenson was himself always on the look-out for new coal-fields, and eventually became a large coal-owner in the neighbourhood of Chesterfield, through discovering new beds of that mineral while constructing the Midland Railway. As early as 1824 we find, from a letter written by him to Mr. Sandars, of Liverpool, handed to us by Robert Stephenson, that he was actively speculating on the subject of the strata underlying the line of the then proposed Liverpool and Manchester Railway. "On my way to Bolton," said he, "and whilst at Bolton, I collected a great deal of useful information respecting the coal-fields in that neighbourhood. It is my opinion that coal will be found under Chat Moss. I think there will be none under Kirkby Moss, but immediately on the south-east point of Mossbro Road, from where the railroad crosses, I think it will be found; and I believe the coal-field will pass up, even under Knowsley Hall, and continue through the whole of that high country by Prescott. But I should not advise any purchase to be made of coal-fields until a closer investigation is made, even though you were certain of the Act passing." We are not aware whether these speculations have been verified or not.

of the land. He accordingly requested his father to come over to Snibston and look at the property, which he did; and after a careful inspection of the ground, he arrived at the same conclusion as his son.

The large manufacturing town of Leicester, about fourteen miles distant, had up to that time been exclusively supplied with coal brought by canal from Derbyshire; and Mr. Stephenson saw that the railway under construction, from Swannington to Leicester, would furnish him with a ready market for any coals which he might find at Snibston. Having induced two of his Liverpool friends to join him in the venture, the Snibston estate was purchased in 1831: and shortly after, Stephenson removed his home from Liverpool to Alton Grange, for the purpose of superintending the sinking of the pit. He travelled thither by gig with his wife,— his favourite horse " Bobby " performing the journey by easy stages.

Sinking operations were immediately commenced, and proceeded satisfactorily until the old enemy, water, burst in upon the workmen, and threatened to drown them out. But by means of efficient pumping-engines, and the skilful casing of the shaft with segments of cast-iron —a process called " tubbing,"[1] which Mr. Stephenson was the first to adopt in the Midland Counties—it was eventually made water-tight, and the sinking proceeded. When a depth of 166 feet had been reached, a still more formidable difficulty presented itself—one which had baffled former sinkers in the neighbourhood, and deterred them from further operations. This was a remarkable bed of whinstone or greenstone, which had originally been poured out as a sheet of burning lava over the

---

[1] Tubbing is now adopted in many cases as a substitute for brick-walling. The tubbing consists of short portions of cast-iron cylinder fixed in segments. Each weighs about 4½ cwt., is about three or four feet long, and about three-eighths of an inch thick. These pieces are fitted closely together, length under length, and form an impermeable wall along the sides of the pit.

denuded surface of the coal measures; indeed it was afterwards found that it had turned to cinders one part of the seam of coal with which it had come in contact. The appearance of this bed of solid rock was so unusual a circumstance in coal mining, that some experienced sinkers urged Stephenson to proceed no further, believing the occurrence of the dyke at that point to be altogether fatal to his enterprise. But, with his faith still firm in the existence of coal underneath, he fell back upon his old motto of " Persevere ! " He determined to go on boring; and down through the solid rock he went until, twenty-two feet lower, he came upon the coal measures. In the mean time, however, lest the boring at that point should prove unsuccessful, he had commenced sinking another pair of shafts[1] about a quarter of a mile west of the " fault;" and after about nine months' labour he reached the principal seam, called the " main coal."

The works were then opened out on a large scale, and Mr. Stephenson had the pleasure and good fortune to send the first train of main coal to Leicester by railway. The price was immediately reduced there to about 8s. a ton, effecting a pecuniary saving to the inhabitants of the town of about 40,000l. per annum, or equivalent to the whole amount then collected in Government taxes and local rates, besides giving an impetus to the manufacturing prosperity of the place, which has continued down to the present day. The correct and scientific principles upon which the mining operations at Snibston were conducted offered a salutary example to the neighbouring colliery owners. The numerous improvements there introduced were freely exhibited to all, and they were afterwards reproduced in many forms all over the Midland Counties, greatly to the advantage of the mining interests.

---

[1] Mr. Stephenson was strongly in favour of working and ventilating coal-mines by means of more shafts than one. He considered the provision of at least a second shaft essential for the safety of the persons working in the pit, in the event of the occurrence of any of the numerous accidents incident to coal-mining.

At the same time Mr. Stephenson endeavoured to extend the benefit of railways throughout the district in which he now resided. He suggested to Lord Stamford the importance of constructing a branch line from the Leicester and Swannington Railway through his property, principally for the purpose of opening out his fine granite quarries at Groby The valuable advice was taken by Lord Stamford, and Mr. Stephenson laid out the line for him and superintended the works gratuitously. Another improvement which he effected for Lord Talbot proved of even greater pecuniary value. He contrived for his Lordship, with no slight difficulty, a plan for "tubbing off" the fresh water from the salt at his mines near Tamworth, which enabled the salt-works there to be subsequently carried on to a great profit, which had not before been practicable. Mr. Stephenson was less successful in his endeavours to induce the late Marquis of Hastings to consent to the Birmingham and Derby Railway, of which he was the engineer, passing through the mineral district of Ashby-de-la-Zouch. The Marquis was the principal owner of the colliery property in the neighbourhood, and Mr. Stephenson calculated upon his Lordship's influence in support of a scheme so certain to increase the value of his estate. But the Marquis, like many others of his class, did not yet detect the great advantages of railways, and he threatened his determined opposition if the Derby line were attempted to be brought through his coal-field. The line was consequently taken further to the west, by way of Burton; and thus Ashby for a time lost the benefits of railway communication. Twenty years elapsed before Mr. Stephenson's designs for its accommodation were carried into effect.

Nor was Mr. Stephenson less attentive to the comfort and well-being of those immediately dependent upon him—the workpeople of the Snibston colliery and their families. Unlike many of those large employers who

have " sprung from the ranks," he was one of the kindest
and most indulgent of masters.   He would have a fair
day's work for a fair day's wages ; but he never forgot
that the employer had his duties as well as his rights.
First of all, he attended to the proper home accommodation of his workpeople.   He erected a village of comfortable cottages, each provided with a snug little garden.
He was also instrumental in erecting a church adjacent
to the works, as well as Church schools for the education
of the colliers' children ; and with that broad catholicity
of sentiment which distinguished him, he further provided a chapel and a school-house for the use of the
Dissenting portion of the colliers and their families—an
example of benevolent liberality which was not without
a salutary influence upon the neighbouring employers.

STEPHENSON'S HOUSE AT ALTON GRANGE

Robert Stephenson

Engraved by W. Holl, after a photograph by Claudet.

Published by John Murray, Albemarle Street, 1862.

CHAPTER XV.

ROBERT STEPHENSON CONSTRUCTS THE LONDON AND BIRMINGHAM
RAILWAY.

OF the numerous extensive projects which followed close
upon the completion of the Liverpool and Manchester
line, and the locomotive triumph at Rainhill, that of
a railway between London and Birmingham was the
most·important.   The scheme originated at the latter
place in 1830.   Two committees were formed, and two
plans were proposed.   One was of a line to London by
way of Oxford, and the other by way of Coventry.
There was at that early period less of the fighting spirit
amongst railway projectors which unhappily prevailed
at a later date.   The simple object of the promoters of
both schemes being to secure the advantages of railway
communication with the metropolis, they wisely determined to combine their strength to secure it.   They
then resolved to call George Stephenson to their aid,
and requested him to advise them as to the two schemes
which were before them.   After a careful examination
of the country, Mr. Stephenson reported in favour of
the Coventry route, when the Lancashire gentlemen,
who were the principal subscribers to the project, having
every confidence in his judgment, supported his decision, and the line recommended by him was adopted
accordingly.

At the meeting of the promoters held at Birmingham
to determine on the appointment of the engineer for the
railway, there was a strong party in favour of associating
with Mr. Stephenson a gentleman with whom he had
been brought into serious collision in the course of the

Liverpool and Manchester undertaking. When the offer was made to him that he should be joint engineer with the other, he requested leave to retire and consider the proposal with his son. The two walked into St. Philip's churchyard, which adjoined the place of meeting, and debated the proposal. The father was in favour of accepting it. His struggle . heretofore had been so hard, that he could not bear the thought of missing so promising an opportunity of professional advancement. But the son, foreseeing the jealousies and heartburnings which the joint engineership would most probably create, recommended to his father the answer which Mr. Bradshaw gave, when shares in the Liverpool and Manchester line were offered to the Duke of Bridgewater's Trustees —"All or none!" "Well, I believe you are right," said Mr. Stephenson; and returning to the Committee, he announced to them his decision. "Then 'all' be it!" replied the Chairman; and he was at once appointed the engineer of the London and Birmingham Railway in conjunction with his son.

The line, as originally laid out, was to have had its London terminus at Maiden Lane, King's Cross, the site of the present Great Northern Station: it passed through Cashiobury and Grove Parks, the seats of Lord Essex and Lord Clarendon, and along the Hemel Hempstead and Little Goddesden valleys, in Hertfordshire. This latter portion of the project excited a vehement opposition on the part of the landowners, who formed a powerful confederacy against the bill. The principal parties who took an active part in the opposition were Lady Bridgewater and her trustees, Lord Essex, and Sir Astley Cooper, supported by the Grand Junction Canal Company. By their influence the landowners throughout the counties of Hertford and Buckingham organised themselves to oppose the measure. The time for preparing the plans to be deposited with the several clerks of the peace, as required by the standing orders of Parliament, being very limited, the necessary documents were prepared in great haste, and were deposited in such an imperfect state as to give just grounds for presuming that they would not pass the ordeal of the Standing Orders Committee. It was also thought that alterations might be made in some parts of the railway which would remove the objections of the principal landowners, and it was therefore determined to postpone the application to Parliament until the following session.

In the mean time the opponents of the bill out of doors were not idle. Numerous pamphlets were published, calling on the public to "beware of the bubbles," and holding up the promoters of railways to ridicule. They were compared to St. John Long and similar quacks, and pronounced fitter for Bedlam than to be left at large. The canal proprietors, landowners, and road trustees, made common cause in decrying and opposing the project. The failure of railways was confidently predicted—indeed, it was elaborately attempted to be proved that they had failed;[1] and it was industriously spread abroad that the locomotive engines, having been found useless and highly dangerous on the Liverpool and Manchester line, were immediately to be

---

[1] In a book published in 1834, entitled 'Railroad Impositions Detected,' by Richard Cort, son of the inventor of the iron-puddling process, the "Bubble Railway Speculations" of the time were strongly inveighed against. The writer proved incontrovertibly, to his own satisfaction, that the Liverpool and Manchester line had not, during the time it had been at work, made so much as one per cent. profit, and that it must soon cease to pay any dividend whatever, and involve its proprietors in hopeless ruin. With canals and common roads, however, the case was altogether different. "Long before any more new lines can be constructed," said the writer, "inland navigation will be still further improved, and steam-carriages will be in the field many times more profitable than railways ever can be, and eventually quite as expeditious." And again:—"As an additional comfort to shareholders in the London and Birmingham Railway, it should be observed that in less than twelve months from the passing of the Bill for the Granite Road from London to Birmingham, now actually planning side by side of that unfortunate speculation, the stone tramway will be ready to receive steam-carriages, to enable them to run quite as fast as the iron railway-coaches. If this be true, who will subscribe one farthing to the Birmingham railway?"

abandoned in favour of horses—a rumour which the directors of the Company considered it necessary publicly to contradict.

Public meetings were held in all the counties through which the line would pass between London and Birmingham, at which the project was denounced, and strong resolutions were passed against it. The county meetings of Northampton[1] were held at Towcester; of Bedford at Leighton Buzzard; of Buckingham at Stony Stratford; of Hertford at Watford and Great Berkhampstead; and of Middlesex, in Exeter Hall, London. It

---

[1] The opposition of the town of Northampton, above referred to (p. 293), was generally understood at the time to have had the effect of compelling the engineer to deviate the line so as to avoid that place, and to render necessary the construction of the Kilsby Tunnel. This had been often stated without contradiction, and was repeated in the first edition of this work, published in 1857. That statement having come under the notice of Mr. W. T. Higgins, Mayor of Northampton at the time, he addressed a letter to the 'Times,' dated September 19th, of that year, enclosing the copy of a resolution passed at a public meeting of the inhabitants held in November, 1830—"That it is the opinion of this meeting that it is highly desirable that such railway should approach as near to the town of Northampton as possible." On this the author wrote to Robert Stephenson for further information, and the following was his reply, dated 30th September, 1857:—"It may be quite true what the Mayor of Northampton says, but it certainly does not convey the whole truth. Meetings were held in almost every town on the line, both for and against the railway, but Northampton distinguished itself by being rather more furious than other places in opposition to railways, and begged that the line might be kept away from them. It is true that the low level of Northampton presented a very great objection to the line approaching it nearer than it does; but I had a strong leaning for that direction, because it would have admitted of the line approaching the Kilsby ridge up the Althorp valley in a favourable manner. I was anxious to go in that direction for another reason, viz., that the line would have reached a point better calculated than Rugby for commanding the midland and northern counties. If you look at the map, you will easily see the bearing of this view. The line by Banbury and Warwick I soon abandoned, in consequence of feeling the absolute importance of enabling the London and Birmingham to command the midland counties and the districts now traversed by the North Midland. Nothing saved a direct line to Manchester in 1845, but the general position of the London and Birmingham, and especially the bending northwards and passing through Rugby, instead of bending southwards, and passing through Banbury and Warwick, which latter course was strongly urged upon me by some of the most influential Birmingham people. Few persons have any notion of how completely the whole system of our railways has been influenced by the bend northwards at Rugby, to which I have referred. Scarcely a single line that now exists to the north of that point would have been made as it now is, but for the determination I then formed as to the direction in which the railway should be constructed."

was insisted at those meetings that there was no necessity whatever for accelerating the existing communications, there being already abundant means of conveyance for travellers by the coaches daily travelling through the district at ten miles an hour, whilst there was water-carriage for heavy goods to a much greater extent than had ever been required.   Deputations from the promoters of the railway attended some of these meetings for the purpose of stating their case, but the landowners would not permit them to be heard.   The Earls of Clarendon and Essex were the most powerful opponents of the measure, and the other landed proprietors followed in their wake.   The attempt was made to conciliate these landlords by explanations, but all such efforts proved futile.

"I remember," said Robert Stephenson, describing the opposition, "that we called one day on Sir Astley Cooper, the eminent surgeon, in the hope of overcoming his aversion to the railway.   He was one of our most inveterate and influential opponents.   His house was at Hemel Hempstead, and the line was so laid out as to pass through part of his property.   We found a courtly, fine-looking old gentleman, of very stately manners, who received us kindly and heard all we had to say in favour of the project.   But he was quite inflexible in his opposition to it.   No deviation or improvement that we could suggest had the slightest effect in conciliating him.   He was opposed to railways generally, and to this in particular.   'Your scheme,' said he, 'is preposterous in the extreme.   It is of so extravagant a character, as to be positively absurd.   Then look at the recklessness of your proceedings!   You are proposing to cut up our estates in all directions for the purpose of making an unnecessary road.   Do you think for one moment of the destruction of property involved by it?   Why, gentlemen, if this sort of thing be permitted to go on, you will in a very few years *destroy the noblesse!*'

We left the honourable baronet without having produced the slightest effect upon him, excepting perhaps, it might be, increased exasperation against our scheme. I could not help observing to my companions as we left the house, ' Well, it is really provoking to find one who has been made a " Sir " for cutting that wen out of George the Fourth's neck, charging us with contemplating the destruction of the *noblesse*, because we propose to confer upon him the benefits of a railroad.' "

Such being the opposition of the owners of land, it was with the greatest difficulty that an accurate survey of the line could be made. At one point the vigilance of the landowners and their servants was such, that the surveyors were effectually prevented taking the levels by the light of day; and it was only at length accomplished at night by means of dark lanterns. There was one clergyman, who made such alarming demonstrations of his opposition, that the extraordinary expedient was resorted to of surveying his property during the time he was engaged in the pulpit. This was managed by having a strong force of surveyors in readiness to commence their operations, who entered the clergyman's grounds on one side the moment they saw him fairly off them on the other. By a well organised and systematic arrangement each man concluded his allotted task just as the reverend gentleman concluded his sermon; so that, before he left the church, the deed was done, and the sinners had all decamped. Similar opposition was offered at many other points, but ineffectually. The laborious application of Robert Stephenson was such, that in examining the country to ascertain the best line, he walked the whole distance between London and Birmingham upwards of twenty times. He was ably supported by his staff of surveyors under the direction of Mr. Gooch, whose united perseverance eventually overcame all obstacles; and by the end of 1831 the requisite plans were deposited preparatory to an application being made to Parliament in the ensuing session.

The principal alterations made in the new line were at the London end; the terminus being changed from Maiden Lane to a large piece of open land adjoining the Regent's Canal—the site of the present London and North-Western Goods Station; and also at Watford, where the direction of the line was altered so as entirely to avoid the parks of Lords Essex and Clarendon. This latter diversion, however, inflicted on the public the inconvenience of the Watford Tunnel, about a mile in length, and on the company a largely increased outlay for its construction. The Hemel Hempstead and Goddesden valleys were also avoided, and the line proceeded by the towns of Berkhampstead and Tring.

It was expected that these alterations would have the effect of mitigating, if not of entirely averting, the powerful opposition of the landowners; but it was found, on the contrary, to become more violent than ever, although the grounds of complaint in regard to their parks and residences had been almost entirely removed. The most exaggerated alarms continued to be entertained, especially by those who had never seen a railway; and although there were a few country gentlemen who took a different view of the subject, when the bill for the altered line was introduced into Parliament in the session of 1832, the owners of nearly seven-eighths of the land required for the railway were returned as dissentients. It was, however, a noticeable fact, that Lords Derby and Sefton, who had so vehemently opposed the Liverpool Railway in all its stages, were found among the assentients to the London and Birmingham line. The scheme had, it is true, many warm friends and supporters, but these were principally confined to classes possessing more intelligence than influence. Indeed, the change which was rapidly taking place in public opinion on the subject of rail-

ways induced the promoters to anticipate a favourable
issue to their application, notwithstanding the hostility
of the landowners. They also drew a favourable
augury from the fact that the Grand Junction Canal
Company, although still opposing the measure as strenu-
ously as ever, so far as the influence of its proprietors
collectively and individually extended, and watching all
the proceedings of the bill with a jealous eye, did not
openly appear in the ranks of its opponents, and, what
was of still greater significance, did not open their
purse-strings to supply funds for the opposition.

When the bill went before the Committee of the Commons, a formidable array of evidence was produced.
All the railway experience of the day was brought to
bear in support of the measure, and all that interested
opposition could do was set in motion against it. The
necessity for an improved mode of communication
between London and Birmingham was clearly demonstrated ; and the engineering evidence was regarded as
quite satisfactory. So strong an impression was made
upon the Committee, that the result was no longer doubtful so far as the Commons were concerned ; but it was
considered very desirable that the case should be fully
brought out in evidence for the information of the public,
and the whole of the witnesses in support of the bill,
about a hundred in number, were examined at great
length. The opponents confined themselves principally
to cross-examination, without producing direct evidence
of their own ; reserving their main opposition for the
House of Lords, where they knew that their strength
lay. Not a single fact was proved against the utility
of the measure, and the bill passed the Committee, and
afterwards the third reading in the Commons, by large
majorities.

It was then sent to the House of Lords, and went into
Committee, when a similar mass of testimony was again
gone through during seven days. An overwhelming

case was made out as before ; though an attempt was
made to break down the evidence of the witnesses on
cross-examination.   The feasibility of the route was
questioned, and the greatest conceivable difficulties were
suggested.   Their lordships seemed to take quite a
paternal interest in the protection of the public against
possible loss by the formation of the line.   The Committee required that the promoters should prove the
traffic to be brought upon the railway, and that the
profits derived from the working should pay a dividend of from six to eight per cent. upon the money
invested.   A few years after, the policy of Parliament completely changed in this respect.   When the
landed interest found railway companies paying from
six to ten times the marketable value of the land taken,
they were ready to grant duplicate lines through the
same districts, without proving any traffic whatever.

It soon became evident, after the proceedings had
been opened before the Committee of the Lords, that
the fate of the bill had been determined before a word
of the evidence had been heard.   At that time the
committees were open to all peers ; and the promoters
of the bill found, to their dismay, many of the lords
who were avowed opponents of the measure as landowners, sitting as judges to decide its fate.   Their
principal object seemed to be, to bring the proceedings
to a termination as quickly as possible.   An attempt at
negociation was made in the course of the proceedings
in committee, but failed, and the bill was thrown out,
on the motion of Earl Brownlow, one of Lady Bridgewater's trustees ; but though carried by a large majority,
the vote was far from unanimous.

As the result had been foreseen, measures were
taken to neutralise the effect of this decision as
regarded future operations.   Not less than 32,000*l.*
had been expended in preliminary and parliamentary
expenses up to this stage ; but the promoters deter-

mined not to look back, and forthwith made arrangements for prosecuting the bill in a future session. A meeting of the friends of the measure was held in London, attended by members of both Houses of Parliament, and by leading bankers and merchants; and a series of resolutions was passed, declaring their conviction of the necessity for the railway, and deprecating the opposition by which it had been encountered. Lord Wharncliffe, who had acted as the chairman of the Lords' Committee, attributed the failure of the bill entirely to the landowners; and Mr. Glyn subsequently declared that they had tried to smother the bill by the high price which they demanded for their property. It was determined to reintroduce the bill in the following session (1833), and measures were taken to prosecute it vigorously. Strange to say, the bill on this occasion passed both Houses silently and almost without opposition. The mystery was afterwards solved by the appearance of a circular issued by the directors of the company, in which it was stated, that they had opened " negotiations " with the most influential of their opponents ; that " these measures had been successful to a greater extent than they had ventured to anticipate ; and the most active and formidable had been conciliated." An instructive commentary on the mode by which these noble lords and influential landed proprietors had been " conciliated," is found in the simple fact that the estimate for land was nearly trebled, and that the owners were paid about 750,000*l.* for what had been originally estimated at 250,000*l.* The total expenses of carrying the bill through Parliament amounted to the frightful sum of 72,868*l.*

The landowners having thus been " conciliated," the promoters of the measure were at length permitted to proceed with the formation of their great highway, and allowed to benefit the country by carrying out one of the grandest public works that has ever been executed

in England, the utility of which may almost be pronounced unparalleled. Eighty miles of the railway
were shortly under construction; the works were let
(within the estimates) to contractors, who were necessarily for the most part new to such work. The business
of railway construction was not then well understood.
There were no leviathans among contractors as now,
able to undertake the formation of a line of railway
hundreds of miles in length; they were for the most
part men of small capital and slender experience. Their
tools and machinery were imperfect; they did not
understand the economy of time and piece labour; the
workmen, as well as their masters, had still to learn
their trade; and every movement of an engineer was
attended with outlays, which were the inevitable result
of a new system of things, but which each succeeding
day's experience tended to diminish.

The difficulties encountered by the Messrs. Stephenson
in constructing the line were thus very great; but the
most formidable of them originated in the character of
the works themselves. Extensive tunnels had to be
driven through unknown strata, and miles of underground excavation had to be carried out in order to
form a level road from valley to valley under the intervening ridges. This kind of work was the newest of
all to the contractors of that day. The experience of
the Messrs. Stephenson in the collieries of the North,
made them, of all living engineers, the best fitted to
grapple with such difficulties; yet even they, with all
their practical knowledge, could scarcely have foreseen
or anticipated the serious obstacles they were called
upon to encounter and overcome in executing the formidable cuttings, embankments, and tunnels of the
London and Birmingham Railway. It would be an
uninteresting, as it would be a fruitless task, to attempt
to describe these works in detail; but a general outline
of their extraordinary character and extent may not be
out of place.

The length of railway to be constructed between London and Birmingham was 112½ miles. The line crossed a series of low-lying districts separated from each other by considerable ridges of hills; and it was the object of the engineer to cross the valleys at as high an elevation, and the hills at as low a one, as possible. The high ground was therefore cut down and the "stuff" led into embankments, in some places of great height and extent, so as to form a road upon as level a plane as was considered practicable for the working of the locomotive engine. In some places, the high grounds were passed in open cuttings, as at the Oxhey summit near Harrow, Dudswell, Tring, Denbigh Hall, and Blisworth; whilst in others it was necessary to bore through them in tunnels with deep cuttings at either end, as at Primrose Hill, Watford, and Kilsby.

The most formidable excavations on the line are those at Tring, Denbigh Hall, and Blisworth. The Tring cutting is an immense chasm across the great chalk ridge of Ivinghoe. It is two miles and a half long, and for a quarter of a mile is fifty-seven feet deep. A million and a half cubic yards of chalk and earth were taken out of this cutting by means of horse-runs, and deposited in spoil banks; besides the immense quantity run into the embankment north of the cutting, forming a solid mound nearly six miles long and about thirty feet high. Passing over the Denbigh Hall cutting, and the Wolverton embankment of a mile and a half in length across the valley of the Ouse, we come to the excavation at Blisworth, a brief description of which will give the reader an idea of one of the most difficult kinds of railway work.

The Blisworth Cutting is one of the longest and deepest

BLISWORTH CUTTING.   [By Percival Skelton.]

grooves ever cut in the solid earth.   It is a mile and a half long, in some places sixty-five feet deep, passing through earth, stiff clay, and hard rock.   Not less than a million cubic yards of these materials were dug, quarried, and blasted out of it.   One-third of the cutting was stone, and beneath the stone lay a thick bed of clay, under which were found beds of loose shale so full of water that almost constant pumping was necessary at many points to enable the works to proceed.   For a year and a half the contractor went on fruitlessly contending with these difficulties, and at length he was compelled to abandon the adventure. The engineer then took the works in hand for the Company, and they were vigorously proceeded with. Steam-engines were set to work to pump out the water; two locomotives were put on, one at either end of the cutting, to drag away the excavated rock and clay; and eight hundred men and boys were employed along the work, in digging, wheeling, and blasting, besides a large number of horses. Some idea of the extent of the blasting operations may be formed from the fact that twenty-five barrels of gunpowder were exploded weekly; the total quantity used in forming this one cutting being about three thousand barrels. Considerable difficulty was experienced in supporting the bed of rock cut through, which overlaid the clay and shale along either side of the cutting. It was found necessary to hold it up by strong retaining walls, to prevent the clay bed from bulging out, and these walls were further supported by a strong invert,—that is, an arch placed in an inverted position under the road,—thus binding together the walls on both sides. Behind the retaining walls, a drift or horizontal drain was provided to enable the water to run off, and occasional openings were left in the walls themselves for the same purpose. The work was at length brought to a successful completion, but the extraordinary difficulties encountered in forming the cutting had the effect of greatly increasing the cost of this portion of the railway.

The tunnels on the line are eight in number, their total length being 7336 yards. The first high ground encountered was Primrose Hill, where the stiff London clay was passed through for a distance of about 1164 yards. The clay was close, compact, and dry, more difficult to work than stone itself. It was entirely free from water; but the absorbing properties of the clay were such that when exposed to the air it swelled out rapidly. Hence an unusual thickness of brick lining was found necessary; and the engineer afterwards informed the author that for some time he entertained an apprehension lest the pressure should force in the brickwork altogether, as afterwards happened in the case of the short Preston Brook tunnel upon the Grand Junction Railway, constructed by his father. He stated that the pressure behind the brickwork was such, that it made the face of the bricks to fly off in minute chips, which covered his clothes whilst he was inspecting the work. The materials used in the building were, however, of excellent quality; and the work was happily brought to a completion without any accident.

At Watford the chalk ridge was penetrated by a tunnel about 1800 yards long; and at Northchurch, Lindslade, and Stowe Hill, there were other tunnels of minor extent. But the chief difficulty of the undertaking was the execution of that under the Kilsby ridge. Though not the largest, this is in many respects one of the most interesting works of the kind in this country. It is about two thousand four hundred yards long, and runs at an average depth of about a hundred and sixty feet below the surface. The ridge under which it extends is of considerable extent, the famous battle of Naseby having been fought upon one of the spurs of the same high ground some seven miles to the eastward.

Previous to the letting of the contract, the character of the underground soil was fairly tested by trial shafts, which indicated that it consisted of shale of the lower oolite, and it was let accordingly. But the works had scarcely been commenced when it was discovered that at an interval between the two trial-shafts which had been sunk about two hundred yards from the south end of the tunnel, there existed an extensive quicksand under a bed of clay forty feet thick, which the borings had escaped in the most singular manner. At the bottom of one of these shafts the excavation and building of the

LINE OF THE SHAFTS OVER KILSBY TUNNEL. [By Percival Skelton.]

tunnel were proceeding, when the roof at one part
suddenly gave way, a deluge of water burst in, and the
party of workmen with the utmost difficulty escaped
with their lives. They were only saved by means of a
raft, on which they were towed by one of the engineers
swimming with the rope in his mouth to the lower end
of the shaft, out of which they were safely lifted to the
daylight. The works were of course at that point
immediately stopped. The contractor, who had undertaken the construction of the tunnel, was so overwhelmed
by the calamity, that, though he was relieved by the
Company from his engagement, he took to his bed and
shortly after died. Pumping-engines were then erected
for the purpose of draining off the water, but for a long
time it prevailed, and sometimes even rose in the shaft.
The question arose, whether in the face of so formidable
a difficulty, the works should be proceeded with or
abandoned. Robert Stephenson sent over to Alton

Grange for his father, and the two took serious counsel together. George was in favour of pumping out the water from the top by powerful engines erected over each shaft, until the water was fairly mastered. Robert concurred in that view, and although other engineers who were consulted pronounced strongly against the practicability of the scheme and advised the abandonment of the enterprise, the directors authorised him to proceed; and powerful steam-engines were ordered to be constructed and delivered without loss of time.

In the mean time, Robert suggested to his father the expediency of running a drift along the heading from the south end of the tunnel, with the view of draining off the water in that way. George said he thought it would scarcely answer, but that it was worth a trial, at all events until the pumping-engines were got ready. Robert accordingly gave orders for the drift to be proceeded with; and the workmen had nearly reached the sand bed, when one day that the engineer, his assistants, and the workmen were clustered about its open entrance, they heard a sudden roar as of distant thunder. It was hoped that the water had burst in—for all the workmen were out of the drift,—and that the sand bed would now drain itself off in a natural way. Instead of which, very little water made its appearance; and on examining the inner end of the drift, it was found that the loud noise had been caused by the sudden discharge into it of an immense mass of sand, which had completely choked up the passage, and prevented the water from flowing away.

The engineer now found that there was nothing for it but sinking numerous additional shafts over the line of the tunnel at the points at which it crossed the quicksand, and endeavouring to master the water by sheer force of engines and pumps. The engines, when at length erected, possessed an aggregate power of 160 horses; and they went on pumping for eight successive months,

emptying out an almost incredible quantity of water.
It was found that the water, with which the bed of sand
extending over many miles was charged, was to a certain
degree held back by the particles of the sand itself, and
that it could only percolate through at a certain average
rate. It appeared in its flow to take a slanting direction
to the suction of the pumps, the angle of inclination
depending upon the coarseness or fineness of the sand,
and regulating the time of the flow. Hence the distri-
bution of the pumping power at short intervals along.
the line of the tunnel had a much greater effect than the
concentration of that power at any one spot. It soon
appeared that the water had found its master. Protected by the pumps, which cleared a space for engineering operations—in the midst, as it were, of two almost
perpendicular walls of water and sand on either side—
the workmen proceeded with the building of the tunnel at
numerous points. Every exertion was used to wall in the
dangerous parts as quickly as possible ; the excavators and
bricklayers labouring night and day until the work was
finished. Even while under the protection of the immense pumping power above described, it often happened
that the bricks were scarcely covered with cement ready
for the setting, ere they were washed quite clean by the
streams of water which poured down overhead. The
men were accordingly under the necessity of holding
over their work large whisks of straw and other appliances to protect the bricks and cement at the moment
of setting.

The quantity of water pumped out of the sand bed
during eight months of incessant pumping, averaged
two thousand gallons per minute, raised from an average
depth of 120 feet. It is difficult to form an adequate
idea of the bulk of the water thus raised, but it may be
stated that if allowed to flow for three hours only, it
would fill a lake one acre square to the depth of one
foot, and if allowed to flow for one entire day it would

fill the lake to over eight feet in depth, or sufficient to
float vessels of a hundred tons' burthen. The water
pumped out of the tunnel while the work was in progress would be nearly equivalent to the contents of the
Thames at high water, between London and Woolwich.
It is a curious circumstance, that notwithstanding the
quantity of water thus removed, the level of the surface
in the tunnel was only lowered about two and a half to
three inches per week, proving the vast area of the
quicksand, which probably extended along the entire
ridge of land under which the railway passed.

The cost of the line was greatly increased by these difficulties encountered at Kilsby. The original estimate for
the tunnel was only 99,000*l.*; but before it was finished
it had cost more than 100*l.* per lineal yard forward, or
a total of nearly 300,000*l.* The expenditure on the
other parts of the line also greatly exceeded the amount
first set down by the engineer; and before the works
were finished, it was more than doubled. The land cost
three times more than the estimate; and the claims for
compensation were enormous. Although the contracts
were let within the estimates, very few of the contractors
were able to complete them without the assistance of
the Company, and many became bankrupt. Speaking
of the difficulties encountered during the construction of
the line, Robert Stephenson afterwards observed to us :—
" After the works were let, wages rose, the prices of
materials of all kinds rose, and the contractors, many of
whom were men of comparatively small capital, were
thrown on their beam ends. Their calculations as to
expenses and profits were completely upset. Let me
just go over the list. There was Jackson, who took the
Primrose Hill contract — he failed. Then there was
the next length—Nowells; then Copeland and Harding; north of them Townsend, who had the Tring
cutting; next Stoke Hammond; then Lyers; then
Hughes : I think all of these broke down, or at least

were helped through by the directors. Then there was
that terrible contract of the Kilsby tunnel, which broke
the Nowells, and killed one of them. The contractors
to the north of Kilsby were more fortunate, though
some of them pulled through only with the greatest
difficulty. Of the eighteen contracts in which the
line was originally let, only seven were completed by
the original contractors. Eleven firms were ruined
by their contracts, which were relet to others at advanced
prices, or were carried on and finished by the Company.
The principal cause of increase in the expense, however,
was the enlargement of the stations. It appeared that
we had greatly under-estimated the traffic, and it accordingly became necessary to spend more and more money
for its accommodation, until I think I am within the
mark when I say that the expenditure on this account
alone exceeded by eight or ten fold the amount of the
Parliamentary estimate."

The magnitude of the works, which were unpre-
cedented in England, was one of the most remarkable
features in the undertaking. The following striking
comparison has been made between this railway and
one of the greatest works of ancient times. The Great
Pyramid of Egypt was, according to Diodorus Siculus,
constructed by three hundred thousand—according to
Herodotus, by one hundred thousand—men. It required
for its execution twenty years, and the labour expended
upon it has been estimated as equivalent to lifting
15,733,000,000 of cubic feet of stone one foot high.
Whereas, if the labour expended in constructing the
London and Birmingham Railway be in like manner
reduced to one common denomination, the result is
25,000,000,000 of cubic feet *more* than was lifted for the
Great Pyramid; and yet the English work was performed by about 20,000 men in less than five years.
And whilst the Egyptian work was executed by a
powerful monarch concentrating upon it the labour and

capital of a great nation, the English railway was constructed, in the face of every conceivable obstruction and difficulty, by a company of private individuals out of their own resources, without the aid of Government or the contribution of one farthing of public money.

The labourers who executed these formidable works were in many respects a remarkable class. The "railway navvies,[1]" as they were called, were men drawn by the attraction of good wages from all parts of the kingdom; and they were ready for any sort of hard work. Many of the labourers employed on the Liverpool line were Irish; others were from the Northumberland and Durham railways, where they had been accustomed to similar work; and some of the best came from the fen districts of Lincoln and Cambridge, where they had been trained to execute works of excavation and embankment. These old practitioners formed a nucleus of skilled manipulation and aptitude, which rendered them of indispensable utility in the immense undertakings of the period. Their expertness in all sorts of earthwork, in embanking, boring, and well-sinking —their practical knowledge of the nature of soils and rocks, the tenacity of clays, and the porosity of certain stratifications — were very great; and, rough-looking though they were, many of them were as important in their own department as the contractor or the engineer.

During the railway-making period the navvy wandered about from one public work to another—apparently belonging to no country and having no home. He usually wore a white felt hat with the brim turned up, a velveteen or jean square-tailed coat, a scarlet plush waistcoat with little black spots, and a bright-coloured kerchief round his herculean neck, when, as often

---

[1] The word "navvie," or "navigator," is supposed to have originated in the fact of many of these labourers having been originally employed in making the navigations, or canals, the construction of which immediately preceded the railway era.

happened, it was not left entirely bare. His corduroy breeches were retained in position by a leathern strap round the waist, and were tied and buttoned at the knee, displaying beneath a solid calf and foot encased in strong high-laced boots. Joining together in a "butty gang," some ten or twelve of these men would take a contract to cut out and remove so much "dirt"—as they denominated earth-cutting—fixing their price according to the character of the "stuff," and the distance to which it had to be wheeled and tipped. The contract taken, every man put himself to his mettle: if any was found skulking, or not putting forth his full working power, he was ejected from the gang. Their powers of endurance were extraordinary. In times of emergency they would work for twelve and even sixteen hours, with only short intervals for meals. The quantity of flesh-meat which they consumed was something enormous; but it was to their bones and muscles what coke is to the locomotive—the means of keeping up the steam. They displayed great pluck, and seemed to disregard peril. Indeed the most dangerous sort of labour — such as working horse-barrow runs, in which accidents are of constant occurrence—has always been most in request amongst them, the danger seeming to be one of its chief recommendations.

Working together, eating, drinking, and sleeping together, and daily exposed to the same influences, these railway labourers soon presented a distinct and well-defined character, strongly marking them from the population of the districts in which they laboured. Reckless alike of their lives as of their earnings, the navvies worked hard and lived hard. For their lodging, a hut of turf would content them; and, in their hours of leisure, the meanest public-house would serve for their parlour. Unburdened, as they usually were, by domestic ties, unsoftened by family affection, and without much moral or religious training, the navvies came to be distinguished by a sort of savage manners, which contrasted strangely with those of the surrounding population. Yet, ignorant and violent though they might be, they were usually good-hearted fellows in the main —frank and open-handed with their comrades, and ready to share their last penny with those in distress. Their pay-nights were often a saturnalia of riot and disorder, dreaded by the inhabitants of the villages along the line of works. The irruption of such men into the quiet hamlet of Kilsby must, indeed, have produced a very startling effect on the recluse inhabitants of the place. Robert Stephenson used to tell a story of the clergyman of the parish waiting upon the foreman of one of the gangs to expostulate with him as to the shocking impropriety of his men working during Sunday. But the head navvy merely hitched up his trowsers, and said, " Why, Soondays hain't cropt out here yet !" In short, the navvies were little better than heathens, and the village of Kilsby was not restored to its wonted quiet until the tunnel-works were finished, and the engines and scaffoldings removed, leaving only the immense masses of *débris* around the line of shafts which extend along the top of the tunnel.

In illustration of the extraordinary working energy and powers of endurance of the English navvies, we may mention that when railway-making extended to France, the English contractors for the works took with them gangs of English navvies, with the usual plant, which included wheelbarrows. These the English navvy was accustomed to run out continuously, loaded with some three or four hundredweight of stuff, piled so high that he could barely see, over the summit of the load, the gang-board along which he wheeled his barrow, whereas the French navvy was contented with half the weight. Indeed, the French navvies on one occasion struck work because of the size of the English barrows, and there was an *émeute* on the Rouen Railway, which was

only quelled by the aid of the military. The consequence was that the big barrows were abandoned to the
English workmen, who earned nearly double the wages
of the Frenchmen. The manner in which they stood to
their work was matter of great surprise and wonder-
ment to the French countrypeople, who came crowding
round them in their blouses, and, after gazing admiringly
at their expert handling of the pick and mattock, and
the immense loads of " dirt" which they wheeled out,
would exclaim to each other, " *Mon Dieu, voila! voila
ces Anglais, comme ils travaillent!* "

KILSBY TUNNEL (NORTH END).

CHAPTER XVI.

Manchester and Leeds, and Midland Railways — Stephenson's
Life at Alton — Visit to Belgium — General Extension
of Railways and their Results.

While the London and Birmingham Railway was
under construction, George Stephenson continued to
reside at Alton Grange. Though he took an active
interest in the progress of the works, and made frequent
visits of inspection at the more important points, he
left the practical part of the business in the hands of
his son. He was himself fully occupied in laying
out and constructing numerous lines in the north of
England, for the purpose of opening up communications
between the more important towns, as well as between
them and the metropolis.

The rapidity with which railways were carried out,
when the spirit of the country became roused, was indeed
remarkable. This was doubtless in some measure owing
to the increased force of the current of speculation at
the time, but chiefly to the desire which the public began
to entertain for the general extension of the system. It
was even proposed to fill up the canals, and convert
them into railways. The new roads became the topic
of conversation in all circles; they were felt to give a
new value to time; their vast capabilities for "business" peculiarly recommended them to the trading
classes; whilst the friends of "progress" dilated on
the great benefits they would eventually confer upon
mankind at large. It began to be seen that Edward
Pease had not been exaggerating when he said, "Let
the country but make the railroads, and the railroads

will make the country!" They also came to be regarded as inviting objects of investment to the thrifty,
and a safe outlet for the accumulations of inert men of
capital. Thus new avenues of iron road were soon in
course of construction in all directions, branching north,
south, east, and west, so that the country promised in a
wonderfully short space of time to become wrapped in
one vast network of iron.

In 1836 the Grand Junction Railway was under
construction between Warrington and Birmingham—the
northern part by Mr. Stephenson, and the southern by

THE DUTTON VIADUCT.

Mr. Rastrick. The works on that line were of the
usual kind—heavy cuttings, long embankments, and
numerous viaducts; but none of these are worthy of
any special description. Perhaps the finest piece of
masonry on the railway is the Dutton Viaduct across
the valley of the Weaver. It consists of twenty arches
of 60 feet span, springing 16 feet from the perpendicular shaft of each pier, and 60 feet in height from
the crown of the arches to the level of the river. The
foundations of the piers were built on piles driven 20

feet deep. The structure has a solid and majestic appearance, and is perhaps the finest of George Stephenson's
viaducts. Although designed by him, it was carried
out by Mr. Locke, on the latter succeeding Mr. Stephenson as engineer to the Grand Junction Railway.

The Manchester and Leeds line was in progress at
the same time—an important railway connecting the
principal manufacturing towns of Yorkshire and Lancashire. An attempt was made to obtain the Act as early
as the year 1831; but its promoters were defeated by
the powerful opposition of the landowners aided by the
canal companies, and the project was not revived for
several years. Mr. Stephenson, having carefully examined the district, had in his own mind settled the
proper direction of the line, and decided that no other
was practicable, without the objectionable expedient of
a tunnel three and a-half miles in length under Black-
stone Edge, and the additional disadvantage of bad
gradients. The line as laid out by him was somewhat
circuitous, and the works were heavy ; but on the whole
the gradients were favourable, and it had the advantage
of passing through a district full of manufacturing
towns and villages, the teeming hives of population,
industry, and enterprise. The Act authorising the construction of the railway was obtained in the session of
1836; it was greatly amended in the succeeding year,
and the first ground was broken on the 18th of August,
1837.

An incident occurred while the second Manchester
and Leeds Bill was before the Committee of the Lords,
which is worthy of passing notice in this place, as illustrative of George Stephenson's character. The line
which was authorised by Parliament in 1836 had been
hastily surveyed within a period of less than six weeks,
and before it received the royal assent Mr. Stephenson
became convinced that many important improvements
might be made in it, and communicated his views to

the directors. They determined, however, to obtain the Act, although conscious at the time that they would have to go for a second and improved line in the following year. The second Bill passed the Commons in 1837 without difficulty, and was expected in like manner to receive the sanction of the Lords' Committee. Quite unexpectedly, however, Lord Wharncliffe, who was interested in the Manchester and Sheffield line, which passed through his colliery property in the south of Yorkshire, and conceived that the new Manchester and Leeds line might have some damaging effect upon it, appeared as an opponent of the Bill. He was himself a member of the Committee, and adopted the unusual course of rising to his feet, and making a set speech against the Bill while Mr. Stephenson was under examination. After pointing out that the Bill applied for and obtained in the preceding session was one that the promoters had no intention of carrying out, that they had secured it only for the purpose of obtaining possession of the ground and reducing the number of the opponents to their present application, and that in fact they had been practising a deception upon the House, his Lordship turned full upon the witness, and, addressing him, said, " I ask you, sir, do you call that conduct *honest?*" Mr. Stephenson, his voice trembling with emotion, replied, " Yes, my Lord, I *do* call it honest. And I will ask your Lordship, whom I served for many years as your enginewright at the Killingworth collieries, did you ever know me to do anything that was not strictly honourable? You know what the collieries were when I went there, and you know what they were when I left them. Did you ever hear that I was found wanting when honest services were wanted, or when duty called me? Let your Lordship but fairly consider the circumstances of the case, and I feel persuaded you will admit that my conduct has been equally honest throughout in this matter." He then briefly but clearly stated the history of the application to Parliament for the Act, which was so satisfactory to the Committee that they passed the preamble of the Bill without further objection. Lord Wharncliffe requested that the Committee would permit his observations, together with Mr. Stephenson's reply, to be erased from the record of the evidence, which, as an acknowledgment of his error, was permitted. Lord Kenyon and several other members of the Committee afterwards came up to Mr. Stephenson, shook him by the hand, and congratulated him on the manly way in which he had vindicated himself in the course of the inquiry.

In conducting this project to an issue, Mr. Stephenson had much opposition and many prejudices to encounter. Predictions were confidently made in many quarters that the line could never succeed. It was declared that the utmost engineering skill could not construct a railway through such a country of hills and hard rocks; and it was maintained that, even if the railway were practicable, it could only be formed at an altogether ruinous cost to the proprietors.

During the progress of the works, as the Summit Tunnel, near Littleborough, was approaching completion, the rumour was spread abroad in Manchester that the tunnel had fallen in and buried a number of the workmen. The last arch had been keyed in, and the work was all but finished, when the accident occurred which was thus exaggerated by the lying tongue of rumour. An invert had given way through the irregular pressure of the surrounding earth and rock at a part of the tunnel where a "fault" had occurred in the strata. A party of the directors accompanied the engineer to inspect the scene of the accident. They entered the tunnel's mouth preceded by upwards of fifty navvies, each bearing a torch.

After walking a distance of about half a mile, the inspecting party arrived at the scene of the "frightful

ENTRANCE TO THE SUMMIT TUNNEL, LITTLEBOROUGH.

[By Percival Skelton.]

accident," about which so much alarm had been spread.
All that was visible was a certain unevenness of the
ground, which had been forced up by the invert under
it giving way; thus the ballast had been loosened, the
drain running along the centre of the road had been
displaced, and small pools of water stood about. But
the whole of the walls and the roof were still as perfect
as at any other part of the tunnel. Mr. Stephenson
explained the cause of the accident: the blue shale, he
said, through which the excavation passed at that point,
was considered so hard and firm, as to render it unnecessary to build the invert very strong there. But
shale is always a deceptive material. Subjected to the
influence of the atmosphere, it gives but a treacherous
support. In this case, falling away like quicklime, it
had left the lip of the invert alone to support the
pressure of the arch above, and hence its springing
inwards and upwards. Mr. Stephenson directed the

attention of the visitors to the completeness of the arch overhead, where not the slightest fracture or yielding could be detected. Speaking of the work, in the course of the same day, he said, " I will stake my character, my head, if that tunnel ever give way, so as to cause danger to any of the public passing through it. Taking it as a whole, I don't think there is such another piece of work in the world. It is the greatest work that has yet been done of this kind, and there has been less repairing than is usual,—though an engineer might well be beaten in his calculations, for he cannot beforehand see into those little fractured parts of the earth he may meet with." As Mr. Stephenson had promised, the invert was put in; and the tunnel was made perfectly safe.

The construction of this subterranean road employed the labour of above a thousand men for nearly four years. Besides excavating the arch out of the solid rock, they used 23,000,000 of bricks, and 8000 tons of Roman cement in the building of the tunnel. Thirteen stationary engines, and about 100 horses, were also employed in drawing the earth and stone out of the shafts. Its entire length is 2869 yards, or nearly a mile and three-quarters,—exceeding the famous Kilsby Tunnel by 471 yards. Mr. T. L. Gooch was the acting engineer on the line, and was afterwards promoted, at Mr. Stephenson's recommendation, to the post of joint principal engineer, sharing the responsibilities of that office with his chief.

The Midland Railway was a favourite line of Mr. Stephenson's for several reasons. It passed through a rich mining district, in which it opened up many valuable coal-fields, and it formed part of the great main line of communication between London and Edinburgh. The line was originally projected by gentlemen interested in the London and Birmingham Railway. Their intention was to extend that line from Rugby to Leeds; but, finding themselves anticipated in

part by the projection of the Midland Counties Railway from Rugby to Derby, they confined themselves to the district between Derby and Leeds; and in 1835, a Company was formed to construct the North Midland line, with George Stephenson for its engineer. The Act was obtained in 1836, and the first ground was broken in February, 1837.

Although the Midland Railway was only one of the many great works of the same kind executed at that time, it was almost enough of itself to be the achievement of a life. Compare it, for example, with Napoleon's military road over the Simplon, and it will at once be seen how greatly it excels that work, not only in the constructive skill displayed in it, but also in its cost and magnitude, and the amount of labour employed in its formation. The road of the Simplon is 45 miles in length; the North Midland Railway 72½ miles. The former has 50 bridges and 5 tunnels, measuring together 1338 feet in length; the latter has 200 bridges and 7 tunnels, measuring together 11,400 feet, or about 2¼ miles. The former cost about 720,000*l.* sterling, the latter above 3,000,000*l.* Napoleon's grand military road was constructed in six years, at the public cost of the two great

kingdoms of France and Italy; while Stephenson's
railway was formed in about three years, by a company of private merchants and capitalists out of their
own funds, and under their own superintendence.

It is scarcely necessary that we should give any
account in detail of the North Midland works. The
making of one tunnel so much resembles the making of
another,—the building of bridges and viaducts, no
matter how extensive, so much resembles the building
of others,—the cutting out of " dirt," the blasting of
rocks, and the wheeling of excavation into embankments, is so much a matter of mere time and hard work,
—that it is quite unnecessary for us to detain the reader
by any attempt at their description. Of course there
were the usual difficulties to encounter and overcome,—
but the railway engineer regarded these as mere matters
of course, and would probably have been disappointed if
they had not presented themselves. On the Midland, as
on other lines, water was the great enemy to be fought
against,—water in the Claycross and other tunnels,—
water in the boggy or sandy foundations of bridges,—
and water in cuttings and embankments. As an illustration of the difficulties of bridge building, we may
mention the case of the five arch bridge over the Derwent,
where it took two years' work, night and day, to get in
the foundations of the piers alone. Another curious
illustration of the mischief done by water in cuttings
may be briefly mentioned. At a part of the North
Midland Line, near Ambergate, it was necessary to pass
along a hillside in a cutting a few yards deep. As the
cutting proceeded, a seam of shale was cut across, lying
at an inclination of 6 to 1 ; and shortly after, the water
getting behind the bed of shale, the whole mass of
earth along the hill above began to move down across
the line of excavation. The accident completely upset
the estimates of the contractor, who, instead of fifty
thousand cubic yards, found that he had about five
hundred thousand to remove ; the execution of this

LAND-SLIP ON NORTH MIDLAND LINE, NEAR AMBERGATE

part of the railway occupying fifteen months instead of two.

The Oakenshaw cutting near Wakefield was also of a very formidable character. About six hundred thousand yards of rock shale and bind were quarried out of it, and led to form the adjoining Oakenshaw embankment. The Normanton cutting was almost as heavy, requiring the removal of four hundred thousand yards of the same kind of excavation into embankment and spoil. But the progress of the works on the line was so rapid in 1839, that not less than 450,000 cubic yards of excavation were effected per month.

As a curiosity in construction, we may also mention a very delicate piece of work executed on the same railway at Bullbridge in Derbyshire, where the line at the same point passes *over* a bridge which here spans the river Amber, and *under* the bed of the Cromford Canal. Water, bridge, railway, and canal, were thus piled one above the other, four stories high; such another

BULL-BRIDGE NEAR AMBERGATE.

curious complication probably not existing. In order to prevent the possibility of the waters of the canal breaking in upon the works of the railroad, Mr. Stephenson had an iron tank made, 150 feet long, of the width of the canal, and exactly fitting the bottom. It was brought to the spot in three pieces, which were firmly welded together, and the trough was then floated into its place and sunk; the whole operation being completed without in the least interfering with the navigation of the canal. The railway works underneath were then proceeded with and finished.

Another line of the same series, constructed by Mr. Stephenson, was the York and North Midland, extending from Normanton—a point on the Midland Railway—to York; but it was a line of easy formation, traversing a comparatively level country. The inhabitants of Whitby, as well as York, were busy projecting railways as early as 1832; and in the year following, Whitby succeeded in obtaining a horse line of twenty-

four miles, connecting it with the small market-town of
Pickering.   The York citizens were more ambitious,
and agitated the question of a locomotive line to connect
them with the town of Leeds.   Mr. Stephenson recommended them to connect their line with the Midland at
Normanton, and they adopted his advice.   The Company was formed, the shares were at once subscribed
for, and Stephenson appointed his pupil and assistant,
Mr. Swanwick, to lay out the line in October, 1835.
The Act was obtained in the following year, and the
works were constructed without difficulty.

As the best proof of his conviction that the York and
North Midland would prove a good investment, Mr.
Stephenson invested in it a considerable portion of his
savings, being a subscriber for 420 shares; and he also
took some trouble in persuading several wealthy gentlemen in London and elsewhere to purchase shares in the
concern.   The interest thus taken in the line by the
engineer was on more than one occasion specially
mentioned by Mr. Hudson, then Lord Mayor of York,
as an inducement to other persons of capital to join the
undertaking; and had it not afterwards been encumbered
and overlaid by comparatively useless, and therefore
profitless branches, in the projection of which Mr.
Stephenson had no part, the sanguine expectations
which he early formed of the paying qualities of that
railway would have been more than realised.

There was one branch, however, of the York and
North Midland Line in which he took an anxious
interest, and of which he may be pronounced the
projector—the branch to Scarborough; which proved
to be one of the most profitable parts of the railway.
He was so satisfied of its value, that, at a meeting of
the York and North Midland proprietors, he volunteered
his gratuitous services as engineer until the Company
was formed, in addition to subscribing largely to the
undertaking.   At that meeting he took an opportunity

of referring to the charges brought against engineers of so greatly exceeding the estimates :—" He had had a good deal to do with making out the estimate of the North Midland Railway, and he believed there never was a more honest one. He had always endeavoured to state the truth as far as was in his power. He had known a director, who, when he (Mr. Stephenson) had sent in an estimate, came forward and said, ' I can do it for half the money.' The director's estimate went into Parliament, but it came out his. He could go through the whole list of the undertakings in which he had been engaged, and show that he had never had anything to do with stock-jobbing concerns. He would say that he would not be concerned in any scheme, unless he was satisfied that it would pay the proprietors : and in bringing forward the proposed line to Scarborough, he was satisfied that it would pay, or he would have had nothing to do with it."

During the time that our engineer was engaged in superintending the execution of these great undertakings, he was occupied in surveying other lines of railway in various parts of the country. With that object he visited the neighbourhood of Glasgow, and surveyed several lines there; and he afterwards surveyed routes along the east coast from Newcastle to Edinburgh, with the view of completing the main line of communication with London. When out on foot in the fields, on these occasions, he was ever foremost in the march; and he delighted to test the prowess of his companions by a good jump at any hedge or ditch that lay in their way. His companions noted with surprise his remarkable quickness of observation. Nothing escaped his attention—the trees, the crops, the birds, or the farmer's stock; and he was usually full of lively conversation, everything in nature affording him an opportunity for making some striking remark, or propounding some ingenious theory. When taking a

flying survey of a new line, his keen observation proved
very useful to him, for he rapidly noted the general
configuration of the country, and inferred its geological
structure. He afterwards remarked to a friend, " I
have planned many a railway travelling along in a
postchaise, and following the natural line of the
country." And it was remarkable that his first im-
pressions of the direction to be taken almost invariably
proved the right ones; and there are few of the lines
surveyed and recommended by him which have not
been executed, either during his lifetime or since. As
an illustration of his quick and shrewd observation on
such occasions, we may mention that when employed
to lay out a line to connect Manchester, through Mac-
clesfield, with the Potteries, the gentleman who accompanied him on the journey of inspection cautioned him
to provide large accommodation for carrying off the
water, observing—" You must not judge by the appearance of the brooks; for after heavy rains these hills
pour down volumes of water, of which you can have no
conception." " Pooh! pooh! *don't I see your bridges?* "
replied the engineer. He had noted the details of each
as he passed along.

Among the other projects which occupied his attention
about the same time, were the projected lines between
Chester and Holyhead, between Leeds and Bradford,
and between Lancaster and Maryport by the western
coast. This latter was intended to form part of a west-
coast line to Scotland, Mr. Stephenson favouring it
partly because of the flatness of the gradients, and also
because it could be formed at comparatively small cost,
whilst it would open out a valuable iron-mining district,
from which a large traffic in ironstone was expected.
One of its collateral advantages, in the engineer's
opinion, was, that by forming the railway directly
across Morecambe Bay, on the north-west coast of
Lancashire, a large tract of valuable land might be

reclaimed from the sea, the sale of which would considerably reduce the cost of the works. He estimated that by means of a solid embankment across the bay, not less than forty thousand acres of rich alluvial land would be gained. His scheme was, to carry the road across the ten miles of sands which lie between Poulton, near Lancaster, and Humphrey Head on the opposite coast, forming the line in a segment of a circle of five miles' radius. His plan was to drive in piles across the entire length, forming a solid fence of stone blocks on the land side for the purpose of retaining the sand and silt brought down by the rivers from the interior. The embankment would then be raised from time to time as the deposit accumulated, until the land was filled up to high-water mark; provision being made, by means of sufficient arches, for the flow of the river waters into the bay. The execution of the railway after this plan would, however, have occupied more years than the promoters of the West Coast line were disposed to wait; and eventually Mr. Locke's more direct but uneven line by Shap Fell was adopted. A railway has, however, since been carried across the head of the bay, in a greatly modified form, by the Ulverstone and Lancaster Railway Company; but it is not improbable that Stephenson's larger scheme of reclaiming the vast tract of land now left bare at every receding tide, may yet be carried out.

While occupied in carrying out the great railway undertakings which we have above so briefly described, Mr. Stephenson's home continued, for the greater part of the time, to be at Alton Grange, near Leicester. But he was so much occupied in travelling about from one committee of directors to another—one week in England, another in Scotland, and probably the next in Ireland, —that he often did not see his home for weeks together. He had also to make frequent inspections of the various important and difficult works in progress, especially

on the Midland and Manchester and Leeds lines; besides occasionally going to Newcastle to see how the
locomotive works were going on there. During the
three years ending in 1837—perhaps the busiest years
of his life [1]—he travelled by postchaise alone upwards
of twenty thousand miles, and yet not less than six
months out of the three years were spent in London.
Hence there is comparatively little to record of Mr.
Stephenson's private life at this period; during which
he had scarcely a moment that he could call his own.

His correspondence increased so much, that he found
it necessary to engage a private secretary, who accompanied him on his journeys. He was himself exceedingly averse to writing letters. The comparatively
advanced age at which he learnt the art of writing, and
the nature of his duties while engaged at the Killingworth colliery, precluded that facility in correspondence
which only constant practice can give. He gradually,
however, acquired great facility in dictation, and had
also the power of labouring continuously at this work;
the gentleman who acted as his secretary in the year
1835, having informed us that during his busy season
he one day dictated not fewer than thirty-seven letters,
several of them embodying the results of much close
thinking and calculation. On another occasion, he
dictated reports and letters for twelve continuous hours,
until his secretary was ready to drop off his chair from
sheer exhaustion, and at length he pleaded for a suspension of the labour. This great mass of correspondence,
although closely bearing on the subjects under discussion,
was not, however, of a kind to supply the biographer

---

[1] During this period he was engaged on the North Midland, extending
from Derby to Leeds; the York and
North Midland, from Normanton to
York; the Manchester and Leeds;
the Birmingham and Derby, and the
Sheffield and Rotherham Railways;
the whole of these, of which he was
principal engineer, having been authorised in 1836. In that session
alone, powers were obtained for the
construction of 214 miles of new railways under his direction, at an expenditure of upwards of five millions
sterling.

with matter for quotation, or to give that insight into
the life and character of the writer which the letters of
literary men so often furnish. They were, for the most
part, letters of mere business, relating to works in
progress, parliamentary contests, new surveys, estimates
of cost, and railway policy,—curt, and to the point; in
short, the letters of a man every moment of whose time
was precious. He was also frequently called upon to
inspect and report upon colliery works, salt works, brass
and copper works, and such like, in addition to his own
colliery and railway business. He usually staked out
himself the lines laid out by him, which involved a
good deal of labour since undertaken by assistants.
And occasionally he would run up to London, attending
in person to the preparation and deposit of the plans
and sections of the projected undertakings for which he
was engaged as engineer.

Fortunately Mr. Stephenson possessed a facility of
sleeping, which enabled him to pass through this
enormous amount of fatigue and labour without injury
to his health. He had been trained in a hard school,
and could bear with ease conditions which, to men more
softly nurtured, would have been the extreme of physical
discomfort. Many, many nights he snatched his sleep
while travelling in his chaise; and at break of day he
would be at work, surveying until dark, and this for
weeks in succession. His whole powers seemed to be
under the control of his will, for he could wake at any
hour, and go to work at once. It was difficult for
secretaries and assistants to keep up with such a man.

It is pleasant to record that in the midst of these
engrossing occupations, his heart remained as soft and
loving as ever. In spring-time he would not be
debarred of his boyish pursuit of bird-nesting; but
would go rambling along the hedges spying for nests.
In the autumn he went nutting, and when he could
snatch a few minutes he indulged in his old love of

gardening. His uniform kindness and good temper, and his communicative, intelligent disposition, made him a great favourite with the neighbouring farmers, to whom he would volunteer much valuable advice on agricultural operations, drainage, ploughing, and labour-saving processes. Sometimes he took a long rural ride on his favourite "Bobby," now growing old, but as fond of his master as ever. Towards the end of his life, "Bobby" lived in clover, its master's pet, doing no work; and he died at Tapton in 1845, more than twenty years old.

During one of George's brief sojourns at the Grange, he found time to write to his son a touching account of a pair of robins that had built their nest within one of the empty upper chambers of the house. One day he observed a robin fluttering outside the windows, and beating its wings against the panes, as if eager to gain admission. He went up stairs, and there found, in a retired part of one of the rooms, a robin's nest, with one of the parent birds sitting over three or four young—all dead. The excluded bird outside still beat against the panes; and on the window being let down, it flew into the room, but was so exhausted that it dropped upon the floor. Mr. Stephenson took up the bird, carried it down stairs, and had it warmed and fed. The poor robin revived, and for a time was one of his pets. But it shortly died too, as if unable to recover from the privations it had endured during its three days' fluttering and beating at the windows. It appeared that the room had been unoccupied, and, the sash having been let down, the robins had taken the opportunity of building their nest within it; but the servant having closed the window again, the calamity befel the birds which so strongly excited Mr. Stephenson's sympathies. An incident such as this, trifling though it may seem, gives a true key to the heart of the man.

The amount of his Parliamentary business having

greatly increased with the projection of new lines of railway, Mr. Stephenson found it necessary to set up an office in London in 1836. His first office was at No 9, Duke-street, Westminster, from whence he removed in the following year to 30½, Great George-street. That office was the busy scene of railway politics for several years. There consultations were held, schemes were matured, deputations were received, and many projectors called upon our engineer for the purpose of submitting to him their plans of railways and railway working. His private secretary at the time has informed us that at the end of the first Parliamentary session in which he had been engaged as engineer for more companies than one, it became necessary for him to give instructions as to the preparation of the accounts to be rendered to the respective companies. In the simplicity of his heart, he directed Mr. Binns to take his full time at the rate of ten guineas a day, and charge the railway companies in the proportion in which he had been actually employed in their respective business during each day. When Robert heard of this instruction, he went directly to his father and expostulated with him against this unprofessional course; and, other influences being brought to bear upon him, George at length reluctantly consented to charge as other engineers did, an entire day's fee to each of the Companies for which he was concerned whilst their business was going forward; but he cut down the number of days charged for, and reduced the daily amount from ten to seven guineas.

Besides his journeys at home, Mr. Stephenson was on more than one occasion called abroad on railway business. Thus, at the desire of King Leopold, he made several visits to Belgium to assist the Belgian engineers in laying out the national lines of that kingdom. That enlightened monarch at an early period discerned the powerful instrumentality of railways in developing a country's resources, and he determined at the earliest possible period to adopt them as the great high-roads of the nation. The country, being rich in coal and minerals, had great manufacturing capabilities. It had good ports, fine navigable rivers, abundant canals, and a teeming, industrious population. Leopold perceived that railways were eminently calculated to bring the industry of the country into full play, and to render the riches of the provinces available to the rest of the kingdom. He therefore openly declared himself the promoter of public railways throughout Belgium. A system of lines was projected, at his instance, connecting Brussels with the chief towns and cities of the kingdom; extending from Ostend eastward to the Prussian frontier, and from Antwerp southward to the French frontier.

Mr. Stephenson and his son, the leading railway-engineers of England, were consulted by the King on the best mode of carrying out his important plans, as early as 1835. In the course of that year they visited Belgium, and had several interesting conferences with Leopold and his ministers on the subject of the proposed railways. The King then appointed George Stephenson by royal ordinance a Knight of the Order of Leopold. At the invitation of the monarch, Mr. Stephenson made a second visit to Belgium in 1837, on the occasion of the public opening of the line from Brussels to Ghent. At Brussels there was a public procession, and another at Ghent on the arrival of the train. Stephenson and his party accompanied it to the Public Hall, there to dine with the chief Ministers of State, the municipal authorities, and about five hundred of the principal inhabitants of the city; the English Ambassador being also present. After the King's health and a few others had been drunk, that of Mr. Stephenson was proposed, on which the whole assembly rose up, amidst great excitement and loud applause, and made their way to where he sat, in order to jingle glasses with him, greatly to his own amazement. On the day following, our engineer dined with the King and Queen at their own table at Laaken, by special invitation; afterwards accompanying his Majesty and suite to a public ball given by the municipality of Brussels, in honour of the opening of the line to Ghent, as well as of their distinguished English guest. On entering the room, the general and excited inquiry was, "Which is Stephenson?" The English engineer had not before imagined that he was esteemed to be so great a man.

The London and Birmingham Railway having been completed in September, 1838, after being about five years in progress, the great main system of railway communication between London, Liverpool, and Manchester was then opened to the public. For some months previously, the line had been partially opened, coaches performing the journey between Denbigh Hall (near Wolverton) and Rugby—the works of the Kilsby tunnel being still incomplete. It was already amusing to hear the complaints of the travellers about the slowness of the coaches as compared with the railway, though the coaches travelled at a speed of eleven miles an hour. The comparison of comfort was also greatly to the disparagement of the coaches. Then the railway train could accommodate any quantity, whilst the road conveyances were limited; and when a press of travellers occurred—as on the occasion of the Queen's coronation —the greatest inconvenience was experienced, and as much as 10*l.* was paid for a seat on a donkey-chaise between Rugby and Denbigh. On the opening of the railway throughout, of course all this inconvenience and delay was brought to an end.

Numerous other openings of railways constructed by Mr. Stephenson took place about the same time. The Birmingham and Derby line was opened for traffic in August, 1839; the Sheffield and Rotherham in November, 1839; and in the course of the following

year, the Midland, the York and North Midland, the
Chester and Crewe, the Chester and Birkenhead, the
Manchester and Birmingham, the Manchester and Leeds,
and the Maryport and Carlisle railways, were all publicly opened in whole or in part. Thus 321 miles of
railway (exclusive of the London and Birmingham)
constructed under Mr. Stephenson's superintendence, at
a cost of upwards of eleven millions sterling, were, in
the course of about two years, added to the traffic
accommodation of the country.

The ceremonies which accompanied the public opening
of these lines were often of an interesting character.
The adjoining population held general holiday; bands
played, banners waved, and assembled thousands cheered
the passing trains amidst the occasional booming of
cannon. The proceedings were usually wound up by a
public dinner; and in the course of his speech which
followed, Mr. Stephenson would revert to his favourite
topic—the difficulties which he had early encountered
in the promotion of the railway system, and in establishing the superiority of the locomotive. On such
occasions, Mr. Stephenson always took great pleasure
in alluding to the services rendered to himself and the
public by the young men brought up under his eye
—his pupils at first, and afterwards his assistants. No
great master ever possessed a more devoted band of
assistants and fellow-workers than he did. And, indeed,
it was one of the most marked evidences of his own
admirable tact and judgment that he selected, with such
undeviating correctness, the men best fitted to carry out
his plans. For, the ability to accomplish great things,
to carry grand ideas into practical effect, depends in no
small measure on an intuitive knowledge of character,
which Mr. Stephenson possessed in a remarkable
degree. Thus, on the Liverpool and Manchester line,
he secured the able services of Messrs. Vignolles and
Locke; the latter having been his pupil, and laid down

for him several coal-lines in the North. John Dixon, trained by him on the Stockton and Darlington Railway, afterwards carried out his views on the Canterbury and Whitstable, the Liverpool and Manchester, and the Chester railways. Thomas L. Gooch was his representative in superintending the execution of the formidable works of the Manchester and Leeds line. Swanwick on the North Midland, Birkenshaw on the Birmingham and Derby, and Cabrey on the York and North Midland, seconded him well and ably, and established their own reputation while they increased the engineering fame of their master. All these men, then comparatively young, became, in course of time, engineers of distinction, and were employed to conduct on their own account numerous railway enterprises of great magnitude.

At the dinner at York, which followed the partial opening of the York and North Midland Railway, Mr. Stephenson said, " he was sure they would appreciate his feelings when he told them, that when he first began railway business, his hair was black, although it was now grey; and that he began his life's labour as but a poor ploughboy. About thirty years since, he had applied himself to the study of how to generate high velocities by mechanical means. He thought he had solved that problem; and they had for themselves seen, that day, what perseverance had brought him to. He was, on that occasion, only too happy to have an opportunity of acknowledging that he had, in the latter portion of his career, received much most valuable assistance, particularly from young men brought up in his manufactory. Whenever talent showed itself in a young man, he had always given that talent encouragement where he could, and he would continue to do so."

That this was no-exaggerated statement is amply

proved by many facts which redound to Mr. Stephenson's credit.   He was no niggard of encouragement and
praise when he saw honest industry struggling for a
footing.   Many were the young men whom, in the
course of his useful career, he took by the hand and led
steadily up to honour and emolument, simply because
he had noted their zeal, diligence, and integrity.   One
youth excited his interest while working as a common
carpenter on the Liverpool and Manchester line; and
before many years had passed, he was recognised as an
engineer of distinction.   Another young man he found
industriously working away at his bye-hours, and, admiring his diligence, engaged him for his private secretary, the gentleman shortly after rising to a position of
eminent influence and usefulness.   Indeed, nothing gave
Mr. Stephenson greater pleasure than in this way to
help on any deserving youth who came under his
observation, and, in his own expressive phrase, to
"make a man of him."

The openings of the great main lines of railroad
communication shortly proved the fallaciousness of the
numerous rash prophecies which had been promulgated
by the opponents of railways.   The proprietors of the
canals were astounded by the fact that, notwithstanding
the immense traffic conveyed by rail, their own traffic
and receipts continued to increase ; and that, in common
with other interests, they fully shared in the expansion
of trade and commerce which had been so effectually
promoted by the extension of the railway system.   The
cattle-owners were equally amazed to find the price of
horse-flesh increasing with the extension of railways,
and that the number of coaches running to and from
the new railway-stations gave employment to a greater
number of horses than under the old stage-coach system.
Those who had prophesied the decay of the metropolis,
and the ruin of the suburban cabbage-growers, in conse-

quence of the approach of railways to London,[1] were also disappointed ; for, while the new roads let citizens out of London, they let country-people in. Their action, in this respect, was centripetal as well as centrifugal. Tens of thousands who had never seen the metropolis could now visit it expeditiously and cheaply ; and Londoners who had never visited the country, or but rarely, were enabled, at little cost of time or money, to see green fields and clear blue skies, far from the smoke and bustle of town. If the dear suburban-grown cabbages became depreciated in value, there were truck-loads of fresh grown country cabbages to make amends for the loss : in this case, the " partial evil " was a far more general good. The food of the metropolis became rapidly improved, especially in the supply of wholesome meat and vegetables. And then the price of coals—an article which, in this country, is as indispensable as daily food to all classes—was greatly reduced. What a blessing to the metropolitan poor is described in this single fact !

The prophecies of ruin and disaster to landlords and farmers were equally confounded by the openings of the railways. The agricultural communications, so far from

---

[1] When the first railways were opened in the immediate neighbourhood of the metropolis, they were naturally regarded with great curiosity, and crowds flocked to see them. The Greenwich Railway was opened in 1836, and was for some time one of the principal shows of London. When the first locomotive was run upon it, a large sum was taken for admissions of persons to witness the sight. Half-a-guinea was charged for reserved seats. When the passenger-trains began to run, a regular band of musicians was engaged to play in front of the station to attract customers. The line was also used as a show-ground for new inventions — a singular machine of Lord Dundonald's, called the Scorpion, 86 feet long, having for some time been a principal attraction. It seems to have been apprehended that the engines would be apt to run off the line at night, unless they had the advantage of lights to enable them to see their way; and lamps were accordingly placed at intervals of 88 yards along the entire railway. When railways ceased to be a novelty the Greenwich Company paid off their band, took down their lamps, and devoted themselves to the conveyance of the regular traffic, which soon became quite as large as they could conveniently manage.

being "destroyed," as had been predicted, were immensely improved. The farmers were enabled to buy
their coals, lime, and manure for less money, while
they obtained a readier access to the best markets for
their stock and farm-produce. Notwithstanding the
predictions to the contrary, their cows gave milk as
before, their sheep fed and fattened, and even skittish
horses ceased to shy at the passing locomotive. The
smoke of the engines did not obscure the sky, nor
were farmyards burnt up by the fire thrown from the
locomotives. The farming classes were not reduced to
beggary; on the contrary, they soon felt that, so far
from having anything to dread, they had very much
good to expect from the extension of railways.

Landlords also found that they could get higher rents
for farms situated near a railway than at a distance
from one. Hence they became clamorous for " sidings."
They felt it to be a grievance to be placed at a distance
from a station. After a railway had been once opened,
not a landlord would consent to have the line taken
from him. Owners who had fought the promoters before Parliament, and compelled them to pass their
domains at a distance, at a vastly-increased expense
in tunnels and deviations, now petitioned for branches
and nearer station accommodation. Those who held
property near towns, and had extorted large sums as
compensation for the anticipated deterioration in the
value of their building land, found a new demand for
it springing up at greatly advanced prices. Land was
now advertised for sale, with the attraction of being
" near a railway station."

The prediction that, even if railways were made, the
public would not use them, was also completely falsified
by the results. The ordinary mode of fast travelling
for the middle classes had heretofore been by mail-coach
and stage-coach. Those who could not afford to pay

the high prices charged for such conveyances went by
waggon, and the poorer classes trudged on foot. George
Stephenson was wont to say that he hoped to see the
day when it would be cheaper for a poor man to travel
by railway than to walk, and not many years passed before his expectation was fulfilled. In no country in the
world is time worth more money than in England; and
by saving time—the criterion of distance—the railway
proved a great benefactor to men of industry in all classes.

Many deplored the inevitable downfall of the old
stage-coach system. There was to be an end of that
delightful variety of incident usually attendant on a
journey by road. The rapid scamper across a fine
country on the outside of the four-horse " Express," or
" Highflyer;" the seat on the box beside Jehu, or the
equally coveted place near the facetious guard behind;
the journey amid open green fields, through smiling
villages and fine old towns, where the stage stopped to
change horses and the passengers to dine—was all very
delightful in its way; and many regretted that this old-
fashioned and pleasant style of travelling was about to
pass away. But it had its dark side also. Any one
who remembers the journey by stage from London to
Manchester or York, will associate it with recollections
and sensations of not unmixed delight. To be perched
for twenty hours, exposed to all weathers, on the outside
of a coach, trying in vain to find a soft seat—sitting
now with the face to the wind, rain, or sun, and now
with the back—without any shelter such as the com-
monest penny-a-mile parliamentary train now daily
provides—was a miserable undertaking, looked forward
to with horror by many whose business required them
to travel frequently between the provinces and the
metropolis. Nor were the inside passengers more agreeably accommodated. To be closely packed up in a little,
inconvenient, straight-backed vehicle, where the cramped
limbs could not be in the least extended, nor the wearied

frame indulge in any change of posture, was felt by many to be a terrible thing.   Then there were the constantly-recurring demands, not always couched in the politest terms, for an allowance to the driver every two or three stages, and to the guard every six or eight; and if the gratuity did not equal their expectations, growling and open abuse were not unusual.   These *désagrémens*, together with the exactions practised on travellers by innkeepers, seriously detracted from the romance of stage-coach travelling, and there was a general disposition on the part of the public to change the system for a better.

The extent to which the new passenger railways were at once made use of proved that this better system had been discovered.   Notwithstanding the reduction of the coach fares on many of the roads to one-third of their previous rate, people preferred travelling by the railway.   They saved in time; and they saved in money, taking the whole expenses into account.   In point of comfort there could be no doubt as to the infinite superiority of the railway carriage.   But there remained the question of safety, which had been a great bugbear with the early opponents of railways, and was made the most of by the coach-proprietors to deter travellers from using them.   It was predicted that trains of passengers would be blown to pieces, and that none but fools would entrust their persons to the conduct of an explosive machine such as the locomotive.   It appeared, however, that during the first eight years not fewer than five millions of passengers had been conveyed along the Liverpool and Manchester Railway, and of this vast number only two persons had lost their lives by accident.   During the same period, the loss of life by the upsetting of stage-coaches had been immensely greater in proportion.   The public were not slow, therefore, to detect the fact that travelling by railways was greatly safer than travelling by common road; and in all districts penetrated by railways the coaches were very shortly taken off for want of support.

George Stephenson himself had a narrow escape in one of the stage-coach accidents so common twenty years ago, but which are already almost forgotten. While the Birmingham line was under construction, he had occasion to travel from Ashby-de-la-Zouch to London by coach. He was an inside passenger with an elderly lady, and the outsides were pretty numerous. When within ten miles of Dunstable, he felt, from the rolling of the coach, that one of the linchpins securing the wheels had given way, and that the vehicle must upset. He endeavoured to fix himself in his seat, holding on firmly by the arm-straps, so that he might save himself on whichever side the coach fell. It soon toppled over, and fell crash upon the road, amidst the shrieks of his fellow-passengers and the smashing of glass. He immediately pulled himself up by the arm-strap above him, let down the coach window, and climbed out. The coachman and passengers lay scattered about on the road, stunned, and some of them bleeding, while the horses were plunging in their harness. Taking out his pocket-knife, he at once cut the traces, and set the horses free. He then went to the help of the passengers, who were all more or less hurt. The guard had his arm broken, and the driver was seriously cut and contused. A scream from one of his fellow-passenger "insides" here attracted his attention : it proceeded from the elderly lady, whom he had before observed to be decorated with one of the enormous bonnets in fashion at the time. Opening the coach-door, he lifted the lady out, and her principal lamentation was that her large bonnet had been crushed beyond remedy! Mr. Stephenson then proceeded to the nearest village for help, and saw the passengers provided with proper assistance before he himself went forward on his journey.

It was some time before the more opulent classes, who could afford to post to town in aristocratic style, became reconciled to railway travelling.   The old families did not relish the idea of being conveyed in a train of passengers of all ranks and conditions, in which the shopkeeper and the peasant were carried along at the same speed as the duke and the baron—the only difference being in price.   It was another deplorable illustration of the levelling tendencies of the age.[1]   It put an end to that gradation of rank in travelling which was one of the few things left by which the nobleman could be distinguished from the Manchester manufacturer and bagman.   But to younger sons of noble families the convenience and cheapness of the railway did not fail to recommend itself.   One of these, whose eldest brother had just succeeded to an earldom, said one day to a railway manager : " I like railways—they just suit young fellows like me with 'nothing per annum paid quarterly.'   You know, we can't afford to post, and it used to be deuced annoying to me, as I was jogging along on the box-seat of the stage-coach, to see the little Earl go by drawn by his four posters, and just look up at me and give me a nod.   But now, with railways, it's different.   It's true, he may take a first-class ticket, while I can only afford a second-class one, but *we both go the same pace.*"

For a time, however, many of the old families sent forward their servants and luggage by railroad, and

[1] At a meeting of the Chesterfield Mechanics' Institute, at which Mr. Stephenson was present, one of the speakers said of him, " Known as he is wherever steam and iron have opened the swift lines of communication to our countrymen, and regarded by all as the Father of Railways, he might be called, in the most honourable acceptation of the term, *the first and greatest leveller of the age.*"   Mr. Stephenson joined heartily in the laugh which followed this description of himself.   Sir Humphry Davy was once similarly characterised; but the remark was somewhat differently appreciated.   When travelling on the Continent, a distinguished person about a foreign Court inquired who and what he was, never having heard of his scientific fame.   Upon being told that his discoveries had " *revolutionised chemistry,*" the courtier promptly replied, "I hate all revolutionists ; his presence will not be acceptable here."

condemned themselves to jog along the old highway in the accustomed family chariot, dragged by country post-horses. But the superior comfort of the railway shortly recommended itself to even the oldest families; posting went out of date; post-horses were with difficulty to be had along even the great high-roads; and nobles and servants, manufacturers and peasants, alike shared in the comfort, the convenience, and the despatch of railway travelling. The late Dr. Arnold, of Rugby, regarded the opening of the London and Birmingham line as another great step accomplished in the march of civilisation. "I rejoice to see it," he said, as he stood on one of the bridges over the railway, and watched the train flashing along under him, and away through the distant hedgerows—"I rejoice to see it, and to think that feudality is gone for ever: it is so great a blessing to think that any one evil is really extinct."

It was long before the late Duke of Wellington would trust himself behind a locomotive. The fatal accident to Mr. Huskisson, which had happened before his eyes, contributed to prejudice him strongly against railways, and it was not until the year 1843 that he performed his first trip on the South-Western Railway, in attendance upon her Majesty. Prince Albert had for some time been accustomed to travel by railway alone, but in 1842 the Queen began to make use of the same mode of conveyance between Windsor and London. Even Colonel Sibthorpe was eventually compelled to acknowledge its utility. For a time he continued to post to and from the country as before. Then he compromised the matter by taking a railway ticket for the long journey, and posting only for a stage or two nearest town; until, at length, he undisguisedly committed himself, like other people, to the express train, and performed the journey throughout upon what he had formerly denounced as "the infernal railroad."

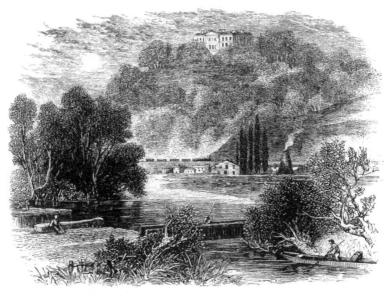

TAPTON HOUSE. [By Percival Skelton]

CHAPTER XVII.

GEORGE STEPHENSON'S COAL MINES — APPEARS AT MECHANICS' INSTITUTES — HIS OPINION ON RAILWAY SPEEDS — ATMOSPHERIC SYSTEM — RAILWAY MANIA — VISITS TO BELGIUM AND SPAIN.

WHILE Mr. Stephenson was engaged in carrying on the works of the Midland Railway in the neighbourhood of Chesterfield, several seams of coal were cut through in the Claycross Tunnel, and it occurred to him that if mines were opened out there, the railway would provide the means of a ready sale for the article in the midland counties, and as far south as even the metropolis itself.

At a time when everybody else was sceptical as to the possibility of coals being carried from the midland counties to London, and sold there at a price to compete with those which were seaborne, he declared his firm conviction that the time was fast approaching when the London market would be regularly supplied with north-country coals led by railway. One of the greatest advantages of railways, in his opinion, was that they would bring iron and coal, the staple products of the country, to the doors of all England. " The strength of Britain," he would say, " lies in her iron and coal beds; and the locomotive is destined, above all other agencies, to bring it forth. The Lord Chancellor now sits upon a bag of wool; but wool has long ceased to be emblematical of the staple commodity of England. He ought rather to sit upon a bag of coals, though it might not prove quite so comfortable a seat. Then think of the Lord Chancellor being addressed as the noble and learned lord *on the coal-sack!* I am afraid it wouldn't answer, after all."

To one gentleman he said : " We want from the coal-mining, the iron-producing and manufacturing districts, a great railway for the carriage of these valuable products. We want, if I may so say, a stream of steam running directly through the country, from the North to London, and from other similar districts to London. Speed is not so much an object as utility and cheapness. It will not do to mix up the heavy merchandize and coal trains with the passenger trains. Coal and most kinds of goods can wait; but passengers will not. A less perfect road and less expensive works will do well enough for coal trains, if run at a low speed; and if the line be flat, it is not of much consequence whether it be direct or not. Whenever you put passenger trains on a line, all the other trains must be run at high speeds to keep out of their way. But coal trains run at high speeds pull the road to pieces, besides causing large expenditure in locomotive power; and I doubt very much whether they will pay after all; but a succession of long coal trains, if run at from ten to fourteen miles an hour, would pay very well. Thus the Stockton and Darlington Company made a larger profit when running coal at low speeds at a halfpenny a ton per mile, than

they have been able to do since they put on their fast
passenger trains, when everything .must needs be run
faster, and a much larger proportion of the gross receipts
is absorbed by working expenses."

In advocating these views, Mr. Stephenson was
considerably ahead of his time; and although he did
not live to see his anticipations fully realised as to
the supply of the London coal-market, he was nevertheless the first to point out, and to some extent to
prove, the practicability of establishing a profitable coal
trade by railway between the northern counties and the
metropolis. So long, however, as the traffic was conducted on main passenger lines at comparatively high
speeds, it was found that the expenditure on tear and
wear of road and locomotive power,—not to mention the
increased risk of carrying on the first-class passenger
traffic with which it was mixed up,—necessarily left a
very small margin of profit; and hence Mr. Stephenson
was in the habit of urging the propriety of constructing
a railway which should be exclusively devoted to goods
and mineral traffic run at low speeds as the only condition on which a large railway traffic of that sort could
be profitably conducted.[1]

Having induced some of his Liverpool friends to join
him in a coal-mining adventure at Chesterfield, a lease
was taken of the Claycross estate, then for sale, and
operations were shortly after begun. At a subsequent
period Mr. Stephenson extended his coal - mining
operations in the same neighbourhood; and in 1841 he
himself entered into a contract with owners of land in
the townships of Tapton, Brimington, and Newbold, for
the working of the coal thereunder; and pits were
opened on the Tapton estate on an extensive scale.

---

[1] A railway of this description has recently been projected from Askern, in Yorkshire, to March, in Cambridge-shire, where it falls into the Eastern Counties Railway, to be devoted mainly to coal traffic, thus carrying out to a great extent George Stephenson's favourite idea.

About the same time he erected great lime-works, close
to the Ambergate station of the Midland Railway, from
which, when in full operation, he was able to turn out
upwards of two hundred tons a day. The limestone
was brought on a tramway from the village of Crich,
about two or three miles distant from the kilns, the coal
with which to burn it being supplied from his adjoining
Claycross colliery. The works were on a scale such
as had not before been attempted by any private
individual engaged in a similar trade ; and we believe
they proved very successful.

LIME WORKS AT AMBERGATE. [By Percival Skelton.]

Tapton House was included in the lease of one of the
collieries, and as it was conveniently situated—being, as
it were, a central point on the Midland Railway, from
which he could readily proceed north or south, on his
journeys of inspection of the various lines then under
construction in the midland and northern counties,—
he took up his residence there, and it continued his
home until the close of his life.

Tapton House is a large roomy brick mansion,
beautifully situated amidst woods, upon a commanding

eminence, about a mile to the north-east of the town of
Chesterfield. Green fields dotted with fine trees slope
away from the house in all directions. The surrounding
country is undulating and highly picturesque. North
and south the eye ranges over a vast extent of lovely
scenery; and on the west, looking over the town of
Chesterfield, with its church and crooked spire, the
extensive range of the Derbyshire hills bounds the
distance. The Midland Railway skirts the western edge
of the park in a deep rock cutting, and the shrill whistle
of the locomotive sounds near at hand as the trains
speed past. The gardens and pleasure-grounds adjoining
the house were in a very neglected state when Mr.
Stephenson first went to Tapton; and he promised
himself, when he had secured rest and leisure from
business, that he would put a new face upon both. The
first improvement he made was cutting a woodland
footpath up the hill-side, by which he at the same time
added a beautiful feature to the park, and secured a shorter
road to the Chesterfield station. But it was some years
before he found time to carry into effect his contemplated
improvements in the adjoining gardens and pleasure-
grounds. He had so long been accustomed to laborious
pursuits, and felt himself still so full of work, that he
could not at once settle down into the habit of quietly
enjoying the fruits of his industry.

He had no difficulty in usefully employing his time.
Besides directing the mining operations at Claycross,
the establishment of the lime-kilns at Ambergate, and
the construction of the extensive railways still in progress, he occasionally paid visits to Newcastle, where
his locomotive manufactory was now in full work,
and the proprietors were reaping the advantages of
his early foresight in an abundant measure of prosperity. One of his most interesting visits to the place
was in 1838, on the occasion of the meeting of the
British Association there, when he acted as one of the

Vice-Presidents in the section of Mechanical Science. Extraordinary changes had occurred in his own fortunes, as well as in the face of the country, since he had first appeared before a scientific body in Newcastle—the members of the Literary and Philosophical Institute—to submit his safety-lamp for their examination. Twenty-three years had passed over his head, full of honest work, of manful struggle; and the humble "colliery enginewright of the name of Stephenson" had achieved an almost world-wide reputation as a public benefactor. His fellow-townsmen, therefore, could not hesitate to recognise his merits and do honour to his name. During the sittings of the Association, Mr. Stephenson took the opportunity of paying a visit to Killingworth, accompanied by some of the distinguished *savans* whom he numbered amongst his friends. He there pointed out to them, with a degree of honest pride, the cottage in which he had lived for so many years, showed what parts of it had been his own handiwork, and told them the story of the sun-dial over the door, describing the study and the labour it had cost him and his son to calculate its dimensions, and fix it in its place. The dial had been serenely numbering the hours through the busy years that had elapsed since that humble dwelling had been his home; during which the Killingworth locomotive had become a great working power, and its contriver had established the railway system, which was now rapidly becoming extended in all parts of the world.

About the same time, his services were very much in request at the meetings of Mechanics' Institutes held throughout the northern counties. From an early period in his history, he had taken an active interest in these valuable institutions. While residing at Newcastle in 1824, shortly after his locomotive foundry had been started in Forth-street, he presided at a public meeting held in that town for the purpose of establishing a

Mechanics' Institute. The meeting was held; but as
George Stephenson was a man comparatively unknown
even in Newcastle at that time, his name failed to
secure "an influential attendance." Among those who
addressed the meeting on the occasion, was Joseph
Locke, then his pupil, and afterwards his rival as an
engineer. The local papers scarcely noticed the proceedings; yet the Mechanics' Institute was founded,
and struggled into existence. Years passed, and it was
now felt to be an honour to secure Mr. Stephenson's
presence at any public meetings held for the promotion of popular education. Among the Mechanics' Institutes in his immediate neighbourhood at Tapton, were
those of Belper and Chesterfield; and at their soirées
he was a frequent and a welcome visitor. On these
occasions he loved to tell his auditors of the difficulties
which had early beset him through want of knowledge,
and of the means by which he had overcome them. His
grand text was—PERSEVERE; and there was manhood
in the very word.

On more than one occasion, the author had the
pleasure of listening to George Stephenson's homely
but forcible addresses at the annual soirées of the Leeds
Mechanics' Institute. He was always an immense
favourite with his audiences there. His personal
appearance was greatly in his favour. A handsome,
ruddy, expressive face, lit up by bright dark-blue eyes,
prepared one for his earnest words when he stood up
to speak and the cheers had subsided which invariably
hailed his rising. He was not glib, but he was very
impressive. And who, so well as he, could serve as a
guide to the working man in his endeavours after
higher knowledge? His early life had been all struggle
—encounter with difficulty—groping in the dark after
greater light, but always earnestly and perseveringly.
His words were therefore all the more weighty, since
he spoke from the fulness of his own experience. On

one occasion he said—" He had commenced his career·
on a lower level than any man present there. He
made that remark for the purpose of encouraging young
mechanics to do as he had done—TO PERSEVERE. And
he would tell them that the humblest amongst them
occupied a much more advantageous position than he
had done on commencing his life of labour. They had
teachers who, going before them, had left their great
discoveries as a legacy and a guide; and their works
were now accessible to all, in such institutions as that
which he addressed. But he remembered the time
when there were none thus to guide and instruct the
young mechanic. With a free access to scientific books,
he knew, from his own experience, that they could be
saved much unnecessary toil and expenditure of mental
capital. Many ingenious young mechanics, if they
failed to profit by the teaching· of those who had
preceded them, might often be induced to believe that
they had hit upon some discovery in mechanics ; and
when they had gone on spending both time and money,
they would only arrive at the unpleasant discovery that
what they had cherished as an original invention ·had
been known many years before, and was to be found
recorded in scientific works." And again—" The man
who wished to rise in his trade or profession must never
see any insurmountable difficulties before him. Obstacles
might appear to be such ; but they must be thrown
overboard or conquered. This was the course which
he had himself pursued." These characteristic sentiments clearly illustrate the man, and show the fibre of
which he was made ; and we need scarcely say that they
served to give new life and hope to all who listened
to him.

Nor did he remain a mere inactive spectator of the
improvements in railway working which increasing
experience from day to day suggested. He continued
to contrive improvements in the locomotive, and to

mature his invention of the carriage-brake. When examined before the Select Committee on Railways in 1841, his mind seems principally to have been impressed with the necessity which existed for adopting a system of self-acting brakes; stating that, in his opinión, this was the most important arrangement that could be provided for increasing the safety of railway travelling. "I believe," he said, "that if self-acting brakes were put upon every carriage, scarcely any accident could take place." His plan consisted in employing the momentum of the running train to throw his proposed brakes into action, immediately on the moving power of the engine being checked. He would also have these brakes under the control of the guard, by means of a connecting line running along the whole length of the train, by which they should at once be thrown out of gear when necessary [1] At the same time he suggested, as an additional means of safety, that the signals of the line should be self-acting, and worked by the locomotives as they passed along the railway. He considered the adoption of this plan of so much importance, that, with a view to the public safety, he would even have it enforced upon railway companies by the legislature. At the same time he was of opinion that it was the interest of the companies themselves to adopt the plan, as it would save great tear and wear of engines, carriages, tenders, and brake-vans, besides greatly diminishing the risk of accidents upon railways.

While before the same Committee, he took the opportunity of stating his views with reference to railway speed, about which wild ideas were then afloat—one gentleman of celebrity having publicly expressed the opinion that a speed of a hundred miles an hour was practicable in railway travelling! Not many years had passed since

---

[1] A full description, with plans, of Mr. Stephenson's self-acting brake, since revived in a modified form by M. Guérin, is given in the 'Practical Mechanics' Journal,' vol. i. p. 53.

George Stephenson had been pronounced insane for stating his conviction that twelve miles an hour could be performed by the locomotive; but now that he had established the fact, and greatly exceeded that speed, he was thought behind the age because he recommended the rate to be limited to forty miles an hour. He said: "I do not like either forty or fifty miles an hour upon any line—I think it is an unnecessary speed; and if there is danger upon a railway, it is high velocity that creates it. I should say no railway ought to exceed forty miles an hour on the most favourable gradient; but upon a curved line the speed ought not to exceed twenty-four or twenty-five miles an hour." He had, indeed, constructed for the Great Western Railway an engine capable of running fifty miles an hour with a load, and eighty miles without one. But he never was in favour of a hurricane speed of this sort, believing it could only be accomplished at an unnecessary increase both of danger and expense.

"It is true," he observed on other occasions,[1] "I have said the locomotive engine *might* be made to travel a hundred miles an hour; but I always put a qualification on this, namely, as to what speed would best suit the public. The public may, however, be unreasonable; and fifty or sixty miles an hour *is* an unreasonable speed. Long before railway travelling became general, I said to my friends that there was no limit to the speed of the locomotive, *provided the works could be made to stand*. But there are limits to the strength of iron, whether it be manufactured into rails or locomotives; and there is a point at which both rails and tyres must break. Every increase of speed, by increasing the strain upon the road and the rolling stock, brings us nearer to that point. At thirty miles a slighter road will do, and less perfect

---

[1] It may be mentioned that these views were communicated to the author by Robert Stephenson, and noted down in his presence.

rolling stock may be run upon it with safety. But if
you increase the speed by say ten miles, then everything
must be greatly strengthened. You must have heavier
engines, heavier and better-fastened rails, and all your
working expenses will be immensely increased. I think
I know enough of mechanics to know where to stop.
I know that a pound will weigh a pound, and that more
should not be put upon an iron rail than it will bear.
If you could ensure perfect iron, perfect rails, and perfect
locomotives, I grant fifty miles an hour or more might
be run with safety on a level railway. But then you
must not forget that iron, even the best, will 'tire,'
and with constant use will become more and more liable
to break at the weakest point—perhaps where there is
a secret flaw that the eye cannot detect. Then look at
the rubbishy rails now manufactured on the contract
system—some of them little better than cast metal :
indeed, I have seen rails break merely on being thrown
from the truck on to the ground. How is it possible
for such rails to stand a twenty or thirty ton engine
dashing over them at the speed of fifty miles an hour ?
No, no," he would conclude, " I am in favour of low
speeds because they are safe, and because they are
economical ; and you may rely upon it that, beyond a
certain point, with every increase of speed there is an
increase in the element of danger."

When railways became the subject of popular discussion, many new and unsound theories were started
with reference to them, which Stephenson opposed as calculated, in his opinion, to bring discredit on the locomotive system. One of these was with reference to what
were called "undulating lines." Among others, Dr.
Lardner, who had originally been somewhat sceptical
about the powers of the locomotive, now promulgated
the idea that a railway constructed with rising and falling
gradients would be practically as easy to work as a line
perfectly level. Mr. Badnell went even beyond him, for

he held that an undulating railway was much better than a level one for purposes of working.[1]   For a time, this theory found favour, and the " undulating system " was extensively adopted; but Mr. Stephenson never ceased to inveigh against it; and experience has amply proved that his judgment was correct.   His practice, from the beginning of his career until the end of it, was to secure a road as nearly as possible on a level, following the course of the valleys and the natural line of the country; preferring to go round a hill rather than to tunnel under it or carry his railway over it, and often making a considerable circuit to secure good, workable gradients. He studied to lay out his lines so that long trains of minerals and merchandise, as well as passengers, might be hauled along them at the least possible expenditure of locomotive power.   He had long before ascertained, by careful experiments at Killingworth, that the engine expends half of its full power in overcoming a rising gradient of 1 in 260, which is about 20 feet in the mile; and that when the gradient is so steep as 1 in 100, not less than three-fourths of its propelling power is sacrificed in ascending the acclivity.   He never forgot the valuable practical lesson taught him by those early trials made and registered in the company of Nicholas Wood, long before the advantages of railways had been recognized.   He saw clearly that the longer flat line must eventually prove superior to the shorter line of steep gradients as respected its paying qualities.   He urged that, after all, the power of the locomotive was but limited; and, although he and his son had done more than any other men to increase its working power, it provoked him to find that every improvement made in it was neutralised by the steep gradients which the new school of engineers were setting

---

[1] ' Treatise on Railway Improvements.' By Mr. Richard Badnell, C.E.

it to overcome. On one occasion, when Robert Stephenson stated before a Parliamentary Committee that
every successive improvement in the locomotive was
being rendered virtually nugatory by the difficult and
almost impracticable gradients proposed on many of the
new lines, his father, on his leaving the witness-box,
went up to him, and said, "Robert, you never spoke
truer words than those in all your life."

To this it must be added, that in urging these views
Mr. Stephenson was strongly influenced by commercial considerations. He had no desire to build up his
reputation at the expense of railway shareholders,
nor to obtain engineering *éclat* by making "ducks
and drakes" of their money. He was persuaded that,
in order to secure the practical success of railways, they
must be so laid out as not only to prove of decided
public utility, but also to be worked economically and
to the advantage of their proprietors. They were not
government roads, but private ventures—in fact, commercial speculations. He therefore endeavoured to
render them financially profitable; and he repeatedly
declared that if he did not believe they could be "made
to pay," he would have nothing to do with them.[1] He
was not influenced by the sordid consideration of what
he could *make* out of any company that employed
him; indeed, in many cases he voluntarily gave up his
claim to remuneration where the promoters of schemes
which he thought praiseworthy had suffered serious loss.

---

[1] He frequently refused to act as
the engineer for lines which he thought
would not prove remunerative, or
when he considered the estimates too
low. Thus, when giving evidence on
the Great Western Bill, Mr. Stephenson said, "I made out an estimate for
the Hartlepool Railway, which they
returned on account of its being too
high, but I declined going to Parliament with a lower estimate." Another engineer was employed. Then,
again, "I was consulted about a line
from Edinburgh to Glasgow. The
directors chalked out a line and sent it
to me, and I told them I could not
support it in that case." Hence the
employment of another engineer to
carry out the line which Mr. Stephenson could not conscientiously advocate.

Thus, when the first application was made to Parliament for the Chester and Birkenhead Railway Bill, the promoters were defeated. They repeated their application, on the understanding that in event of their succeeding, the engineer and surveyor were to be paid their costs in respect of the defeated measure. The Bill was successful, and to several parties their costs were paid. Mr. Stephenson's amounted to 800*l.*, and he very nobly said, "You have had an expensive career in Parliament; you have had a great struggle; you are a young Company; you cannot afford to pay me this amount of money; I will reduce it to 200*l.*, and I will not ask you for that 200*l.* until your shares are at 20*l.* premium; for whatever may be the reverses you will go through, I am satisfied I shall live to see the day when your shares will be at 20*l.* premium, and when I can legally and honourably claim that 200*l.*"[1] We may add that the shares did eventually rise to the premium specified, and the engineer was no loser by his generous conduct in the transaction.

Another novelty of the time, with which George Stephenson had to contend, was the substitution of atmospheric pressure for locomotive steam-power in the working of railways. The idea of obtaining motion by means of atmospheric pressure is said to have originated with Papin, the French philosopher, more than a century and a half ago; but it slept until revived in 1810 by Mr. Medhurst, who published a pamphlet to prove the practicability of carrying letters and goods by air. In 1824, Mr. Vallance of Brighton took out a patent for projecting passengers through a tube large enough to contain a train of carriages; the tube being previously exhausted of its atmospheric air. The same idea was afterwards taken up, in 1835, by Mr. Pinkus, an inge-

[1] Speech of Wm. Jackson, Esq., M.P., at the meeting of the Chester and | Birkenhead Railway Company, held at Liverpool, October, 1845.

nious American. Scientific gentlemen, Dr. Lardner and Mr. Clegg amongst others, advocated the plan; and an association was formed to carry it into effect. Shares were created, and 18,000*l.* raised; and a model apparatus was . exhibited in London. Mr. Vignolles took his friend Mr. Stephenson to see the model; and after carefully examining it, he observed emphatically, "*It won't do:* it is only the fixed engines and ropes over again, in another form; and, to tell you the truth, I don't think this rope of wind will answer so well as the rope of wire did." He did not think the principle would stand the test of practice, and he objected to the mode of applying the principle. After all, it was only a modification of the stationary-engine plan; and every day's experience was proving that fixed engines could not compete with locomotives in point of efficiency and economy. He stood by the locomotive engine; and subsequent experience proved that he was right.

Messrs. Clegg and Samuda afterwards, in 1840, patented their plan of an atmospheric railway; and they publicly tested its working on an unfinished portion of the West London Railway. The results of the experiment were so satisfactory, that the directors of the Dublin and Kingstown line adopted it between Kingstown and Dalkey. The London and Croydon Company also adopted the atmospheric principle; and their line was opened in 1845. The ordinary mode of applying the power was to lay between the line of rails a pipe, in which a large piston was inserted, and attached by a shaft to the framework of a carriage. The propelling power was the ordinary pressure of the atmosphere acting against the piston in the tube on one side, a vacuum being created in the tube on the other side of the piston by the working of a stationary engine. Great was the popularity of the atmospheric system; and still George Stephenson said, "It won't do: it's but

a gimcrack." Engineers of distinction said he was
prejudiced, and that he looked upon the locomotive as
a pet child of his own. "Wait a little," he replied,
"and you will see that I am right."

Mr. Brunel, Mr. Cubitt, Mr. Vignolles, Mr. James
Walker, Dr. Lardner, and many others equally distinguished, strongly approved of the atmospheric railway;
and it was generally supposed that the locomotive
system was about to be snuffed out. "Not so fast,"
said Stephenson. "Let us wait to see if it will pay."
He never believed it would. It was ingenious, clever,
scientific, and all that; but railways were commercial
enterprises, not toys; and if the atmospheric railway
could not work to a profit, it would not do. Considered
in this light, he even went so far as to call it "a great
humbug."

No one can say that the atmospheric railway had not
a fair trial. The Government engineer, General Pasley,
did for it what had never been done for the locomotive
—he reported in its favour, whereas a former Government engineer had inferentially reported against the
use of locomotive power on railways. The House of
Commons also had reported in favour of the use of the
steam-engine on common roads; yet the railway locomotive had vitality enough in it to live through all.
"Nothing will beat it," said George Stephenson, "for
efficiency in all weathers, for economy in drawing loads
of average weight, and for power and speed as occasion
may require."

The atmospheric system was fairly and fully tried,
and it was found wanting. It was admitted to be an
exceedingly elegant mode of applying power; its devices
were very skilful, and its mechanism was most ingenious.
But it was costly, irregular in action, and, in particular
kinds of weather, not to be depended upon. At best,
it was but a modification of the stationary-engine system,
and experience proved it to be so expensive that it was

shortly after entirely abandoned in favour of locomotive power.[1]

One of the remarkable results of the system of railway locomotion which George Stephenson had by his persevering labours mainly contributed to establish, was the outbreak of the railway mania towards the close of his professional career. The success of the first main lines of railway naturally led to their extension into many new districts; but a strongly speculative tendency soon began to display itself, which contained in it the elements of great danger.[2] In the sessions of 1836 and 1837, seventy-six Acts were obtained, authorising the construction of 1458 miles of new railway at an expenditure of 25,680,000l.; and by the end of 1837 notices were given of seventy-five more Bills, to authorise the formation of 1230 additional miles of railway at an estimated cost of about 19,000,000l. This was more than the means of the country could fairly bear. The shares of many companies went to a discount; and a collapse took place, which, together with the restrictions imposed by Parliament on the obtaining of new Acts, had the effect, for a time, of placing a wholesome restraint on further speculation. During the sessions of

---

[1] The question of the specific merits of the atmospheric as compared with the fixed engine and locomotive systems, will be found fully discussed in Robert Stephenson's able 'Report on the Atmospheric Railway System,' 1844, in which he gives the result of numerous observations and experiments made by him on the Kingstown Atmospheric Railway, with the object of ascertaining whether the new power would be applicable for the working of the Chester and Holyhead Railway, then under construction. His opinion was decidedly against the atmospheric railway system.

[2] The traffic cases got up by the professional advocates of some of the Bills applied for in 1836 and 1837 were of the most fallacious character, as has been proved by the actual results. Traffic-taking was one of the arts by which extraordinary profits were then proved. Thus in 1836, the traffic case of the Eastern Counties Railway showed that there would be a clear profit on the outlay of 23½ per cent.! the York and North Midland, of 13½; and the London and Cambridge, of 14½ per cent. During the session of 1837 the traffic-takers grew bolder, and reached their highest flights. Thus, the promoters of the Sheffield and Manchester Bill "proved" a traffic which was to yield a net profit of 18½ per cent. on the outlay. One of the fortunate shareholders in the company, in a letter to the 'Railway Magazine,' even went so far beyond the traffic-taker, as to calculate on a dividend of 80 per cent.!

1838 and 1839 only five new railway companies obtained Acts of incorporation.   In 1840, not a single Act was passed; and in 1841, only a branch of 5¾ miles in length was authorized, which was not carried out. The next two sessions were equally quiet; and it was not until 1844, that the tide of railway enterprise suddenly rose again, and in the following year burst all bounds, breaking out in the wildest fury of speculation.

The extension of railways had, up to the year 1844, been mainly effected by men of the commercial classes, and the shareholders in them principally belonged to the manufacturing districts,—the capitalists of the metropolis as yet holding aloof, and prophesying disaster to all concerned in railway projects.[1]   The Stock Exchange looked askance upon them, and it was with difficulty that respectable brokers could be found to do business in the shares.   But when the lugubrious anticipations of the City men were found to be so entirely falsified by the results—when, after the lapse of years, it was ascertained that railway traffic rapidly increased and dividends steadily improved—a change came over the spirit of the London capitalists.   They then invested largely in railways, the shares in which became a leading branch of business on the Stock Exchange, and the prices of some rose to nearly double their original value.

A stimulus was thus given to the projection of further lines, the shares in most of which came out at a premium, and became the subject of immediate traffic.   A reckless spirit of gambling set in, which completely changed the character and objects of railway enterprise.   The public outside the Stock Exchange became also infected, and many persons utterly ignorant of railways, knowing

---

[1] The leading "City men" looked with great suspicion on the first railway projects, having no faith in their success.   In 1835, the solicitorship of the Brighton Railway (then projected) was offered to a city firm of high standing, and refused,—one of the partners assigning as a reason, that "the coaches would drive the railway trains off the road in a month!"

and caring nothing about their national uses, but hunger-
ing and thirsting after premiums, rushed eagerly into
the vortex. They applied for allotments, and subscribed
for shares in lines, of the engineering character or
probable traffic of which they knew nothing. Provided
they could but obtain allotments which they could sell
at a premium, and put the profit—in many cases the
only capital they possessed[1]—into their pocket, it was
enough for them. The mania was not confined to the
precincts of the Stock Exchange, but infected all ranks.
It embraced merchants and manufacturers, gentry and
shopkeepers, clerks in public offices, and loungers at the
clubs. Noble lords were pointed at as " stags ; " there
were even clergymen who were characterised as " bulls ; "
and amiable ladies who had the reputation of " bears,"
in the share-markets. The few quiet men who remained
uninfluenced by the speculation of the time were, in not
a few cases, even reproached for doing injustice to their
families, in declining to help themselves from the stores
of wealth that were poured out on all sides.

Folly and knavery were, for a time, completely in the
ascendant. The sharpers of society were let loose, and
jobbers and schemers became more and more plentiful.
They threw out railway schemes as lures to catch the
unwary. They fed the mania with a constant succession
of new projects. The railway papers became loaded
with their advertisements. The post-office was scarcely
able to distribute the multitude of prospectuses and
circulars which they issued. For a time their popularity was immense. They rose like froth into the
upper heights of society, and the flunkey FitzPlushe, by
virtue of his supposed wealth, sat amongst peers and
was idolised. Then was the harvest-time of scheming

---

[1] The Marquis of Clanricarde brought
under the notice of the House of Lords,
in 1845, that one Charles Guernsey,
the son of a charwoman, and a clerk
in a broker's office, at 12s. à week,
had his name down as a subscriber for
shares in the London and York line,
for 52,000l. Doubtless he had been
made useful for the purpose by the
brokers, his employers.

lawyers, parliamentary agents, engineers, surveyors, and traffic-takers, who were alike ready to take up any railway scheme however desperate, and to prove any amount of traffic even where none existed. The traffic in the credulity of their dupes was, however, the great fact that mainly concerned them, and of the profitable character of which there could be no doubt.

Parliament, whose previous conduct in connection with railway legislation was so open to reprehension, interposed no check—attempted no remedy. On the contrary, it helped to intensify the evils arising from this unseemly state of things. Many of its members were themselves involved in the mania, and as much interested in its continuance as the vulgar herd of money-grubbers. The railway prospectuses now issued —unlike the original Liverpool and Manchester, and London and Birmingham schemes — were headed by peers, baronets, landed proprietors, and strings of M.P's. Thus, it was found in 1845 that no fewer than 157 members of Parliament were on the lists of new companies as subscribers for sums ranging from 291,000*l.* downwards! The projectors of new lines even came to boast of their parliamentary strength, and of the number of votes which they could command in "the House." At all events, it is matter of fact, that many utterly ruinous branches and extensions projected during the mania, calculated only to benefit the inhabitants of a few miserable boroughs accidentally omitted from Schedule A, were authorised in the memorable sessions of 1844 and 1845.

Mr. Stephenson was anxiously entreated to lend his name to prospectuses during the railway mania; but he invariably refused. He held aloof from the headlong folly of the hour; and endeavoured to check it, but in vain. Had he been less scrupulous, and given his countenance to the numerous projects about which he was consulted, he might, without any trouble, have thus secured enormous gains; but he had no desire to accumulate a fortune without labour and without honour. He himself never speculated in shares. When he was satisfied as to the merits of any undertaking, he subscribed for a certain amount of capital in it, and held on, neither buying nor selling. At a dinner of the Leeds and Bradford directors at Ben Rydding in October, 1844, before the mania had reached its height, he warned those present against the prevalent disposition towards railway speculation. It was, he said, like walking upon a piece of ice with shallows and deeps; the shallows were frozen over, and they would carry, but it required great caution to get over the deeps. He was satisfied that in the course of the next year many would step on places not strong enough to carry them, and would get into the deeps; they would be taking shares, and afterwards be unable to pay the calls upon them. Yorkshiremen were reckoned clever men, and his advice to them was, to stick together and promote communication in their own neighbourhood,—not to go abroad with their speculations. If any had done so, he advised them to get their money back as fast as they could, for if they did not they would not get it at all. He informed the company, at the same time, of his earliest holding of railway shares; it was in the Stockton and Darlington Railway, and the number he held was *three*— "a very large capital for him to possess at the time." But a Stockton friend was anxious to possess a share, and he sold him *one* at a premium of 33*s.*; he supposed he had been about the first man in England to sell a railway share at a premium.

During 1845, his son's offices in Great George-street, Westminster, were crowded with persons of various conditions seeking interviews, presenting very much the appearance of the levee of a minister of state. The burly figure of Mr. Hudson, the " Railway King," surrounded by an admiring group of followers, was often to be seen there; and a still more interesting person, in the estimation of many, was George Stephenson, dressed in black, his coat of somewhat old-fashioned cut, with square pockets in the tails. He wore a white neckcloth, and a large bunch of seals was suspended from his watch-ribbon. Altogether, he presented an appearance of health, intelligence, and good humour, that rejoiced one to look upon in that sordid, selfish, and eventually ruinous saturnalia of railway speculation.

Being still the consulting engineer of several of the older companies, he necessarily appeared before Parliament in support of their branches and extensions. In 1845 his name was associated with that of his son as the engineer of the Southport and Preston Junction. In the same session he gave evidence in favour of the Syston and Peterborough branch of the Midland Railway; but his principal attention was confined to the promotion of the line from Newcastle to Berwick, in which he had never ceased to take the deepest interest. At the same time he was engaged in examining and reporting upon certain foreign lines of considerable importance.

Powers were granted by Parliament, in 1845, to construct not less than 2883 miles of new railways in Britain, at an expenditure of about forty-four millions sterling! Yet the mania was not appeased; for in the following session of 1846, applications were made to Parliament for powers to raise 389,000,000*l.* sterling for the construction of further lines; and powers were actually conceded for forming 4790 miles (including 60 miles of tunnels), at a cost of about 120,000,000*l.* sterling.[1] During this session, Mr. Stephenson appeared as engineer for only one new line,—the Buxton, Maccles-

---

[2] On the 17th November, 1845, Mr. Spackman published a list of the lines *projected* (many of which were not afterwards prosecuted), from which it appeared that there were then 620 new railway projects before the public, requiring a capital of 563,203,000*l.*

field, Congleton, and Crewe Railway—a line in which,
as a coal-owner, he was personally interested;—and of
three branch-lines in connexion with existing companies
for which he had long acted as engineer. At the same
period, all the leading professional men were fully
occupied, some of them appearing as consulting engineers
for upwards of thirty lines each !

One of the features of the mania was the rage for
" direct lines " which everywhere displayed itself. There
were "Direct Manchester," "Direct Exeter," "Direct
York," and, indeed, new direct lines between most of
the large towns. The Marquis of Bristol, speaking in
favour of the " Direct Norwich and London " project, at
a public meeting at Haverhill, said, "If necessary, they
might *make a tunnel beneath his very drawing-room,* rather
than be defeated in their undertaking ! "   And the Rev.
F. Litchfield, at a meeting in Banbury, on the subject of
a line to that town, said "He had laid down for himself
a limit to his approbation of railways,—at least of such
as approached the neighbourhood with which he was
connected,—and that limit was, that he did not wish
them to approach any nearer to him than *to run through
his bedroom, with the bedposts for a station !* "   How
different was the spirit which influenced these noble
lords and gentlemen but a few years before !

The course adopted by Parliament in dealing with
the multitude of railway bills applied for during the
prevalence of the mania, was as irrational as it proved
to be unfortunate. The want of foresight displayed by
both Houses in obstructing the railway system so long
as it was based upon sound commercial principles, was
only equalled by the fatal facility with which they now
granted railway projects based only upon the wildest
speculation. Parliament interposed no check, laid down
no principle, furnished no guidance, for the conduct of
railway projectors; but left every company to select its
own locality, determine its own line, and fix its own

gauge. No regard was paid to the claims of existing companies, which had already expended so large an amount in the formation of useful railways; and speculators were left at full liberty to project and carry out lines almost parallel with theirs.

The House of Commons became thoroughly influenced by the prevailing excitement. Even the Board of Trade began to favour the views of the fast school of engineers. In their "Report on the Lines projected in the Manchester and Leeds District," they promulgated some remarkable views respecting gradients, declaring themselves in favour of the "undulating system." They there stated that lines of an undulating character "which have gradients of 1 in 70 or 1 in 80 distributed over them in short lengths, may be positively *better* lines, *i. e., more susceptible of cheap and expeditious working*, than others which have nothing steeper than 1 in 100 or 1 in 120!" They concluded by reporting in favour of the line which exhibited the worst gradients and the sharpest curves, chiefly on the ground that it could be constructed for less money.

Sir Robert Peel took occasion, when speaking in favour of the continuance of the Railways Department of the Board of Trade, to advert to this Report in the House of Commons on the 4th of March following, as containing " a novel and highly important view on the subject of gradients, which, he was certain, never could have been taken by any Committee of the House of Commons, however intelligent;" and he might have added, that the more intelligent, the less likely they were to arrive at any such conclusion. When Mr. Stephenson saw this report of the Premier's speech in the newspapers of the following morning, he went forthwith to his son, and asked him to write a letter to Sir Robert Peel on the subject. He saw clearly that if these views were adopted, the utility and economy of railways would be seriously curtailed. " These members

of Parliament," said he, " are now as much disposed to
exaggerate the powers of the locomotive, as they were
to under-estimate them but a few years ago." Robert
accordingly wrote a letter for his father's signature,
embodying the views which he so strongly entertained as to the importance of flat gradients, and
referring to the experiments conducted by him many
years before, in proof of the great loss of working power
which was incurred on a line of steep as compared with
easy gradients. It was clear, from the tone of Sir
Robert Peel's speech in a subsequent debate, that he
had carefully read and considered Mr. Stephenson's
practical observations on the subject; though it did not
appear that he had come to any definite conclusion
thereon, further than that he strongly approved of the
Trent Valley Railway, by which Tamworth would be
placed upon a direct main line of communication.

The result of the labours of Parliament was a tissue
of legislative bungling, involving enormous loss to the
public. Railway Bills were granted in heaps. Two
hundred and seventy-two additional Acts were passed in
1846. Some authorised the construction of lines running
almost parallel to existing railways, in order to afford
the public "the benefits of unrestricted competition."
Locomotive and atmospheric lines, broad-gauge and
narrow-gauge lines, were granted without hesitation.
Committees decided without judgment and without
discrimination; it was a scramble for Bills, in which
the most unscrupulous were the most successful. As an
illustration of the legislative folly of the period, Mr.
Robert Stephenson, speaking at Toronto, in Upper
Canada, some years later, adduced the following instances:—"There was one district through which it
was proposed to run two lines, and there was no other
difficulty between them than the simple rivalry that, if
one got a charter, the other might also. But here,
where the Committee might have given both, they gave

neither.   In another instance, two lines were projected
through a barren country, and the Committee gave the
one which afforded the least accommodation to the public.
In another, where two lines were projected to run,
merely to shorten the time by a few minutes, leading
through a mountainous country, the Committee gave
both.   So that, where the Committee might have given
both, they gave neither, and where they should have
given neither, they gave both."

Amongst the many ill effects of the mania, one of the
worst was that it introduced a low tone of morality into
railway transactions.   The bad spirit which had been
evoked by it unhappily extended to the commercial
classes, and many of the most flagrant swindles of recent
times had their origin in the year 1845.   Those who
had suddenly gained large sums without labour, and
also without honour, were too ready to enter upon
courses of the wildest extravagance ; and a false style
of living shortly arose, the poisonous influence of which
extended through all classes.   Men began to look upon
railways as instruments to job with.   Persons, sometimes possessing information respecting railways, but
more frequently possessing none, got upon boards for
the purpose of promoting their individual objects, often
in a very unscrupulous manner; landowners, to promote branch lines through their property ; speculators
in shares, to trade upon the exclusive information which
they obtained; whilst some directors were appointed
through the influence mainly of solicitors, contractors,
or engineers, who used them as tools to serve their own
ends.   In this way the unfortunate proprietors were, in
many cases, betrayed, and their property was shame-
fully squandered, much to the discredit of the railway
system.

One of the most prominent celebrities of the mania
was George Hudson, of York.   He was a man of some
local repute in that city when the line between Leeds

and York was projected. His views as to railways were then extremely moderate, and his main object in joining the undertaking was to secure for York the advantages of the best railway communication. The Company was not very prosperous at first, and during the years 1840 and 1841 the shares had greatly sunk in value. Mr. Alderman Meek, the first chairman, having retired, Mr. Hudson was elected in his stead, and he very shortly contrived to pay improved dividends to the proprietors, who asked no questions. Desiring to extend the field of his operations, he proceeded to lease the Leeds and Selby Railway at five per cent. That line had hitherto been a losing concern; so its owners readily struck a bargain with Mr. Hudson, and sounded his praises in all directions. He increased the dividends on the York and North Midland shares to ten per cent., and began to be cited as the model of a railway chairman.

He next interested himself in the North Midland Railway, where he appeared in the character of a reformer of abuses. The North Midland shares also had gone to a great discount, and the shareholders were very willing to give Mr. Hudson an opportunity of reforming their railway. They elected him a director. His bustling, pushing, persevering character gave him an influential position at the board, and he soon pushed the old directors from their stools. He laboured hard, at much personal inconvenience, to help the concern out of its difficulties, and he succeeded. The new directors recognised his power, and elected him their chairman.

Railway affairs revived in 1842, and public confidence in them as profitable investments was gradually increasing. Mr. Hudson had the benefit of this growing prosperity. The dividends in his lines improved, and the shares rose in value. The Lord Mayor of York began to be quoted as one of the most capable of railway directors. Stimulated by his success and encou-

raged by his followers, he struck out or supported many
new projects—a line to Scarborough, a line to Bradford,
lines in the Midland districts, and lines to connect York
with Newcastle and Edinburgh.   He was elected chairman of the Newcastle and Darlington Railway ; and
when—in order to complete the continuity of the main
line of communication—it was found necessary to secure
the Durham junction, which was an important link in
the chain, he and Mr. Stephenson boldly purchased that
railway between them, at the price of 88,500*l.*   It was
an exceedingly fortunate purchase for the Company, to
whom it was worth double the money.   The act, though
not strictly legal, proved successful, and was much
lauded.   Thus encouraged, Mr. Hudson proceeded to
buy the Brandling Junction line for 500,000*l.*, in his
own name—an operation at the time regarded as equally
favourable, though he was afterwards charged with
appropriating 1600 of the shares created for the purchase, when worth 21*l.* premium each.   The Great
North of England line being completed, Mr. Hudson
had thus secured the entire line of communication from
York to Newcastle, and the route was opened to the
public in June, 1844.   On that occasion Newcastle
eulogised Mr. Hudson in its choicest local eloquence,
and he was pronounced to be the greatest benefactor
the district had ever known.

The adulation which now followed Mr. Hudson would
have intoxicated a stronger and more self-denying man.
He was pronounced the man of the age, and hailed as
" the Railway King."   The grand test by which the
shareholders judged him was the dividends that he paid,
although subsequent events proved that these dividends
were in many cases delusive, intended only " to make
things pleasant."   The policy, however, had its effect.
The shares in all the lines of which he was chairman
went to a premium, and then arose the temptation to
create new shares in branch and extension lines, often

worthless, which were issued at a premium also. Thus he
shortly found himself chairman of nearly 600 miles of railways, extending from Rugby to Newcastle, and at the
head of numerous new projects, by means of which
paper wealth could be created, as it were, at pleasure.
He held in his own hands almost the entire administra-
tive power of the companies over which he presided :
he was chairman, board, manager, and all. His ad-
mirers for the time, inspired sometimes by gratitude for
past favours, but oftener by the expectation of favours
to come, supported him in all his measures. At the
meetings of the companies, if any suspicious inquirer
ventured to put a question about the accounts, he was
summarily put down by the chair, and hissed by the
proprietors. Mr. Hudson was voted praises, testimo-
nials, and surplus shares, alike liberally ; and scarcely a
word against him could find a hearing. He was equally
popular outside the circle of railway proprietors. His
entertainments at Albert Gate. were crowded by sycophants, many of them titled ; and he went his round of
visits among the peerage like a prince.

Of course Mr. Hudson was a great authority on railway questions in Parliament, to which the burgesses of
Sunderland had sent him. His experience of railways,
still little understood, though the subject of so much
legislation, gave value and weight to his opinions, and
in many respects he was a useful member. During the
first years of his membership he was chiefly occupied in
passing the railway bills in which he was more particularly interested ; and in the session of 1845, when he
was at the height of his power, it was triumphantly said
of him, that " he walked quietly through Parliament
with some sixteen railway bills under his arm." One
of these bills, however, was the subject of a very severe
contest—we mean that empowering the construction of
the railway from Newcastle to Berwick. It was almost
the only bill in which George Stephenson was that year

concerned. Mr. Hudson displayed great energy in supporting the measure, and he worked hard to ensure its success both in and out of Parliament; but he himself attributed the chief merit to Mr. Stephenson. He accordingly suggested to the shareholders that they should present him with some fitting testimonial in recognition of his valuable services. Indeed, a Stephenson testimonial had long been spoken of, and a committee was formed for the purpose of raising subscriptions as early as the year 1839. Mr. Hudson now revived the subject, and successively appealed to the Newcastle and Darlington, the Midland, and the York and North Midland Companies, who unanimously adopted the resolutions which he proposed to them amidst "loud applause;" but there the matter ended.

The Hudson Testimonial was a much more taking thing; for Mr. Hudson had it in his power to allot shares (selling at a premium) to the subscribers to his testimonial. But Mr. Stephenson pretended to fill no man's pocket with premiums; he was no creator of shares, and could not therefore work upon shareholders' gratitude for "favours to come." The proposed testimonial to him accordingly ended with resolutions and speeches. The York, Newcastle, and Berwick Board—in other words, Mr. Hudson—did indeed mark their sense of the "great obligations" which they were under to Mr. Stephenson for helping to carry their bill through Parliament, by making him an allotment of thirty of the new shares authorised by the Act. But, as afterwards appeared, the chairman had at the same time appropriated to himself not fewer than 10,894 of the same shares, the premiums on which were then worth, in the market, about 145,000*l*. This shabby manner of acknowledging the gratitude of the Company to their engineer, was strongly resented by Mr. Stephenson at the time, and a coolness took place between him and Mr. Hudson which was never wholly removed—though they afterwards shook hands, and Mr. Stephenson declared that all was forgotten.

Mr. Hudson's brief reign soon drew to a close. The speculation of 1845 was followed by a sudden reaction. Shares went down faster than they had gone up; the holders of them hastened to sell in order to avoid payment of the calls, and many found themselves ruined. Then came repentance, and a sudden return to virtue. The betting-man who, temporarily abandoning the turf for the share-market, had played his heaviest stake and lost; the merchant who had left his business, and the doctor who had neglected his patients, to gamble in railway stock, and been ruined; the penniless knaves and schemers, who had speculated so recklessly and gained so little; the titled and fashionable people, who had bowed themselves so low before the idol of the day, and found themselves deceived and "done;" the credulous small capitalists, who, dazzled by premiums, had invested their all in railway shares, and now saw themselves stripped of everything—were grievously enraged, and looked about them for a victim. In this temper were shareholders, when, at a railway meeting in York, some pertinent questions were put to the Railway King. His replies were not satisfactory, and the questions were pushed home. Mr. Hudson became confused. Angry voices rose in the meeting. A committee of investigation was appointed. The golden calf was found to be of brass, and hurled down; Hudson's own toadies and sycophants eagerly joining the chorus of popular indignation. Similar proceedings shortly after occurred at the meetings of other companies, and the bubbles having by that time burst, the Railway Mania came to a sudden and ignominious end.

While the mania was at its height in England, railways were also being extended abroad, and George Stephenson was requested on several occasions to give the benefit of his advice to the directors of foreign undertakings.   One of the most agreeable of these excursions was to Belgium in 1845, in company with his friends Mr. Sopwith and Mr. Starbuck.   His special object was to examine the proposed line of the Sambre and Meuse Railway, for which a concession had been granted by the Belgian legislature.   Arrived on the ground, he went carefully over the entire length of the proposed line, to Couvins, the Forest of Ardennes, and Rocroi, across the French frontier; examining the bearings of the coal-field, the slate and marble quarries, and the numerous iron mines in existence between the Sambre and the Meuse, as well as carefully exploring the ravines which extended through the district, in order to satisfy himself that the best possible route had been selected.   Mr. Stephenson was delighted with the novelty of the journey, the beauty of the scenery, and the industry of the population.   His companions were entertained by his ample and varied stores of practical information on all subjects, and his conversation was full of reminiscences of his youth, on which he always delighted to dwell when in the society of his more inti-mate friends.   The journey was varied by a visit to the coal-mines near Jemappe, where Stephenson exa-mined with interest the mode adopted by the Belgian miners of draining the pits, inspecting their engines and brakeing machines, so familiar to him in early life.

The engineers of Belgium took the opportunity of Mr. Stephenson's visit to their country to invite him to a magnificent banquet at Brussels.   The Public Hall, in which they entertained him, was gaily decorated with flags, prominent amongst which was the Union Jack, in honour of their distinguished guest.   A handsome marble pedestal, ornamented with his bust crowned with laurels, occupied one end of the room.   The chair was occupied by M. Massui, the Chief Director of the National Railways of Belgium; and the most eminent scientific men of the kingdom were present.   Their

reception of "the Father of railways" was of the
most enthusiastic description.   Mr. Stephenson was
greatly pleased with the entertainment.   Not the
least interesting incident of the evening was his
observing, when the dinner was about half over, a
model of a locomotive engine placed upon the centre
table, under a triumphal arch.   Turning suddenly to
his friend Sopwith, he exclaimed, " Do you see the
' Rocket ? ' "   It was indeed the model of that celebrated engine ; and Mr. Stephenson prized the compliment thus paid him, perhaps more than all the
encomiums of the evening.

The next day (April 5th) King Leopold invited him
to a private interview at the palace.   Accompanied by
Mr. Sopwith, he proceeded to Laaken, and was very
cordially received by His Majesty.   Nothing was more
remarkable in Mr. Stephenson than his extreme ease
and self-possession in the presence of distinguished and
highly-educated persons.   The king immediately entered
into familiar conversation with him, discussing the railway project which had been the object of Mr. Stephenson's visit to Belgium, and then the structure of the
Belgian coal-fields,— his Majesty expressing his sense
of the great importance of economy in a fuel which had
become indispensable to the comfort and well-being of
society, which was the basis of all manufactures, and the
vital power of railway locomotion.   The subject was
always a favourite one with Mr. Stephenson, and,
encouraged by the king, he proceeded to describe to
him the geological structure of Belgium, the original
formation of coal, its subsequent elevation by volcanic
forces, and the vast amount of denudation.   In describing the coal-beds he used his hat as a sort of model to
illustrate his meaning ; and the eyes of the king were
fixed upon it as he proceeded with his interesting description.   The conversation then passed to the rise and
progress of trade and manufactures,— Mr. Stephenson

pointing out how closely they everywhere followed the coal, being mainly dependent upon it, as it were, for their very existence.

The king seemed greatly pleased with the interview, and at its close expressed himself obliged by the interesting information which Mr. Stephenson had communicated. Shaking hands cordially with both the gentlemen, and wishing them success in all their important undertakings, he bade them adieu. As they were leaving the palace Mr. Stephenson, bethinking him of the model by which he had just been illustrating the Belgian coal-fields, said to his friend, "By the bye, Sopwith, I was afraid the king would see the inside of my hat; it's a shocking bad one!". Little could George Stephenson, when brakesman at a coal-pit, have dreamt that, in the course of his life, he should be admitted to an interview with a monarch, and describe to him the manner in which the geological foundations of his kingdom had been laid!

Mr. Stephenson paid a second visit to Belgium in the course of the same year, on the business of the West Flanders Railway; and he had scarcely returned from it ere he made arrangements to proceed to Spain, for purpose of examining and reporting upon a scheme then on foot for constructing "the Royal North of Spain Railway." A concession had been made by the Spanish Government of a line of railway from Madrid to the Bay of Biscay, and a numerous staff of engineers was engaged in surveying the proposed line. The directors of the Company had declined making the necessary deposits until more favourable terms had been secured; and Sir Joshua Walmsley, on their part, was about to visit Spain and press the Government on the subject. Mr. Stephenson, whom he consulted, was alive to the difficulties of the office which Sir Joshua was induced to undertake, and offered to be his companion and adviser on the occasion,—

declining to receive any recompense beyond the simple expenses of the journey. He could only arrange to be absent for six weeks, and set out from England about the middle of September, 1845.

The party was joined at Paris by Mr. Mackenzie, the contractor for the Orleans and Tours Railway, then in course of construction, who took them over the works, and accompanied them as far as Tours. Sir Joshua Walmsley was struck during the journey by Mr. Stephenson's close and accurate observation. Of course he was fully alive to any important engineering works which came in his way. Thus, in crossing the river Dordogne, on the road to Bordeaux, he was struck with the construction of . the stupendous chain-bridge which had recently been . erected there. Not satisfied with his first inspection, he walked back and again crossed the bridge. On reaching the shore he said : " This bridge cannot stand ; it is impossible that it can sustain any unusual weight. Supposing a large body of troops to march over it, there would be so much oscillation as to cause the greatest danger ; in fact it could not stand." He determined to write to the public authorities, warning them on the subject ; which he did. His judgment proved to be quite correct, for only a few years after, no improvement having been made in the bridge, a body of troops marching over it under the precise circumstances which he had imagined, the chains broke, the men were precipitated into the river, and many lives were lost.

They soon reached the great chain of the Pyrenees, and crossed over into Spain. It was on a Sunday evening, after a long day's toilsome journey through the mountains, that the party suddenly found themselves in one of those beautiful secluded valleys lying amidst the Western Pyrenees. A small hamlet lay before them, consisting of some thirty or forty houses and a fine old church. The sun was low on the horizon, and,

under the wide porch, beneath the shadow of the
church, were seated nearly all the inhabitants of the
place.   They were dressed in their holiday attire.
The bright bits of red and amber colour in the dresses
of the women, and the gay sashes of the men, formed
a striking picture, on which the travellers gazed in
silent admiration.   It was something entirely novel
and unexpected.   Beside the villagers sat two venerable old men, whose canonical hats indicated their
quality as village pastors.   Two groups of young
women and children were dancing outside the porch
to the accompaniment of a simple pipe; and within a
hundred yards of them, some of the youths of the
village were disporting themselves in athletic exercises; the whole being carried on beneath the fostering
care of the old church, and with the sanction of its
ministers.   It was a beautiful scene, and deeply moved
the travellers as they approached the principal group.
The villagers greeted them courteously, supplied their
present wants, and pressed upon them some fine melons,
brought from their adjoining gardens.   Mr. Stephenson
used afterwards to look back upon that simple scene,
and speak of it as one of the most charming pastorals he
had ever witnessed.

They shortly reached the site of the proposed railway,
passing through Irun, St. Sebastian, St. Andero, and
Bilbao, at which places they met deputations of the
principal inhabitants who were interested in the subject of their journey.   At Raynosa Mr. Stephenson
carefully examined the mountain passes and ravines
through which a railway could be made.   He rose
at break of day, and surveyed until the darkness set
in; and frequently his resting-place at night was the
floor of some miserable hovel.   He was thus laboriously
occupied for ten days, after which he proceeded across
the province of Old Castile towards Madrid, surveying
as he went.   The proposed plan included the purchase

of the Castile canal; and that property was also surveyed.
He next proceeded to El Escorial, situated at the foot
of the Guadarama mountains, through which he found
that it would be necessary to construct two formidable
tunnels; added to which he ascertained that the country
between El Escorial and Madrid was of a very difficult
and expensive character to work through. Taking these
circumstances into account, and looking at the expected
traffic on the proposed line, Sir Joshua Walmsley, acting
under the advice of Mr. Stephenson, offered to construct
the line from Madrid to the Bay of Biscay, only on condition that the requisite land was given the Company
for the purpose; that they should be allowed every
facility for cutting such timber belonging the Crown
as might be required for the purposes of the railway;
and also that the materials required from abroad for the
construction of the line should be admitted free of duty.
In return for these concessions the Company offered to
clothe and feed several thousands of convicts while
engaged in the execution of the earthworks. General
Narvaez, afterwards Duke of Valencia, received Sir
Joshua Walmsley and Mr. Stephenson on the subject
of their proposition, and expressed his willingness to
close with them; but it was necessary that other influential parties should give their concurrence before the
scheme could be carried into effect. The deputation
waited ten days to receive the answer of the Spanish
government; but no answer of any kind was vouchsafed.
The authorities, indeed, invited them to be present
at a Spanish bull-fight, but that was not quite the business Mr. Stephenson had gone all the way to Spain to
transact; and the offer was politely declined. The
result was, that Mr. Stephenson dissuaded his friend
from making the necessary deposit at Madrid. Besides, he had by this time formed an unfavourable
opinion of the entire project, and considered that the
traffic would not amount to one-eighth of the estimate.

Mr. Stephenson was now anxious to be in England. During the journey from Madrid he often spoke with affection of friends and relatives; and when apparently absorbed by other matters, he would revert to what he thought might then be passing at home. Few incidents worthy of notice occurred on the journey homeward, but one may be mentioned. While travelling in an open conveyance between Madrid and Vittoria, the driver urged his mules down hill at a dangerous pace. He was requested to slacken speed; but suspecting his passengers to be afraid, he only flogged the brutes into a still more furious gallop. Observing this, Mr. Stephenson coolly said, "Let us try him on the other tack; tell him to show us the fastest pace at which Spanish mules can go." The rogue of a driver, when he found his tricks of no avail, pulled up and proceeded at a moderate rate for the rest of his journey.

Urgent business required Mr. Stephenson's presence in London on the last day of November. They travelled, therefore almost continuously, day and night; and the fatigue consequent on the journey, added to the privations voluntarily endured by the engineer while carrying on the survey among the Spanish mountains, began to tell seriously on his health. By the time he reached Paris he was evidently ill, but he nevertheless determined on proceeding. He reached Havre in time for the Southampton boat; but when on board, pleurisy developed itself, and it was necessary to bleed him freely. During the voyage, he spent his time chiefly in dictating letters and reports to Sir Joshua Walmsley, who never left him, and whose kindness on the occasion he gratefully remembered. His friend was struck by the clearness of his dictated composition, which exhibited a vigour and condensation which to him seemed marvellous. After a few weeks' rest at home, Mr. Stephenson gradually recovered, though his health remained severely shaken.

NEWCASTLE, FROM THE HIGH LEVEL BRIDGE. [By R. P. Leitch.]

CHAPTER XVIII.

ROBERT STEPHENSON'S CAREER—THE STEPHENSONS AND BRUNEL—
EAST COAST ROUTE TO SCOTLAND — ROYAL BORDER BRIDGE,
BERWICK—HIGH LEVEL BRIDGE, NEWCASTLE.

THE career of George Stephenson was drawing to a
close. He had for some time been gradually retiring
from the more active pursuit of railway engineering,
and confining himself to the promotion of only a
few undertakings in which he took a more than ordinary personal interest. In 1840, when the extensive
main lines in the Midland districts had been finished
and opened for traffic, he publicly expressed his intention
of withdrawing from the profession. He had reached
sixty, and, having spent the greater part of his life in
very hard work, he naturally desired rest and retirement
in his old age. There was the less necessity for his continuing "in harness," as Robert Stephenson was now in
full career as a leading railway engineer, and his father
had pleasure in handing over to him, with the sanction
of the companies concerned, nearly all the railway
appointments which he held.

Robert Stephenson amply repaid his father's care. The sound education of which he had laid the foundations at Bruce's school at Newcastle, improved by his subsequent culture at Edinburgh College, but more than all by his father's example in application, industry, and thoroughness in all that he undertook, told powerfully in the formation of his character, not less than in the discipline of his intellect. His father had early implanted in him habits of mental activity, familiarized him with the laws of mechanics, and carefully trained and stimulated his inventive faculties, the first great fruits of which, as we have seen, were exhibited in the triumph of the "Rocket" at Rainhill. "I am fully conscious in my own mind," said the son, at a meeting of the Mechanical Engineers at Newcastle, in 1858, "how greatly my civil engineering has been regulated and influenced by the mechanical knowledge which I derived directly from my father; and the more my experience has advanced, the more convinced I have become that it is necessary to educate an engineer in the workshop. That is, emphatically, the education which will render the engineer most intelligent, most useful, and the fullest of resources in times of difficulty."

Robert Stephenson was but twenty-six years old when the performances of the "Rocket" established the practicability of steam locomotion on railways. He was shortly after appointed engineer of the Leicester and Swannington Railway; after which, at his father's request, he was made joint engineer with himself in the engineering of the London and Birmingham Railway, and the execution of that line was afterwards almost entirely entrusted to him. The stability and excellence of the works of that railway, the difficulties which had been successfully overcome in the course of its construction, and the judgment which was displayed by Robert Stephenson throughout the whole conduct of the undertaking to its completion, established his reputation as an

engineer; and his father could now look with confidence
and with pride upon his son's achievements. From that
time forward, father and son worked together as one
man, each jealous of the other's honour; and on the
father's retirement, it was generally recognized that, in
the sphere of railways, Robert Stephenson was the
foremost man, the safest guide, and the most active
worker.

Robert Stephenson was subsequently appointed engineer of the Eastern Counties, the Northern and Eastern,
the Blackwall, and many other railways in the midland
and southern districts. When the speculation of 1844
set in, his services were, of course, greatly in request.
Thus, in one session we find him engaged as engineer
for not fewer than thirty-three new schemes. Projectors
thought themselves fortunate who could secure his name,
and he had only to propose his terms to obtain them.
The work which he performed at this period of his life
was indeed enormous, and his income was large beyond
any previous instance of engineering gain. But much
of his labour was heavy hackwork, of a very uninteresting character. During the sittings of the committees
of Parliament, almost every moment of his time was
occupied in consultations, and in preparing evidence or
in giving it. The crowded, low-roofed committee-
rooms of the old Houses of Parliament were altogether
inadequate to accommodate the rush of perspiring projectors for bills, and even the lobbies were sometimes
choked with them. To have borne that noisome atmosphere and heat would have tested the constitutions of
salamanders, and engineers were only human. With
brains kept in a state of excitement during the entire
day, no wonder their nervous systems became unstrung.
Their only chance of refreshment was during an occasional rush to the bun and sandwich stand in the lobby,
though sometimes even that recourse failed them.
Then, with mind and body jaded—perhaps after under-

going a series of consultations upon many bills after
the rising of the committees—the exhausted engineers
would seek to stimulate nature by a late, perhaps a
heavy, dinner.   What chance had any ordinary constitution of surviving such an ordeal ?   The consequence
was, that stomach, brain, and liver were alike irretrievably injured; and hence the men who bore the
heat and brunt of those struggles—Stephenson, Brunel,
Locke, and Errington—have already all died, comparatively young men.

In mentioning the name of Brunel, we are reminded
of him as the principal rival and competitor of Robert
Stephenson.   Both were the sons of distinguished men,
and both inherited the fame and followed in the foot-
steps of their fathers.   The Stephensons were inventive,
practical, and sagacious; the Brunels ingenious, ima-
ginative, and daring.   The former were as thoroughly
English in their characteristics as the latter perhaps
were as thoroughly French.   The fathers and the sons
were alike successful in their works, though not in the
same degree.   Measured by practical and profitable
results, the Stephensons were unquestionably the safer
men to follow.

Robert Stephenson and Isambard Kingdom Brunel
were destined often to come into collision in the course
of their professional life.   Their respective railway districts "marched" with each other, and it became their
business to invade or defend those districts, according as
the policy of their respective boards might direct.   The
gauge fixed by Mr. Brunel for the Great Western
Railway, so entirely different from that adopted by the
Stephensons on the Northern and Midland lines,[1] was

---

[1] The original width of the coal
tramroads in the North virtually determined the British gauge.   It was
the width of the ordinary road-track,
—not fixed after any scientific theory,
but adopted simply because its use had
already been established.   George Stephenson introduced it without alteration on the Liverpool and Manchester
Railway; and the lines subsequently
formed in the same district were laid
down of the same width.   Mr. Ste-

from the first a great cause of contention. But Mr. Brunel had always an aversion to follow any man's lead; and that another engineer had fixed the gauge of a railway, or built a bridge, or designed an engine, in one way, was of itself often a sufficient reason with him for adopting an altogether different course. Robert Stephenson, on his part, though less bold, was more practical, preferring to follow the old routes, and to tread in the safe steps of his father.

Mr. Brunel, however, determined that the Great Western should be a giant's road, and that travelling should be conducted upon it at double speed. His ambition was to make the *best* road that imagination could devise; whereas the main object of the Stephensons, both father and son, was to make a road that would *pay*. Although, tried by the Stephenson test, Brunel's magnificent road was a failure so far as the shareholders in the Great Western Company were concerned, the stimulus which his ambitious designs gave to mechanical invention at the time proved a general good. The narrow-gauge engineers exerted themselves to quicken

---

phenson from the first anticipated the general extension of railways throughout England; and one of the ideas with which he started was, the essential importance of preserving such a uniformity as would admit of perfect communication between them. When consulted about the gauge of the Canterbury and Whitstable, and Leicester and Swannington Railways, he said, "Make them of the same width: though they may be a long way apart now, depend upon it they will be joined together some day." All the railways, therefore, laid down by himself and his assistants in the neighbourhood of Manchester, extending from thence to London on the south, and to Leeds on the east, were constructed on the Liverpool and Manchester, or narrow gauge. Besides the Great Western Railway, where the gauge adopted was seven feet, the only other line on which a

broader gauge than four feet eight and a-half inches was adopted was the Eastern Counties, where it was five feet, Mr. Braithwaite, the engineer, being of opinion that an increase of three and a-half inches in the width of his line would give him better space for the machinery of the locomotive. But when the northern and eastern extension of the same line was formed, which was to work into the narrow-gauge system of the Midland Railway, Mr. Robert Stephenson, its new engineer, strongly recommended the directors of the Eastern Counties line to alter their gauge accordingly, for the purpose of securing uniformity; and they adopted his recommendation. Mr. Braithwaite himself afterwards justified the wisdom of this step, and stated that he considered the narrow gauge "infinitely superior to any other," more especially for passenger traffic.

their locomotives to the utmost; they were improved
and re-improved; their machinery was simplified and
perfected; outside cylinders gave place to inside; the
steadier and more rapid and effective action of the
engine was secured; and in a few years the highest
speed on the narrow-gauge lines went up from thirty to
about fifty miles an hour.   For this rapidity of progress
we are in no small degree indebted to the stimulus imparted to the narrow-gauge engineers by Mr. Brunel.
And it is well for a country that it should possess men
such as he, ready to dare the untried, and to venture
boldly into new paths.   Individuals may suffer from the
cost of the experiments, but the nation, which is an
aggregate of individuals, gains, and so does the world
at large.

It was one of the characteristics of Brunel to *believe*
in the success of the schemes for which he was professionally engaged as engineer; and he proved this by
investing his savings largely in the Great Western
Railway, in the South Devon atmospheric line, and in
the Great Eastern steamship, with what results are well
known.   Robert Stephenson, on the contrary, with characteristic caution, towards the latter years of his life,
avoided holding unguaranteed railway shares; and
though he might execute magnificent structures, such as
the Victoria Bridge across the St. Lawrence, he was
careful not to embark any portion of his own fortune in
the ordinary capital of these concerns.   In 1845, he
shrewdly foresaw the inevitable crash that was about to
follow the mania of that year; and while shares were
still at a premium he took the opportunity of selling out
all that he had.   He urged his father to do the same
thing, but George's reply was characteristic.   "No,"
said he; "I took my shares for an investment, and not
to speculate with, and I am not going to sell them now
because folks have gone mad about railways."   The
consequence was, that he continued to hold the 60,000*l*.

which he had invested in the shares of various railways until his death, when they were at once sold out by his son, though at a great depreciation on their original cost.

One of the hardest battles fought between the Stephensons and Brunel was for the railway between Newcastle and Berwick, forming part of the great East Coast route to Scotland. As early as 1836, George Stephenson had surveyed two lines to connect Edinburgh with Newcastle : one by Berwick and Dunbar along the coast, and the other, more inland, by Carter Fell, up the vale of the Gala, to the northern capital. Two years later, he made a further examination of the intervening country, and again reported more decidedly than before in favour of the coast line. The inland route, however, was not without its advocates : Stephenson's old friend, Nicholas Wood, heading the opposition to his proposed Coast railway. But both projects lay dormant for several years longer, until the completion of the Midland and other main lines as far north as Newcastle had the effect of again reviving the subject of the extension of the route as far as Edinburgh.

On the 18th of June, 1844, the Newcastle and Darlington line—an important link of the great main highway to the north—was completed and publicly opened, thus connecting the Thames and the Tyne by a continuous line of railway. On that day Mr. Stephenson and a distinguished party of railway men travelled by express train from London to Newcastle in about nine hours. It was a great event, and was worthily celebrated. The population of Newcastle held holiday ; and a banquet given in the Assembly Rooms the same evening assumed the form of an ovation to Mr. Stephenson and his son. Thirty years before, in the capacity of a workman, he had been labouring at the construction of his first locomotive in the immediate neighbourhood. By slow and laborious steps he had worked his way on, dragging the locomotive into notice, and raising himself in public

estimation; and at length he had victoriously established the railway system, and went back amongst his
townsmen to receive their greeting.

After the opening of this railway, the project of
the East Coast line from Newcastle to Berwick was
revived; and George Stephenson, who had already
identified himself with the question, and was intimately
acquainted with every foot of the ground, was again
called upon to assist the promoters with his judgment
and experience. He again recommended as strongly
as before the line he had previously surveyed; and on
its being adopted by the local committee, the necessary
steps were taken to have the scheme brought before
Parliament in the ensuing session. The East Coast
line was not, however, to be allowed to pass without a
fight. On the contrary, it had to encounter as stout
an opposition as Stephenson had ever experienced.

We have already stated that about this time the
plan of substituting atmospheric pressure for locomotive
steam-power in the working of railways, had become
very popular. Many eminent engineers avowedly supported atmospheric in preference to locomotive lines;
and there was a strong party in Parliament, headed by
the Prime Minister, who were much disposed in their
favour. Mr. Brunel warmly espoused the atmospheric
principle, and his persuasive manner, as well as his
admitted scientific ability, unquestionably exercised considerable influence in determining the views of many
leading members of both Houses on the subject. Amongst
others, Lord Howick, one of the members for Northumberland, adopted the new principle, and, possessing great
local influence, he succeeded in forming a powerful
confederacy of the landed gentry in favour of Brunel's
atmospheric railway through that county.

George Stephenson could not brook the idea of seeing
the locomotive, for which he had fought so many stout
battles, pushed to one side, and that in the very county

in which its great powers had been first developed. Nor did he relish the appearance of Mr. Brunel as the engineer of Lord Howick's scheme, in opposition to the line which had occupied his thoughts and been the object of his strenuous advocacy for so many years. When Stephenson first met Brunel in Newcastle, he good-naturedly shook him by the collar, and asked " What business he had north of the Tyne ? " George gave him to understand that they were to have a fair stand-up fight for the ground, and, shaking hands before the battle like Englishmen, they parted in good humour. A public meeting was held at Newcastle in the following December, when, after a full discussion of the merits of the respective plans, Stephenson's line was almost una-nimously adopted as the best.

The rival projects went before Parliament in 1845, and a severe contest ensued. The display of ability and tactics on both sides was great. Mr. Hudson and the Messrs. Stephenson were the soul of the struggle for the locomotive line, and Lord Howick and Mr. Brunel in support of the atmospheric system of working. Robert Stephenson was examined at great length as to the merits of the former, and Brunel at equally great length as to the merits of the latter. Mr. Brunel, in the course of his evidence, said that after numerous experiments, he had arrived at the conclusion that the mechanical contrivance of the atmospheric system was perfectly applicable, and he believed that it would likewise be more economical in most cases than locomotive power. " In short," said he, " rapidity, comfort, safety, and economy, are its chief recommendations."

Notwithstanding the promise of Mr. Sergeant Wrang-ham, the counsel for Lord Howick's scheme, that the Northumberland atmospheric was to be " a *respectable* line, and not one that was to be converted into a road for the accommodation of the coal-owners of the district," the locomotive again triumphed. The Stephenson Coast

Line secured the approval of Parliament; and the shareholders in the Atmospheric Company were happily
prevented investing their capital in what would unquestionably have proved a gigantic blunder.   For, less than
three years later, the whole of the atmospheric tubes which
had been laid down on other lines were pulled up, and the
materials sold—including Mr. Brunel's immense tube on
the South Devon Railway[1]—to make way for the working
of the locomotive engine.   George Stephenson's first
verdict of " It won't do," was thus conclusively confirmed.

Robert Stephenson used afterwards to describe with
great gusto an interview which took place between Lord
Howick and his father, at his office in Great George
Street, during the progress of the bill in Parliament.
His father was in the outer office, where he used to
spend a good deal of his spare time; occasionally taking
a quiet wrestle with a friend when nothing else was
stirring.[2]   On the day in question, George was standing
with his back to the fire, when Lord Howick called to
see Robert.   Oh! thought George, he has come to try
and talk Robert over about that atmospheric gimcrack;
but I'll tackle his Lordship.   " Come in, my Lord," said
he, " Robert's busy; but I'll answer your purpose quite
as well; sit down here, if you please."   George began,
" Now, my Lord, I know very well what you have come
about: it's that atmospheric line in the north; I will
show you in less than five minutes that it can never
answer."   " If Mr. Robert Stephenson is not at liberty,
I can call again," said his Lordship.   " He's certainly

---

[1] During the last half-year of the
atmospheric experiment on this line,
in 1848, the expenditure exceeded the
gross income (26,782*l*.) by 2487*l*., or
about 9¾ per cent. excess of working
expenses beyond gross receipts.

[2] " When my father came about
the office," said Robert, " he sometimes did not well know what to do
with himself. So he used to invite
Bidder to have a wrestle with him,
for old acquaintance' sake. And the
two wrestled together so often, and
had so many ' falls ' (sometimes I
thought they would bring the house
down between them), that they broke
half the chairs in my outer office. I
remember once sending my father in
a joiner's bill of about 2*l*. 10*s*. for
mending broken chairs."

occupied on important business just at present," was
George's answer, " but I can tell you far better than he
can what nonsense the atmospheric system is : Robert's
good-natured, you see, and if your Lordship were to get
alongside of him you might talk him over; so you
have been quite lucky in meeting with me.   Now, just
look at the question of expense,"—and then he proceeded in his strong Doric to explain his views in detail,
until Lord Howick could stand it no longer, and he rose
and walked towards the door.   George followed him
down stairs, to finish his demolition of the atmospheric
system, and his parting words were, " You may take
my word for it, my Lord, it will never answer."   George
afterwards told his son with glee of " the settler " he
had given Lord Howick.

So closely were the Stephensons identified with this
measure, and so great was the personal interest which
they were both known to take in its success, that, on
the news of the triumph of the bill reaching Newcastle,
a sort of general holiday took place, and the workmen
belonging to the Stephenson Locomotive Factory, upwards of eight hundred in number, walked in procession
through the principal streets of the town, accompanied
with music and banners.

It is unnecessary to enter into any description of the
works on the Newcastle and Berwick Railway.   There
are no fewer than a hundred and ten bridges of all sorts
on the line—some under and some over it,—the viaducts
over the Ouseburn, the Wansbeck, and the Coquet, being
of considerable importance.   But by far the most formidable piece of masonry work on this railway, is at its
northern extremity, where it passes across the Tweed
into Scotland, immediately opposite the formerly redoubtable castle of Berwick.   Not many centuries had
passed since the district amidst which this bridge stands
was the scene of almost constant warfare.   Berwick was
regarded as the key of Scotland, and was fiercely fought

for, sometimes held by a Scotch and sometimes by an English garrison. Though strongly fortified, it was repeatedly taken by assault. On its capture by Edward I., Boetius says, 17,000 persons were slain, so that its streets "ran with blood like a river." Within sight of the ramparts, a little to the west, is Halidon Hill, where a famous victory was gained by Edward III., over the Scottish army under Douglas; and there is scarcely a foot of ground in the neighbourhood but has been the scene of contention in days long past. In the reigns of James I. and Charles I., a bridge of fifteen arches was built across the Tweed at Berwick; and in our own day a second railway-bridge of twenty-eight arches was built a little above the old one, but at a much higher level. The bridge built by the Kings, out of the national resources, cost 15,000*l.*, and occupied twenty-four years and four months in the building; the bridge built by the Railway Company, with funds drawn from private resources, cost 120,000*l.*, and was finished in three years and four months from the day of laying the foundation stone.

This important viaduct consists of a series of twenty-eight semicircular arches, each 61 feet 6 inches in span, the greatest height above the bed of the river being 126 feet. The whole is built of ashlar, with a hearting of rubble; excepting the river parts of the arches, which are constructed with bricks laid in cement. The total length of the work is 2160 feet. The foundations of the piers were got in by coffer-dams in the ordinary way, Nasmyth's steam-hammer being extensively used in driving the piles. The bearing piles, from which the foundations of the piers were built up, were each capable of carrying 70 tons. The work was designed by Robert Stephenson, and carried out by George Barclay Bruce, who acted as resident engineer.

Another bridge, of still greater importance, necessary to complete the continuity of the East Coast route, was

THE ROYAL BORDER BRIDGE, BERWICK-UPON-TWEED.

[By R. P. Leitch, after his original Drawing.]

the masterwork erected by Robert Stephenson between the north and south banks of the Tyne at Newcastle, commonly known as the High Level Bridge. Mr. R. W. Brandling—to the public spirit and enterprise of whose family the prosperity of Newcastle has been in no small degree indebted, and who first brought to light the strong original genius of George Stephenson in connexion with the safety-lamp—is entitled to the merit of originating the idea of the High Level Bridge, as it was eventually carried out, with a central terminus for the northern railways in the Castle Garth at Newcastle. He first promulgated the plan in 1841; and in the following year it was resolved that Mr. George Stephenson should be consulted as to the most advisable site for the proposed structure. A prospectus of a High Level Bridge Company was issued in 1843, the names of George Stephenson and George Hudson appearing on the committee of management, Mr. Robert Stephenson being the consulting engineer. The project was eventually taken up by the Newcastle and Darlington Railway Company, and an act for the construction of the bridge was obtained in 1845.

The rapid extension of railways had given an extraordinary stimulus to the art of bridge-building; the number of such structures erected in Great Britain alone, since 1830, having been above twenty-five thousand, or more than all that previously existed in the country. Instead of the erection of a single large bridge constituting, as formerly, an epoch in engineering, hundreds of extensive bridges of novel design were simultaneously constructed. The necessity which existed for carrying rigid roads, capable of bearing heavy railway trains at high speeds, over extensive gaps free of support, rendered it apparent that the methods which had up to that time been employed for bridging space were altogether insufficient. The railway engineer could not, like the ordinary road engineer, divert his road, and make choice of the best point for crossing a river or a valley. He must take such ground as lay in the line of his railway, be it bog, or mud, or shifting sand. Navigable rivers and crowded thoroughfares had to be crossed without interruption to the existing traffic, sometimes by bridges at right angles to the river or road, sometimes by arches more or less oblique. In many cases great difficulty arose from the limited nature of the headway; but, as the level of the original road must generally be preserved, and that of the railway was in a measure fixed and determined, it was necessary to modify the form and structure of the bridge, in almost every case, in order to comply with the public requirements. Novel conditions were met by fresh inventions, and difficulties of the most unusual character were one after another successfully surmounted. In executing these extraordinary works, iron has been throughout the sheet-anchor of the engineer. In its various forms of cast and wrought iron, it offered a valuable resource, where rapidity of execution, great strength, and cheapness of construction in the first instance, were elements of prime importance; and by its skilful use, the railway architect was enabled to achieve results which thirty years ago would scarcely have been thought possible.

In many of the early cast-iron bridges the old form of the arch was adopted, the stability of the structure depending wholly on compression, the only novel feature being the use of iron instead of stone. But in a large proportion of cases, the arch, with the railroad over it, was found inapplicable in consequence of the limited headway which it provided. Hence it early occurred to George Stephenson, when constructing the Liverpool and Manchester Railway, to adopt the simple cast-iron beam for the crossing of several roads and canals along that line—this beam resembling in some measure the lintel of the early temples—the pressure on the abutments being purely vertical. One of the earliest instances of this kind of bridge was that erected over Water Street, Manchester, in 1829; after which, cast-iron girders, with their lower webs considerably larger than their upper, were ordinarily employed where the span was moderate; and wrought-iron tie rods below were added to give increased strength where the span was greater.

The next step was the contrivance of arched beams or bowstring girders, firmly held together by horizontal ties to resist the thrust, instead of abutments. Numerous excellent specimens of this description of bridge were erected by Robert Stephenson on the original London and Birmingham Railway; but by far the grandest work of the kind—perfect as a specimen of modern constructive skill—was the High Level Bridge, which we owe to the genius of the same engineer.

The problem was, to throw a railway bridge across the deep ravine which lies between the towns of Newcastle and Gateshead, at the bottom of which flows the navigable river Tyne. Along and up the sides of the valley —on the Newcastle bank especially—run streets of old-fashioned houses, clustered together in the strange forms peculiar to the older cities. The ravine is of great depth—so deep and so gloomy-looking towards dusk, that local tradition records that when the Duke of Cumberland arrived late in the evening, at the brow of the hill overlooking the Tyne, on his way to Culloden, he exclaimed to his attendants, on looking down into the black gorge before him, "For God's sake, don't think of taking me down that coal-pit at this time of night!" The road down the Gateshead High Street is almost as steep as the roof of a house, and up the Newcastle Side, as the street there is called, it is little better. During many centuries the traffic north and south passed along this dangerous and difficult route, over the old bridge which crosses the river in the bottom of the valley. For some thirty years the Newcastle Corporation

had discussed various methods of improving the communication between the towns. Captain Brown, Telford,
and other engineers, were consulted, and the discussion
might have gone on for thirty years more, but for the
advent of railways, when the skill and enterprise to
which they gave birth speedily solved the difficulty, and
bridged the ravine. The locality adroitly took advantage of the opportunity, and insisted on the provision
of a road for ordinary vehicles and foot-passengers in
addition to the railroad. In this circumstance originated
one of the striking peculiarities of the High Level
Bridge, which serves two purposes, being a railway
above and a carriage roadway underneath.

The breadth of the river at the point of crossing is
515 feet, but the length of the bridge and viaduct
between the Gateshead station and the terminus on the
Newcastle side is about 4000 feet. It springs from
Pipewell Gate Bank, on the south, directly across to
Castle Garth, where, nearly fronting the bridge, stands
the fine old Norman keep of the *New* Castle, now
nearly eight hundred years old, and a little beyond it is
the spire of St. Nicholas Church, with its light and
graceful Gothic crown; the whole forming a grand
architectural group of unusual historic interest. The
bridge passes completely over the roofs of the houses
which fill both sides of the valley; and the extraordinary
height of the upper parapet, which is about 130 feet
above the bed of the river,[1] offers a prospect to the

---

[1] Notwithstanding the extraordinary
height of the bridge, it is remarkable
that several persons have thrown
themselves from it into the river beneath, and survived. One tipsy arti-
san, for a wager of a pot of drink,
jumped from the parapet, and was
picked out of the water alive. Another
person afterwards attempted suicide in
the same manner, and was rescued.
But the most singular accident occurred during the construction of the
bridge, when a shipwright, at work
upon the timber platform, stepping
from the permanent to the temporary
work, set his foot upon a loose plank,
which canted over. Accidentally,
however, a huge nail had been driven
—no one knew why—into the end of
a crossbearer, on which the temporary
platform rested; and this nail-head
catching the leg of the man's fustian
trowsers near the lower hem as he fell,
held him suspended, head downwards,

passing traveller the like of which is perhaps nowhere else to be seen. Far below are the queer chares and closes, the wynds and lanes of old Newcastle; the water is crowded with pudgy, black, coal keels; and, when there is a lull in the great smoke volcanos which usually obscure the sky, the funnels of steamers and the masts of the shipping may be seen far down the river. The old bridge lies so far beneath that the passengers crossing it seem like so many bees passing to and fro.

The first difficulty encountered in building the bridge was in securing a solid foundation for the piers. The dimensions of the piles to be driven were so huge, that the engineer found it necessary to employ some extraordinary means for the purpose. He called Nasmyth's Titanic steam-hammer to his aid—the first occasion, we believe, on which this prodigious power was employed in bridge pile-driving.[1] A temporary staging was erected for the steam-engine and hammer apparatus, which rested on two keels, and, notwithstanding the newness and stiffness of the machinery, the first pile was driven on the 6th of October, 1846, to a depth of 32 feet, in four minutes. Two hammers of 30 cwt. each were kept in regular use, making from 60 to 70 strokes per minute; and the results were astounding to those who had been accustomed to the old style of pile-driving by means of the ordinary pile-frame, consisting of slide, ram, and monkey. By the old

---

swinging to and fro, gazing at the river a hundred feet beneath him. The man's comrades ran to his assistance, and placing a ladder from the lower bridge, they with difficulty rescued him from his perilous position. Being a devout Methodist, the shipwright attributed his preservation to the direct interposition of Providence in his behalf. In the course of about a week, however, a tailor's advertisement appeared in the local papers, containing a letter from the rescued workman himself, in which he gave the sole credit to the trowsers by which he had been suspended. On another tailor publishing his claim to the merit of having made them, a controversy between the tailors ensued, which may possibly remain unsettled to this day.

[1] This work was not executed without dismal forebodings on the part of some of the Gateshead people; one of whom, on hearing the pile-driving machine at work on the foundations, was wont to ejaculate, " There goes another nail in the coffin of Gateshead ! "

system, the pile was driven by a comparatively small mass of iron descending with great velocity from a considerable height—the velocity being in excess and the mass deficient, and calculated, like the momentum of a cannon ball, rather for destructive than impulsive action. In the case of the steam pile-driver, on the contrary, the whole weight of a heavy mass is delivered rapidly upon a driving-block of several tons weight placed directly over the head of the pile, the weight never ceasing, and the blows being repeated at the rate of a blow a second, until the pile is driven home. It is a curious fact, that the rapid strokes of the steam-hammer evolved so much heat, that on many occasions the pile-head burst into flames during the process of driving. The elastic force of steam is the power that lifts the ram, the escape permitting its entire force to fall upon the head of the driving block; while the steam above the piston on the upper part of the cylinder, acting as a buffer or recoil-spring, materially enhances the effect of the downward blow. As soon as one pile was driven, the traveller, hovering overhead, presented another, and down it went into the solid bed of the river, with as much ease as a lady sticks pins into a cushion. By the aid of this formidable machine, what was formerly among the most costly and tedious of engineering operations, was rendered simple, easy, and economical.

When the piles had been driven and the coffer-dams formed and puddled, the water within the enclosed space was pumped out by the aid of powerful engines, so as, if possible, to lay bare the bed of the river. Considerable difficulty was experienced in getting in the foundations of the middle pier, in consequence of the water forcing itself through the quicksand beneath as fast as it was removed. This fruitless labour went on for months, and many expedients were tried. Chalk was thrown in in large quantities, outside the piling, but without effect. Cement concrete was at last put within the coffer-dam, until it set, and the bottom was then found to be secure. A bed of concrete was laid up to the level of the heads of the piles, the foundation course of stone blocks being commenced about two feet below low water, and the building proceeded without further difficulty. It may serve to give an idea of the magnitude of the work, when we state that 400,000 cubic feet of ashlar, rubble, and concrete were worked up in the piers, and 450,000 cubic feet in the land-arches and approaches.

The most novel feature of the structure is the use of cast and wrought iron in forming the double bridge, which admirably combines the two principles of the arch and suspension; the railway being carried over the back of the ribbed arches in the usual manner, while the carriage-road and footpaths, forming a long gallery or aisle, are suspended from these arches by wrought-iron vertical rods, with horizontal tie-bars to resist the thrust. The suspension-bolts are enclosed within spandril pillars of cast iron, which give great stiffness to the superstructure. This system of longitudinal and vertical bracing has been much admired, for it not only accomplishes the primary object of securing rigidity in the roadway, but at the same time, by its graceful arrangement, heightens the beauty of the structure. The arches consist of four main ribs, disposed in pairs, with a clear distance between the two inner arches of 20 feet 4 inches, forming the carriage-road, while between each of the inner and outer ribs there is a space of 6 feet 2 inches, constituting the footpaths. Each arch is cast in five separate lengths or segments, strongly bolted together. The ribs spring from horizontal plates of cast iron, bedded and secured on the stone piers. All the abutting joints were carefully executed by machinery, the fitting being of the most perfect kind. In order to provide for the expansion and contraction of the iron arching, and to preserve the equilibrium of the piers

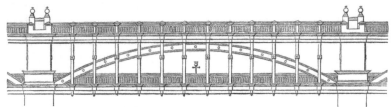

HIGH LEVEL BRIDGE—ELEVATION OF ONE ARCH.

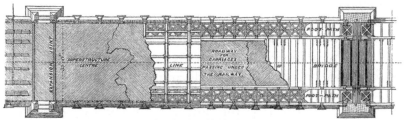

PLAN OF ONE ARCH.

without disturbance or racking of the other parts of the
bridge, it was arranged that the ribs of every two
adjoining arches resting on the same pier should be
secured to the springing-plates by keys and joggles;
whilst on the next piers upon either side the ribs
remained free and were at liberty to expand or contract
according to temperature—a space being left for the
purpose. Hence each arch is complete and independent
in itself, the piers having simply to sustain their vertical
pressure. The arches are six in number, of 125 feet
span each; the two approaches to the bridge being
formed of cast-iron pillars and bearers in keeping with
the arches.

The result is a bridge that for massive solidity may
be pronounced unrivalled. It is perhaps the most
magnificent and striking of all the bridges to which
railways have given birth, and has been worthily styled
" the King of railway structures." It is a monument of
the highest engineering skill of our time, with the
impress of power grandly stamped upon it. It will also
be observed, from Mr. Leitch's masterly drawing, placed
as the frontispiece of this book, that the High Level

Bridge forms a very fine object in a picture of great interest, full of striking architectural variety and beauty. The bridge was opened on the 15th of August, 1849, and a few days after the royal train passed over it, halting for a few minutes to enable her Majesty to survey the wonderful scene below.   In the course of the following year the Queen opened the extensive stone viaduct across the Tweed, above described, by which the last link was completed of the continuous line of railway between London and Edinburgh.   Over the entrance to the Berwick station, occupying the site of the once redoubtable Border fortress, so often the deadly battle-ground of the ancient Scots and English, was erected an arch under which the royal train passed, bearing in large letters of gold the appropriate words, " *The last act of the Union.*"

The warders at Berwick no longer look out from the castle walls to descry the glitter of Southron spears. The bell-tower, from which the alarm was sounded of old, though still standing, is deserted ; the only bell heard within the precincts of the old castle being the railway porter's bell announcing the arrival or the departure of trains.   You see the Scotch Express pass along the bridge and speed southward on the wings of steam. But no alarm spreads along the Border now.   Northumbrian beeves are safe.   Chevy-Chase and Otterburn are quiet sheep pastures.   The only men at arms on the battlements of Alnwick Castle are of stone.   Bamborough Castle has become an asylum for shipwrecked mariners, and the Norman Keep at Newcastle has been converted into a Museum of Antiquities.   The railway has indeed consummated the Union.

CHAPTER XIX.

CHESTER AND HOLYHEAD RAILWAY — MENAI AND CONWAY
BRIDGES.

WE have lastly to describe briefly another great
undertaking, begun by George Stephenson, and taken
up and completed by his son, in the course of which the
latter carried out some of his greatest works—we mean
the Chester and Holyhead Railway, completing the
railway connection with Dublin, as the Newcastle and
Berwick line completed the connection with Edinburgh.
It will thus be seen how closely Telford was followed
by the Stephensons in perfecting the highways of their
respective epochs; the former by means of turnpike
roads, and the latter by means of railways.

George Stephenson surveyed a line from Chester to
Holyhead in 1838, and at the same time reported on the
line through North Wales to Port Dynllaen, proposed
by the Irish Railway Commissioners. His advice was
strongly in favour of adopting the line to Holyhead,
as less costly and presenting better gradients. A public
meeting was held at Chester, in January, 1839, in
support of the latter measure, at which the Marquis of
Westminster, Mr. Wilbraham, and other influential
gentlemen, were present. Mr. Uniacke, the Mayor,
in opening the proceedings, observed, that it clearly
appeared that the rival line through Shrewsbury was
quite impracticable. Mr. Stephenson, he added, was
present in the room, ready to answer any questions
which might be put to him on the subject; and
" it would be better that he should be asked questions
than required to make a speech; for, though a very

good engineer, he was a bad speaker." One of the questions then put to Mr. Stephenson related to the mode by which he proposed to haul the passenger carriages over the Menai Suspension Bridge by horse power; and he was asked whether he knew the pressure the bridge was capable of sustaining. His answer was, that "he had not yet made any calculations; but he proposed getting data which would enable him to arrive at an accurate calculation of the actual strain upon the bridge during the late gale. He had, however, no hesitation in saying that it was more than twenty times as much as the strain of a train of carriages and a locomotive engine. The only reason why he proposed to convey the carriages over by horses, was in order that he might, by distributing the weight, not increase the wavy motion. All the train would be on at once; but distributed. This he thought better than passing them linked together, by a locomotive engine." It will thus be observed that the practicability of throwing a rigid railway bridge across the Straits had not yet been contemplated.

The Dublin Chamber of Commerce passed resolutions in favour of Stephenson's line, after hearing his explanations of its essential features. The project, after undergoing much discussion, was at length embodied in an Act passed in 1844; and the work was brought to a successful completion by his son, with several important modifications, including the grand original feature of the tubular bridges across the Menai Straits and the estuary of the Conway. Excepting these great works, the construction of this line presented no unusual features; though the remarkable terrace cut for the accommodation of the railway under the steep slope of Penmaen Mawr is worthy of a passing notice.

About midway between Conway and Bangor, Penmaen Mawr forms a bold and almost precipitous headland, at

PENMAEN MAWR.   [By Percival Skelton, after his original Drawing.]

the base of which, in rough weather, the ocean dashes
with great fury.   There was not space enough between
the mountain and the strand for the passage of the
railway; hence in some places the rock had to be
blasted to form a terrace, and in others sea walls had
to be built up to the proper level, on which to form
an embankment of sufficient width to enable the road
to be laid. A tunnel 10½ chains in length was cut
through the headland itself; and on its east and west
sides the line was formed by a terrace cut out of the
cliff, and by embankments protected by sea walls; the

terrace being three times interrupted by embankments in its course of about a mile and a quarter. The road lies so close under the steep mountain face, that it was even found necessary at certain places to protect it against possible accidents from falling stones, by means of a covered way. The terrace on the east side of the headland was, however, in some measure protected against the roll of the sea by the mass of stone run out from the tunnel, and forming a deep shingle bank in front of the wall.

The part of the work which lies on the westward of the headland penetrated by the tunnel, was exposed to the full force of the sea; and the formation of the road at that point was attended with great difficulty. While the sea wall was still in progress, its strength was severely tried by a strong north-westerly gale, which blew in October, 1846, with a spring tide of 17 feet. On the following morning it was found that a large portion of the rubble was irreparably injured, and 200 yards of the wall were then replaced by an open viaduct, with the piers placed edgeways to the sea, the openings between them being spanned by ten cast-iron girders each 42 feet long. This accident induced the engineer to alter the contour of the sea wall, so that it should present a diminished resistance to the force of the waves. But the sea repeated its assaults, and made further havoc with the work; entailing heavy expenses and a complete reorganisation of the contract. Increased solidity was then given to the masonry, and the face of the wall underwent further change. At some points outworks were constructed, and piles were driven into the beach about 15 feet from the base of the wall, for the purpose of protecting its foundations and breaking the force of the waves. The work was at length finished after about three years' anxious labour; but Mr. Stephenson confessed that if a long tunnel had been made in the first instance through the solid rock of Penmaen Mawr, a saving of from 25,000*l.* to 30,000*l.* would have been effected.   He also said he had arrived at the conclusion that in railway works engineers should endeavour as far as possible to avoid the necessity of contending with the sea ; [1] but if he were ever again compelled to go within its reach, he would adopt, instead of retaining walls, an open viaduct, placing all the piers edgeways to the force of the sea, and allowing the waves to break upon a natural slope of beach.   He was ready enough to admit the errors he had committed in the original design of this work; but he said he had always gained more information from studying the causes of failures and endeavouring to surmount them, than he had done from easily-won successes.   Whilst many of the latter had been forgotten, the former were indelibly fixed in his memory.

But by far the greatest difficulty which Robert Stephenson had to encounter in executing this railway, was in carrying it across the Straits of Menai and the estuary of the Conway, where, like his predecessor Telford when forming his high road through North Wales, he was under the necessity of resorting to new and altogether untried methods of bridge construction. At Menai the waters of the Irish Sea are perpetually vibrating along the precipitous shores of the Strait ; rising and falling from 20 to 25 feet at each successive tide ; the width and depth of the channel being such as to render it available for navigation by the largest ships. The problem was, to throw a bridge across this wide chasm—a bridge of unusual span and dimensions—of such strength as to be capable of bearing the heaviest loads at high speeds, and at such a uniform height

---

[1] The simple fact that in a heavy storm the force of impact of the waves is from one and a-half to two tons per square foot, must necessarily dictate the greatest possible caution in approaching so formidable an element. Mr. R. Stevenson (Edinburgh) registered a force of three tons per square foot at Skerryvore, during a gale in the Atlantic, when the waves were supposed to run twenty feet high.

throughout as not in any way to interfere with the
navigation of the Strait. From an early period, Mr.
Stephenson had fixed upon the spot where the Britannia
Rock occurs, nearly in the middle of the channel, as the
most eligible point for
crossing ; the water-
width from shore to shore
at high water there being
about 1100 feet.

BRITANNIA BRIDGE

The engineer's first
idea was to construct the
bridge of two cast iron
arches, each of 350 feet
span. There was no
novelty in this idea ; for,
as early as the year 1801,
Mr. Rennie prepared a
design of a cast-iron
bridge across the Strait
at the Swilly rocks, the
great centre arch of
which was to be 450 feet
span ; and at a later
period, in 1810, Telford
submitted a design of a
similar bridge at Inys-y-
Moch, with a single cast-iron arch of 500 feet. But
the same objections which led to the rejection of Rennie's and Telford's designs, proved fatal to Robert
Stephenson's, and his iron-arched railway bridge was
rejected by the Admiralty. The navigation of the
Strait was under no circumstances to be interfered
with ; and even the erection of scaffolding from below,
to support the bridge during construction, was not to
be permitted. The idea of a suspension bridge was
dismissed as inapplicable ; a degree of rigidity and
strength, greater than could be secured by any bridge

constructed on the principle of suspension, being considered indispensable conditions of the proposed structure.

Mr. Stephenson next considered the expediency of erecting a bridge by means of suspended centering, after the ingenious method proposed by Telford in 1810;[1] by which the arching was to be carried out by placing equal and corresponding voussoirs on opposite sides of the pier at the same time, tying them together by horizontal tie-bolts. The arching thus extended outwards from each pier and held in equilibrium, would have been connected at the crown with the extremity of the arch advanced in like manner from the adjoining pier. It was, however, found that this method of construction was not applicable at the Conway; and it was eventually abandoned. Various other plans were suggested; but the whole question remained unsettled even down to the time when the Company went before Parliament, in 1844, for power to construct the proposed bridges. No existing kind of structure seemed to be capable of bearing the fearful extension to which rigid bridges of the necessary spans would be subjected; and some new expedient of engineering therefore became necessary.

Mr. Stephenson was then led to reconsider a design which he had made in 1841 for a road bridge over the river Lea at Ware, with a span of 50 feet,—the conditions only admitting of a platform 18 or 20 inches thick. For this purpose a wrought-iron platform was designed, consisting of a series of simple cells, formed of boiler-plates riveted together with angle-iron. The bridge was not, however, carried out after this design, but was made of separate wrought-iron girders composed of riveted wrought-iron plates.[2] Recurring to his first

---

[1] See 'Lives of the Engineers,' vol. ii. p. 445. It appears that Mr. Fairbairn suggested this idea in his letter to Mr. Stephenson, dated the 3rd June, 1845, accompanied by a draw-

ing. See his 'Account of the Construction of the Britannia and Conway Tubular Bridges, &c.' London, 1849.

[2] Robert Stephenson's narrative of the early history of the design, in

idea of this bridge, Mr. Stephenson thought that a stiff platform might be constructed, with sides of strongly trussed frame-work of wrought-iron, braced together at top and bottom with plates of like material riveted together with angle-iron, after a method adopted by Mr. Rendel in stiffening the suspension bridge at Montrose with wooden trellis-work a few years before; and that such platform might be suspended by strong chains on either side to give it increased security. " It was now," says Mr. Stephenson, " that I came to regard the tubular platform as a beam, and that the chains should be looked upon as auxiliaries." It appeared to him, nevertheless, that without a system of diagonal struts inside, which of course would have prevented the passage of trains *through* it, this kind of structure was ill-suited for maintaining its form, and would be very liable to become lozenge-shaped. Besides, the rectangular figure was deemed objectionable, from the large surface which it presented to the wind.

It then occurred to him that circular or elliptical tubes might better answer the intended purpose; and in March, 1845, he gave instructions to two of his assistants to prepare drawings of such a structure, the tubes being made with a double thickness of plate at top and bottom. The results of the calculations made as to the strength of such a tube, were considered so satisfactory, that Mr. Stephenson says he determined to fall back on a bridge of this description, on the rejection of his design of the two cast-iron arches by the Parliamentary Committee. Indeed, it became evident that a tubular wrought-iron beam was the only structure which combined the necessary strength and stability for a railway, with the conditions deemed

Edwin Clark's ' Britannia and Conway Tubular Bridges,' vol. i. p. 25. London, 1850.

essential for the protection of the navigation. " I stood," says Mr. Stephenson, " on the verge of a responsibility from which, I confess, I had nearly shrunk. The construction of a tubular beam of such gigantic dimensions, on a platform elevated and supported by chains at such a height, did at first present itself as a difficulty of a very formidable nature. Reflection, however, satisfied me that the principles upon which the idea was founded were nothing more. than an extension of those daily in use in the profession of the engineer. The method, moreover, of calculating the strength of the structure which I had adopted, was of the simplest and most elementary character; and whatever might be the form of the tube, the principle on which the calculations were founded was equally applicable, and could not fail to lead to equally accurate results." [1] Mr. Stephenson accordingly announced to the directors of the railway that he was prepared to carry out a bridge of this general description, and they adopted his views, though not without considerable misgivings.

While the engineer's mind was still occupied with the subject, an accident occurred to the *Prince of Wales* iron steamship, at Blackwall, which singularly corroborated his views as to the strength of wrought-iron beams of large dimensions. When this vessel was being launched, the cleet on the bow gave way, in consequence of the bolts breaking, and let the vessel down so that the bilge came in contact with the wharf, and she remained suspended between the water and the wharf for a length of about 110 feet, but without any injury to the plates of the ship; satisfactorily proving the great strength of this form of construction. Thus, Mr. Stephenson became gradually confirmed in his opinion that the most

---

[1] Robert Stephenson's narrative in Clark's ' Britannia and Conway Tubular Bridges,' vol. i. p. 27.

feasible method of bridging the strait at Menai and the river at Conway was by means of a hollow beam of wrought iron.   As the time was approaching for giving evidence before Parliament on the subject, it was necessary for him to settle some definite plan for submission to the committee.   " My late revered father," says he, " having always taken a deep interest in the various proposals which had been considered for carrying a railway across the Menai Straits, requested me to explain fully to him the views which led me to suggest the use of a tube, and also the nature of the calculations I had made in reference to it.   It was during this personal conference that Mr. William Fairbairn accidentally called upon me, to whom I also explained the principles of the structure I had proposed.   He at once acquiesced in their truth, and expressed confidence in the feasibility of my project, giving me at the same time some facts relative to the remarkable strength of iron steamships, and invited me to his works at Millwall, to examine the construction of an iron steamship which was then in progress."   The date of this consultation was early in April, 1845, and Mr. Fairbairn states that, on that occasion, " Mr. Stephenson asked whether such a design was practicable, and whether I could accomplish it : and it was ultimately arranged that the subject should be investigated experimentally, to determine not only the value of Mr. Stephenson's original conception (of a circular or egg-shaped wrought-iron tube, supported by chains), but that of any other tubular form of bridge which might present itself in the prosecution of my researches.   The matter was placed unreservedly in my hands ; the entire conduct of the investigation was entrusted to me ; and, as an experimenter, I was to be left free to exercise my own discretion in the investigation of whatever forms or conditions of the structure might appear to me best calculated to secure a safe

passage across the Straits." [1]   Mr. Fairbairn then proceeded to construct a number of experimental models
for the purpose of testing the strength of tubes of
different forms.   The short period which elapsed, however, before the bill was in committee, did not admit of
much progress being made with those experiments ; but
from the evidence in chief given by Mr. Stephenson on
the subject, on the 5th of May following, it appears that
the idea which prevailed in his mind was that of a
bridge with openings of 450 feet (afterwards increased
to 460 feet) ; with a roadway formed of a hollow
wrought-iron beam, about 25 feet in diameter, presenting a rigid platform, suspended by chains.   At the
same time, he expressed the confident opinion that a
tube of wrought iron would possess sufficient strength
and rigidity to support a railway train running inside
of it without the help of the chains.

While the bill was still in progress, Mr. Fairbairn
proceeded with his experiments.   He first tested tubes
of a cylindrical form, in consequence of the favourable
opinion entertained by Mr. Stephenson of tubes in that
shape, extending them subsequently to those of an elliptical form.[2]   He found tubes thus shaped more or less
defective, and proceeded to test those of a rectangular
kind.   After the bill had received the royal assent on
the 30th of June, 1845, the directors of the company,
with great liberality, voted a sum for the purpose of
enabling the experiments to be prosecuted, and upwards
of 6000l. were thus expended to make the assurance
of their engineer doubly sure.   Mr. Fairbairn's tests

---

[1] 'Account of the Construction of
the Britannia and Conway Tubular
Bridges.' By W. Fairbairn, C.E. London, 1849.

[2] Mr. Stephenson continued to hold
that the elliptical tube was the right
idea, and that sufficient justice had
not been done to it.   A year or two
before his death Mr. Stephenson remarked to the author, that had the
same arrangement for stiffening been
adopted to which the oblong rectangular tubes owe a great part of their
strength, a very different result would
have been obtained.

were of the most elaborate and eventually conclusive
character, bringing to light many new and important
facts of great practical value.   The due proportions and
thicknesses of the top, bottom, and sides of the tubes
were arrived at after a vast number of separate trials ;
one of the results of the experiments being the adoption
of Mr. Fairbairn's invention of rectangular hollow cells
in the top of the beam for the purpose of giving it the
requisite degree of strength.   About the end of August
it was thought desirable to obtain the assistance of a
mathematician, who should prepare a formula by which
the strength of a full-sized tube might be calculated
from the results of the experiments made with tubes of
smaller dimensions.   Professor Hodgkinson was accordingly called in, and he proceeded to verify and confirm the
experiments which Mr. Fairbairn had made, and afterwards reduced them to the required formulæ ; though
Mr. Fairbairn states that they did not appear in time to
be of any practical service in proportioning the parts of
the largest tubes.[1]

Mr. Stephenson's time was so much engrossed with
his extensive engineering business that he was in a
great measure precluded from devoting himself to the
consideration of the practical details, which he felt
were safe in the hands of Mr. Fairbairn—" a gentleman," as he stated to the committee of the Commons, " whose experience was greater than that of any
other man in England."   The results of the experiments were communicated to him from time to time,
and were regarded by him as exceedingly satisfactory
It would appear, however, that while Mr. Fairbairn
urged the sufficient rigidity and strength of the tubes
without the aid of chains, Mr. Stephenson had not quite
made up his mind upon the point.   Mr. Hodgkinson, also,

---

[1] Fairbairn's ' Account,' p. 22.

was strongly inclined to retain them.[1] Mr. Fairbairn
held that it was quite practicable to make the tubes
" sufficiently strong to sustain not only their own weight,
but, in addition to that load, 2000 tons equally distributed over the surface of the platform,—a load ten times
greater than they will ever be called upon to support."

It was thoroughly characteristic of Mr. Stephenson,
and of the caution with which he proceeded in every
step of this great undertaking—probing every inch of
the ground before he set down his foot upon it—that he
should, early in 1846, have appointed his able assistant,
Mr. Edwin Clark, to scrutinise carefully the results of
every experiment, whether made by Mr. Fairbairn or
Mr. Hodgkinson, and subject them to a separate and
independent analysis before finally deciding upon the
form or dimensions of the structure, or upon any mode
of procedure connected with it. That great progress
had been made by the two chief experimenters before
the end of 1846, appears from the papers read by Messrs.
Fairbairn and Hodgkinson before the British Association
at Southampton in September of that year. In the
course of the following month Mr. Stephenson had become fully satisfied that the use of auxiliary chains was

---

[1] The following passage occurs in
Robert Stephenson's Report to the directors of the Chester and Holyhead
Railway, dated the 9th February,
1846 :—" You will observe in Mr.
Fairbairn's remarks, that he contem-
plates the feasibility of stripping the
tube entirely of all the chains that
may be required in the erection of the
bridge; whereas, on the other hand,
Mr. Hodgkinson thinks the chains
will be an essential, or at all events a
useful auxiliary, to give the tube the
requisite strength and rigidity. This,
however, will be determined by the
proposed additional experiments, and
does not interfere with the construction
of the masonry, which is designed so
as to admit of the tube, with or without chains. The application of chains
as an auxiliary has occupied much of
my attention, and I am satisfied that
the ordinary mode of applying them
to suspension bridges is wholly inadmissible in the present instance; if,
therefore, it be hereafter found necessary or desirable to employ them in
conjunction with the tube, another
mode of employing them must be devised, as it is absolutely essential to
attach them in such a manner as to
preclude the possibility of the smallest
oscillation. In the accomplishment
of this I see no difficulty whatever;
and the designs have been arranged
accordingly, in order to avoid any
further delay."

unnecessary, and that the tubular bridge might be made of such strength as to be entirely self-supporting.[1]

While these important discussions were in progress, measures were taken to proceed with the masonry of the bridges simultaneously at Conway and the Menai Strait. The foundation-stone of the Britannia Bridge was laid by Mr. Frank Forster, the resident engineer, on the 10th of April, 1846; and on the 12th of May following that of the Conway Bridge was laid by Mr. A. M. Ross, resident engineer at that part of the works. Suitable platforms and workshops were also erected for proceeding with the punching, fitting, and riveting of the tubes; and when these operations were in full progress, the neighbourhood of the Conway and Britannia Bridges presented scenes of extraordinary bustle and industry. On the 11th of July, 1847, Mr. Clark informs Mr. Stephenson that "the masonry gets on rapidly. The abutments on the Anglesey side resemble the foundations of a great city rather than of a single structure, and nothing appears to stand still here." About 1500 men were employed on the Britannia Bridge alone, and they mostly lived upon the ground in wooden cottages erected for the occasion. The iron plates were brought in ship-loads from Liverpool, Anglesey marble from Penmon, and red sandstone from Runcorn, in Cheshire, as wind and tide, and shipping and convenience, might determine. There was an unremitting clank of hammers, grinding of machinery, and blasting of rock, going

---

[1] In a letter of Mr. Fairbairn to Mr. Stephenson, dated July 18th, 1846, he says:—" To get rid of the chains will be a desideratum; and I have made the tube of such strength, and intend putting it together upon such a principle, as will insure its carrying a dead weight, equally distributed over its hollow surface, of 4000 tons. With a bridge of such powers, what have we to fear? and why, in the name of truth and in the face of conclusive facts, should we hesitate to adopt measures, calculated not only to establish the principle as a triumph of art, but what is of infinitely more importance to the shareholders, the saving of a large sum of money, nearly equal to half the cost of the bridge? I have been ably assisted by Mr. Clark in all these contrivances; but in a matter of such importance we must have your sanction and support."—Mr. Fairbairn's ' Account,' p. 93.

on from morning till night. In fitting the Britannia
tubes together, not less than 2,000,000 of bolts were
riveted, weighing some 900 tons.

The Britannia Bridge consists of two independent
continuous tubular beams, each 1511 feet in length, and
each weighing 4680 tons, independent of the cast-iron
frames inserted at their bearings on the masonry of the
towers. These immense beams are supported at five
places, namely, on the abutments and on three towers,
the central of which is known as the Great Britannia
Tower, 230 feet high, built on a rock in the middle of the
Strait. The side towers are 18 feet less in height than
the central one, and the abutments 35 feet lower than
the side towers. The design of the masonry is such as
to accord with the form of the tubes, being somewhat
of an Egyptian character, massive and gigantic rather
than beautiful, but bearing the unmistakable impress of
power.

The bridge has four spans,—two of 460 feet over the
water, and two of 230 feet over the land. The weight
of the longer spans, at the points where the tubes repose
on the masonry, is not less than 1587 tons. On the
centre tower the tubes lie solid; but on the land towers
and abutments they lie on roller-beds, so as to allow of
expansion and contraction. The road within each tube
is 15 feet wide, and the height varies from 23 feet at
the ends to 30 feet at the centre. To give an idea of
the vast size of the tubes by comparison with other
structures, it may be mentioned that each length constituting the main spans is twice as long as London
Monument is high; and if it could be set on end in
St. Paul's Churchyard, it would reach nearly 100 feet
above the cross.

The Conway Bridge is, in most respects, similar to
the Britannia, consisting of two tubes, of 400 feet span,
placed side by side, each weighing 1180 tons. The
principle adopted in the construction of the tubes, and

CONSTRUCTION OF MAIN BRITANNIA TUBE ON THE STAGING

the mode of floating and raising them, were nearly the
same as at the Britannia Bridge, though the general
arrangement of the plates is in many respects different.

It was determined to construct the shorter outer tubes
of the Britannia Bridge on scaffoldings in the positions
in which they were permanently to remain, and to erect
the larger tubes upon wooden platforms at high-water-
mark on the Caernarvon shore, from whence they were
to be floated in pontoons,—in like manner as Rennie had
floated into their places the centerings of his Waterloo
and other bridges,—and then raised into their proper
places by means of hydraulic power, after a method originally suggested by Mr. Edwin Clark, to whose valuable

history of the construction of the Britannia and Conway
Bridges we would refer the reader for full details
as to the methods of construction employed in these
extraordinary works.

The floating of the tubes on pontoons, from the places
where they had been constructed to the recesses in the
masonry of the towers, up which they were to be hoisted
to the positions they were permanently to occupy, was
an anxious and exciting operation. The first part of this
process was performed at Conway, where Mr. Stephenson directed it in person, assisted by Captain Claxton,
Mr. Brunel, and other engineering friends. On the 6th
March, 1848, the pontoons bearing the first great tube
of the up-line were floated round quietly and majestically into their place between the towers in about twenty
minutes. Unfortunately, one of the sets of pontoons
had become slightly slued by the stream, by which the
Conway end of the tube was prevented from being
brought home; and five anxious days to all concerned
intervened before it could be set in its place. In
the mean time, the presses and raising machinery had
been fitted in the towers above, and the lifting process
was begun on the 8th of April, when the immense mass
was raised 8 feet, at the rate of about 2 inches a minute.
On the 16th, the tube had been raised and finally
lowered into its permanent bed; the rails were laid
through it; and, on the 18th, Mr. Stephenson passed
through with the first locomotive. The second tube was
proceeded with on the removal of the first from the
platform, and was completed and floated in seven
months. The rapidity with which this second tube was
constructed was in no small degree owing to the Jac-
quard punching-machine, contrived for the purpose by
Mr. Roberts, of Manchester. This tube was finally fixed
in its permanent bed on the 2nd of January, 1849.

The floating and fixing of the great Britannia tubes
was a still more formidable enterprise, though the ex-

CONWAY BRIDGE. [By Percival Skelton.]

perience gained at Conway rendered it easy compared
with what it otherwise would have been.   Mr. Stephenson superintended the operation of floating the first in
person, giving the arranged signals from the top of the
tube on which he was mounted, the active part of the
business being performed by a numerous corps of sailors,
under the immediate direction of Captain Claxton.
Thousands of spectators lined the shores of the Strait
on the evening of the 19th of June, 1849.   On the land
attachments being cut, the pontoons began to float off;
but one of the capstans having given way from the too
great strain put upon it, the tube was brought home
again for the night.   By next morning the defective
capstan was restored, and all was in readiness for another
trial.   At half-past seven in the evening the tube was

afloat, and the pontoons swung out into the current like
a monster pendulum, held steady by the shore guide-
lines, but increasing in speed to almost a fearful extent
as they neared their destined place between the piers.
" The success of this operation," says Mr. Clark, " depended mainly on properly striking the ' butt' beneath
the Anglesey tower, on which, as upon a centre, the
tube was to be veered round into its position across the
opening. This position was determined by a 12-inch
line, which was to be paid out to a fixed mark from the
Llanfair capstan. The coils of the rope unfortunately
over-rode each other upon this capstan, so that it could
not be paid out. In resisting the motion of the tube,
the capstan was bodily dragged out of the platform by
the action of the palls, and the tube was in imminent
danger of being carried away by the stream, or the
pontoons crushed upon the rocks. The men at the
capstan were all knocked down, and some of them
thrown into the water, though they made every exertion
to arrest the motion of the capstan-bars. In this dilemma
Mr. Charles Rolfe, who had charge of the capstan, with
great presence of mind, called the visitors on shore to
his assistance; and handing out the spare coil of the
12-inch line into the field at the back of the capstan, it
was carried with great rapidity up the field, and a crowd
of people, men, women, and children, holding on to this
huge cable, arrested the progress of the tube, which was
at length brought safely against the butt and veered
round. The Britannia end was then drawn into the
recess of the masonry by a chain passing through the
tower to a crab on the far side. The violence of the
tide abated, though the wind increased, and the Anglesey
end was drawn into its place beneath the corbelling in
the masonry; and as the tide went down, the pontoons
deposited their valuable cargo on the welcome shelf at
each end. The successful issue was greeted by cannon
from the shore and the hearty cheers of many thousands

of spectators, whose sympathy and anxiety were but too clearly indicated by the unbroken silence with which the whole operation had been accompanied." [1]   By midnight all the pontoons had been got clear of the tube, which now hung suspended over the waters of the Strait by its two ends, which rested upon the edges cut in the rock for the purpose at the base of the Britannia and Anglesey towers respectively, up which the tube had now to be lifted by hydraulic power to its permanent place near the summit.   The accuracy with which the gigantic beam had been constructed may be inferred from the fact that, after passing into its place, a clear space remained between the iron plating and the rock outside of it of only about three-quarters of an inch!

Mr. Stephenson's anxiety was, of course, very great up to the time of performing this trying operation. When he had got the first tube floated at Conway, and saw all safe, he said to Captain Moorsom, "Now I shall go to bed." But the Britannia Bridge was a still more difficult enterprise, and cost him many a sleepless night. Afterwards describing his feelings to his friend Mr. Gooch, he said: "It was a most anxious and harassing time with me. Often at night I would lie tossing about, seeking sleep in vain. The tubes filled my head. I went to bed with them and got up with them. In the grey of the morning, when I looked across the Square, [2] it seemed an immense distance across to the houses on the opposite side. It was nearly the same length as the span of my tubular bridge!" When the first tube had been floated, a friend observed to him, "This great work has made you ten years older." "I have not slept sound," he replied, "for three weeks." Sir F Head, however, relates, that when he revisited the spot on the following morning, he observed, sitting on a

---

[1] 'The Britannia and Conway Tubular Bridges.' By Edwin Clark. Vol. II. p. 683-4.

[2] No. 34, Gloucester Square, Hyde Park, where he lived.

platform overlooking the suspended tube, a gentleman,
reclining entirely by himself, smoking a cigar, and
gazing, as if indolently, at the aërial gallery beneath
him.   It was the engineer himself, contemplating his
new-born child.   He had strolled down from the neighbouring village, after his first sound and refreshing
sleep for weeks, to behold in sunshine and solitude, that
which during a weary period of gestation had been
either mysteriously moving in his brain, or, like a vision
  sometimes of good omen, and sometimes of evil—had,
by night as well as by day, been flitting across his mind.

  The next process was the lifting of the tube into its
place, which was performed very deliberately and cautiously.   It was raised by powerful hydraulic presses,
only a few feet at a time, and carefully under-built, before
being raised to a farther height.   When it had been got
up by successive stages of this kind to about 24 feet, an
extraordinary accident occurred, during Mr. Stephenson's
absence in London, which he afterwards described to the
author in as nearly as possible the following words :—
" In a work of such novelty and magnitude, you may
readily imagine how anxious I was that every possible
contingency should be provided for.   Where one chain
or rope was required, I provided two.   I was not satisfied
with ' enough :' I must have absolute security, as far as
that was possible.   I knew the consequences of failure
would be most disastrous to the Company, and that the
wisest economy was to provide for all contingencies at
whatever cost.   When the first tube at the Britannia
had been successfully floated between the piers ready for
being raised, my young engineers were very much
elated ; and when the hoisting apparatus had been fixed,
they wrote to me, saying,—' We are now all ready for
raising her : we could do it in a day, or in two at the
most.' But my reply was, ' No : you must only raise the
tube inch by inch, and you must build up under it as you
rise.   Every inch must be made good.   Nothing must

be left to chance or good luck.' And fortunate it was
that I insisted upon this cautious course being pursued;
for, one day, while the hydraulic presses were at work,
the bottom of one of them burst clean away! The
crosshead and the chains, weighing more than 50 tons,
descended with a fearful crash upon the press, and the
tube itself fell down upon the packing beneath. Though
the fall of the tube was not more than nine inches, it
crunched solid castings, weighing tons, as if they had
been nuts. The tube itself was slightly strained and
deflected, though it still remained sufficiently serviceable.
But it was a tremendous test to which it was put, for a
weight of upwards of 5000 tons falling even a few
inches must be admitted to be a very serious matter.
That it stood so well was extraordinary. Clark immediately wrote me an account of the circumstance, in
which he said, ' Thank God, you have been so obstinate.
For if this accident had occurred without a bed for the
end of the tube to fall on, the whole would now have
been lying across the bottom of the Straits.' Five
thousand pounds extra expense was caused by this
accident, slight though it might seem. But careful
provision was made against future failure; a new and
improved cylinder was provided; and the work was
very soon advancing satisfactorily towards completion."[1]

---

[1] The hydraulic-presses were of an
extraordinary character. The cylinders of those first constructed were of
wrought-iron (cast-iron being found
altogether useless), not less than 8
inches thick. They were tested by
being subjected to an internal pressure
of 3 or 3¼ tons to the circular inch.
The pressure was such that it squeezed
the fibres of the iron together; so that
after a few tests of this kind the
piston, which at first fitted it quite
closely, was found considerably too
small. " A new piston," says Mr.
Clark, " was then made to suit the enlarged cylinder; and a further enlargement occurring again and again with
subsequent use, the new pistons became as formidable an obstacle as the
cylinders. The wrought-iron cylinder was on the point of being abandoned, when Mr. Amos (the iron
manufacturer), having carefully gauged
the cylinder inside and out, found to
his surprise, that although the internal diameter had increased considerably, the external diameter had retained precisely its original dimensions. He consequently persevered in
the construction of new pistons; and
ultimately found that the cylinder
enlarged no longer, and to this day
it continues in constant use. Layer
after layer having attained additional

When the Queen first visited the Britannia Bridge, on her return from the North in 1852, Robert Stephenson accompanied Her Majesty and Prince Albert over the works, explaining the principles on which the bridge had been built, and the difficulties which had attended its erection. He conducted the Royal party to near the margin of the sea, and, after describing to them the incident of the fall of the tube, and the reason of its preservation, he pointed with pardonable pride to a pile of stones which the workmen had there raised to commemorate the event. While nearly all the other marks of the work during its progress had been obliterated, that cairn had been left standing in commemoration of the caution and foresight of their chief.

The floating and raising of the remaining tubes need not be described in detail. The second was floated on the 3rd December, and set in its permanent place on the 7th January, 1850. The others[1] were floated and raised

---

permanent set, sufficient material was at length brought into play, with sufficient tenacity to withstand the pressure; and thus an obstacle, apparently insurmountable, and which threatened at one time to render much valuable machinery useless, was entirely overcome. The workman may be excused for calling the stretched cylinder stronger than the new one, though it is only stronger as regards the amount of its yielding to a given force."—Clark, vol. I. 306. The hydraulic-presses used in raising the tubes of the Britannia Bridge, it may be remembered, were afterwards used in starting the *Great Eastern* from her berth on the shore at Milwall where she had been built.

[1] While the preparations were in progress for floating the third tube, Mr. Stephenson received a pressing invitation to a public railway celebration at Darlington, in honour of his old friend Edward Pease. His reply, dated the 15th May, 1850, was as follows:—" I am prevented having the pleasure of a visit to Darlington, on the 22nd, owing to that or the fol-

lowing day having been fixed upon for floating the next tube at the Menai Straits; and as this movement depends on the tide, it is, of course, impossible for me to alter the arrangements. I sincerely regret this circumstance, for every early association connected with my profession, would have tended to render my visit a gratifying one. It would, moreover, have given me an opportunity of saying publicly how much the wonderful progress of railways was dependent upon the successful issue of the first great experiment, and how much that issue was influenced by your great discernment, and your confidence in my late revered father. In my remembrance you stand amongst the foremost of his patrons and early advisers; and I know that throughout his life he regarded you as one of his very best friends. One of the things in which he took especial delight, was in frequently and very graphically describing his first visit to Darlington, on foot, to confer with you on the subject of the Stockton and Darlington Railway."

in due course.   On the 5th of March, Mr. Stephenson
put the last rivet in the last tube, and passed through
the completed bridge, accompanied by about a thousand
persons, drawn by three locomotives.   The bridge was
opened for public traffic on the 18th of March.   The
cost of the whole work was 234,450*l.*

BRITANNIA BRIDGE. [By Percival Skelton, after his original Drawing.]

The Britannia Bridge is one of the most remarkable
monuments of the enterprise and skill of the present
century   Robert Stephenson was the master spirit of
the undertaking.   To him belongs the merit of first
seizing the ideal conception of the structure best adapted
to meet the necessities of the case ; and of selecting the

best men to work out his idea, himself watching, controlling, and testing every result, by independent check and counter-check.   And finally, he organised and directed, through his assistants, the vast band of skilled workmen and labourers who were for so many years occupied in carrying his magnificent original conception to a successful practical issue.   As he himself said of the work, —" The true and accurate calculation of all the conditions and elements essential to the safety of the bridge had been a source not only of mental but of bodily toil; including, as it did, a combination of abstract thought and well-considered experiment adequate to the magnitude of the project."

The Britannia Bridge was the result of a vast combination of skill and industry.   But for the perfection of our tools and the ability of our mechanics to use them to the greatest advantage; but for the matured powers of the steam-engine; but for the improvements in the iron manufacture, which enabled blooms to be puddled of sizes before deemed impracticable, and plates and bars of immense size to be rolled and forged; but for these, the Britannia Bridge would have been designed in vain.   Thus, it was not the product of the genius of the railway engineer alone, but of the collective mechanical genius of the English nation.

CONWAY BRIDGE—FLOATING THE FIRST TUBE.

VIEW IN TAPTON GARDENS   [By Percival Skelton.]

CHAPTER XX.

Closing Years of George Stephenson's Life — Illness and
Death — Character — Death of Robert Stephenson.

In describing the completion of the series of great works
detailed in the preceding chapter, we have somewhat
anticipated the closing years of George Stephenson's life.
He could not fail to take an anxious interest in the success of his son's designs, and he accordingly paid many
visits to Conway and to Menai, during the progress of the
works.   He was present on the occasion of the floating
and raising of the first Conway tube, and there witnessed
a clear proof of the soundness of Robert's judgment as
to the efficiency and strength of the tubular bridge, of
which he had at first experienced some doubts; but
before the like test could be applied at the Britannia
Bridge, George Stephenson's mortal anxieties were at
an end, for he had then ceased from all his labours.

Towards the close of his life, George Stephenson almost
entirely withdrew from the active pursuit of his profession as an engineer.   He devoted himself chiefly to his
extensive collieries and lime-works, taking a local in-

terest only in such projected railways as were calculated
to open up new markets for their products.

At home he lived the life of a country· gentleman,
enjoying his garden and grounds, and indulging his
love of nature, which, through all his busy life, had
never left him. It was not until the year 1845 that
he took an active interest in horticultural pursuits.
Then he began to build new melon-houses, pineries,
and vineries, of great extent; and he now seemed as
eager to excel all other growers of exotic plants in his
neighbourhood, as he had been to surpass the villagers
of Killingworth in the production of gigantic cabbages
and cauliflowers some thirty years before. He had
a pine-house built sixty-eight feet in length and a
pinery a hundred and forty feet. Workmen were constantly employed in enlarging them, until at length he
had no fewer than ten glass forcing-houses, heated with
hot water, which he was one of the first in that neighbourhood to make use of for such a purpose. He did
not take so much pleasure in flowers as in fruits. At
one of the county agricultural meetings, he said that
he intended yet to grow pine-apples at Tapton as big
as pumpkins. The only man to whom he would " knock
under " was his friend Paxton, the gardener to the
Duke of Devonshire; and he was so old in the service,
and so skilful, that he could scarcely hope to beat him.
Yet his " Queen " pines did take the first prize at a
competition with the Duke,—though this was not until
shortly after his death, when the plants had become
more fully grown. His grapes also took the first prize
at Rotherham, at a competition open to all England. He
was extremely successful in producing melons, having
invented a method of suspending· them in baskets of
wire gauze, which, by relieving the stalk from tension,
allowed nutrition to proceed more freely, and better
enabled the fruit to grow and ripen. Amongst his
other erections, he built a joiner's shop, where he kept

a workman regularly employed in carrying out his many ingenious contrivances of this sort.

He took much pride also in his growth of cucumbers. He raised them very fine and large, but he could not make them grow straight. Place them as he would, notwithstanding all his propping of them, and humouring them by modifying the application of heat and the admission of light for the purpose of effecting his object, they would still insist on growing crooked in their own way. At last he had a number of glass cylinders made at Newcastle, for the purpose of an experiment; into these the growing cucumbers were inserted, and then he succeeded in growing them perfectly straight. Carrying one of the new products into his house one day, and exhibiting it to a party of visitors, he told them of the expedient he had adopted, and added gleefully, " I think I have bothered them noo !"

Mr. Stephenson also carried on farming operations with some success. He experimented on manure, and fed cattle after methods of his own. He was very particular as to breed and build in stock-breeding. " You see, sir," he said to one gentleman, " I like to see the *coo's* back at a gradient something like this" (drawing an imaginary line with his hand), " and then the ribs or girders will carry more flesh than if they were so—or so." When he attended the county agricultural meetings, which he frequently did, he was accustomed to take part in the discussions, and he brought the same vigorous practical mind to bear upon questions of tillage, drainage, and farm economy, which he had been accustomed to exercise on mechanical and engineering matters. At one of the meetings of the North Derbyshire Agricultural Society, he favoured the assembled farmers with an explanation of his theory of vegetation. The practical conclusion to which it led was, that the agriculturist ought to give as much light and heat to the soil as possible. At the same time he stated his opinion that, in some cold soils, water contributed to promote vegetation, rather than to impede it, as was generally believed; for the water, being exposed to the sun and atmosphere, became specifically warmer than the earth it covered, and when it afterwards irrigated the fields, it communicated this additional heat to the soil which it permeated.

All his early affection for birds and animals revived. He had favourite dogs, and cows, and horses; and again he began to keep rabbits, and to pride himself on the beauty of his breed. There was not a bird's nest upon the grounds that he did not know of; and from day to day he went round watching the progress which the birds made with their building, carefully guarding them from injury. No one was more minutely acquainted with the habits of British birds, the result of a long, loving, and close observation of nature.

At Tapton he remembered the failure of his early experiment in hatching birds' eggs by heat, and he now performed it successfully, being able to secure a proper apparatus for maintaining a uniform temperature. He was also curious about the breeding and fattening of fowls; and when his friend Edward Pease of Darlington visited him at Tapton, he explained a method which he had invented for fattening chickens in half the usual time. The chickens were shut up in boxes, which were so made as to exclude the light. Dividing the day into two or three periods, the birds were shut up at the end of each after a heavy feed, and went to sleep. The plan proved very successful, and Mr. Stephenson jocularly said that if he were to devote himself to chickens he could soon make a little fortune.

Mrs. Stephenson tried to keep bees, but found they would not thrive at Tapton. Many hives perished, and there was no case of success. The cause of failure was a puzzle to the engineer; but one day his acute powers of observation enabled him to unravel it. At the foot of the hill on which Tapton House stands, he saw some bees trying to rise up from amongst the grass, laden with honey and wax. They were already exhausted, as if with long flying; and then it occurred to him that the height at which the house stood above the bees' feeding-ground rendered it difficult for them to reach their hives when heavy laden, and hence they sank exhausted. He afterwards incidentally mentioned the circumstance to Mr. Jesse the naturalist, who concurred in his view as to the cause of failure, and was much struck by the keen observation which had led to its solution.

Mr. Stephenson had none of the in-door habits of the student. He read very little; for reading is a habit which is generally acquired in youth; and his youth and manhood had been for the most part spent in hard work. Books wearied him, and sent him to sleep. Novels excited his feelings too much, and he avoided them, though he would occasionally read through a philosophical book on a subject in which he felt particularly interested. He wrote very few letters with his own hand; nearly all his letters were dictated, and he avoided even dictation when he could. His greatest pleasure was in conversation, from which he gathered most of his imparted information; hence he was always glad in the society of intelligent, conversible persons.

It was his practice, when about to set out on a journey by railway, to walk along the train before it started, and look into the carriages to see if he could find "a conversible face." On one of these occasions, at the Euston Station, he discovered in a carriage a very hand-some, manly, and intelligent face, which he afterwards found was that of the late Lord Denman. He was on his way down to his seat at Stony Middelton, in Derbyshire. Mr. Stephenson entered the carriage, and the two were shortly engaged in interesting conversation. It turned upon chronometry and horology, and the engineer

amazed his lordship by the extent of his knowledge on
the subject, in which he displayed as much minute information, even down to the latest improvements in watchmaking, as if he had been bred a watchmaker and lived
by the trade.   Lord Denman was curious to know how
a man whose time must have been mainly engrossed by
engineering, had gathered so much knowledge on a
subject quite out of his own line, and he asked the
question.   "I learnt clockmaking and watchmaking,"
was the answer, "while a working man at Killingworth,
when I made a little money in my spare hours, by cleaning the pitmen's clocks and watches; and since then I
have kept up my information on the subject."   This led
to further questions, and then Mr. Stephenson told Lord
Denman the interesting story of his life, which held him
entranced during the remainder of the journey.

Many of his friends readily accepted invitations to
Tapton House to enjoy his hospitality, which never
failed.   With them he would "fight his battles o'er
again," reverting often to his battle for the locomotive;
and he was never tired of telling, nor were his auditors
of listening to, the lively anecdotes with which he was
accustomed to illustrate the struggles of his early career.
Whilst walking in the woods or through the grounds,
he would arrest his friends' attention by allusion to
some simple object,—such as a leaf, a blade of grass, a
bit of bark, a nest of birds, or an ant carrying its eggs
across the path,—and descant in glowing terms upon
the creative power of the Divine Mechanician, whose
contrivances were so exhaustless and so wonderful.
This was a theme upon which he was often accustomed to dwell in reverential admiration, when in the
society of his more intimate friends.

One night, when walking under the stars, and gazing
up into the field of suns, each the probable centre of a
system, forming the Milky Way, a friend said to him,
"What an insignificant creature is man in sight of so

immense a creation as that!" "Yes!" was his reply; "but how wonderful a creature also is man, to be able to think and reason, and even in some measure to comprehend works so infinite!"

A microscope, which he had brought down to Tapton, was a source of immense enjoyment to him; and he was never tired of contemplating the minute wonders which it revealed. One evening, when some friends were visiting him, he induced each of them to puncture his skin so as to draw blood, in order that he might examine the globules through the microscope. One of the gentlemen present was a teetotaller, and Mr. Stephenson pronounced his blood to be the most lively of the whole. He had a theory of his own about the movement of the globules in the blood, which has since become familiar. It was, that they were respectively charged with electricity, positive at one end and negative at the other, and that thus they attracted and repelled each other, causing a circulation. No sooner did he observe anything new, than he immediately set about devising a reason for it. His training in mechanics, his practical familiarity with matter in all its forms, and the strong bent of his mind, led him first of all to seek for a mechanical explanation. And yet he was ready to admit that there was a something in the principle of *life*—so mysterious and inexplicable—which baffled mechanics, and seemed to dominate over and control them. He did not care much, either, for abstruse mechanics, but only for the experimental and practical, as is usually the case with those whose knowledge has been self-acquired.

Even at his advanced age, the spirit of frolic had not left him. When proceeding from Chesterfield station to Tapton House with his friends, he would almost invariably challenge them to a race up the steep path, partly formed of stone steps, along the hill side. And he would struggle, as of old, to keep the front place, though

by this time his "wind" had greatly failed. He would occasionally invite an old friend to take a quiet wrestle with him on the lawn, to keep up his skill, and perhaps to try some new "knack" of throwing. In the evening, he would sometimes indulge his visitors by reciting the old pastoral of "Damon and Phyllis," or singing his favourite song of "John Anderson my Joe." But his greatest glory amongst those with whom he was most intimate, was "a crowdie!" "Let's have a crowdie night," he would say; and forthwith a kettle of boiling water was ordered in, with a basin of oatmeal. Taking a large bowl, containing a sufficiency of hot water, and placing it between his knees, he poured in oatmeal with one hand, and stirred the mixture vigorously with the other. When enough meal had been added, and the stirring was completed, the crowdie was made. It was then supped with new milk, and Stephenson generally pronounced it "capital!" It was the diet to which he had been accustomed when a working man, and all the dainties with which he had become familiar in recent years had not spoiled his simple tastes. To enjoy crowdie at his age, besides, indicated that he still possessed that quality on which no doubt much of his practical success in life had depended,—a strong and healthy digestion.

He would also frequently invite to his house the humbler companions of his early life, and take pleasure in talking over old times with them. He never assumed any of the bearings of a great man on such occasions, but treated the visitors with the same friendliness and respect as if they had been his equals, sending them away pleased with themselves and delighted with him. At other times, needy men who had known him in youth would knock at his door, and they were never refused access. But if he had heard of any misconduct on their part, he would rate them soundly. One who knew him intimately in private life has seen him exhorting such backsliders, and denouncing their misconduct and imprudence, with the tears streaming down his cheeks. And he would generally conclude by opening his purse, and giving them the help which they needed "to make a fresh start in the world."

Young men would call upon him for advice or assistance in commencing a professional career. When he noted their industry, prudence, and good sense, he was always ready. But, hating foppery and frippery above all things, he would reprove any tendency to this weakness which he observed in the applicants. One day, a youth desirous of becoming an engineer called upon him, flourishing a gold-headed cane : Mr. Stephenson said, " Put by that stick, my man, and then I will speak to you." To another extensively decorated gentleman, he one day said, " You will, I hope, Mr. ———, excuse me ; I am a plain-spoken person, and am sorry to see a nice-looking and rather clever young man like you disfigured with that fine-patterned waistcoat, and all these chains and fang-dangs. If I, sir, had bothered my head with such things at your age, I would not have been where I am now."

Mr. Stephenson's life at Tapton during his later years was occasionally diversified with a visit to London. His engineering business having become limited, he gene-

rally went there for the purpose of visiting friends, or
" to see what there was fresh going on."   He found a
new race of engineers springing up on all hands—men
who knew him not; and his London journeys gradually
ceased to yield him real pleasure.   A friend used to
take him to the opera, but by the end of the first act,
he was generally observed in a profound slumber.   Yet
on one occasion he enjoyed a visit to the Haymarket,
with a party of friends on his birthday, to see T. P.
Cooke, in "Black-eyed Susan ; "—if that can be called
enjoyment which kept him in a state of tears during
half the performance.   At other times he visited
Newcastle, which always gave him great pleasure.
He would, on such occasions, go out to Killingworth
and seek up old friends, and if the people whom he
knew were too retiring and shrunk into their cottages,
he went and sought them there.   Striking the floor
with his stick, and holding his noble person upright,
he would say, in his own kind way, " Well, and how's
all here to-day ? "   To the last he had always a warm
heart for Newcastle and its neighbourhood.

Sir Robert Peel, on more than one occasion, invited
Mr. Stephenson to his mansion at Drayton, where he
was accustomed to assemble round him men of the
highest distinction in art, science, and legislation,
during the intervals of his parliamentary life.   The
first invitation was respectfully declined.   Sir Robert
invited him a second time, and a second time he
declined : " I have no great ambition," he said, " to
mix in fine company, and perhaps should feel out
of my element amongst such high folks."   But Sir
Robert a third time pressed him to come down to
Tamworth early in January, 1845, when he would
meet Buckland, Follett, and others well known to
both.   " Well, Sir Robert," said he, " I feel your
kindness very much, and can no longer refuse : I will
come down and join your party."

Mr. Stephenson's strong powers of observation, together with his native humour and shrewdness, imparted to his conversation at all times much vigour and originality, and made him, to young and old, a delightful companion.　Though mainly an engineer, he was also a profound thinker on many scientific questions: and there was scarcely a subject of speculation, or a department of recondite science, on which he had not employed his faculties in such a way as to have formed large and original views.　At Drayton, the conversation usually turned upon such topics, and Mr. Stephenson freely joined in it.　On one occasion, an animated discussion took place between himself and Dr. Buckland on one of his favourite theories as to the formation of coal.　But the result was, that Dr. Buckland, a much greater master of tongue-fence than Mr. Stephenson, completely silenced him.　Next morning, before breakfast, when he was walking in the grounds, deeply pondering, Sir William Follett came up and asked what he was thinking about? "Why, Sir William, I am thinking over that argument I had with Buckland last night; I know I am right, and that if I had only the command of words which he has, I'd have beaten him."　"Let me know all about it," said Sir William, "and I'll see what I can do for you."　The two sat down in an arbour, and the astute lawyer made himself thoroughly acquainted with the points of the case; entering into it with all the zeal of an advocate about to plead the dearest interests of his client.　After he had mastered the subject, Sir William rose up, rubbing his hands with glee, and said, "Now I am ready for him."　Sir Robert Peel was made acquainted with the plot, and adroitly introduced the subject of the controversy after dinner.　The result was, that in the argument which followed, the man of science was overcome by the man of law; and Sir William Follett had at all points the mastery over Dr. Buckland.　"What do *you* say, Mr. Stephenson?"

asked Sir Robert, laughing. " Why," said he, " I will
only say this, that of all the powers above and under
the earth, there seems to me to be no power so great as
the gift of the gab."

One day, at dinner, during the same visit, a scientific
lady asked him the question : " Mr. Stephenson, what
do you consider the most powerful force in nature ? "
" Oh ! " said he, in a gallant spirit, " I will soon answer
that question : it is the eye of a woman for the man
who loves her ; for if a woman look with affection on a
young man, and he should go to the uttermost ends of
the earth, the recollection of that look will bring him
back : there is no other force in nature could do that."

One Sunday, when the party had just returned from
church, they were standing together on the terrace near
the Hall, and observed in the distance a railway-train
flashing along, tossing behind its long white plume
of steam. " Now, Buckland," said Stephenson, " I
have a poser for you. Can you tell me what is the
power that is driving that train ? "   " Well," said the
other, " I suppose it is one of your big engines."   " But
what drives the engine ? "   " Oh, very likely a canny
Newcastle driver."   " What do you say to the light of
the sun ? "   " How can that be ? " asked the doctor.
" It is nothing else," said the engineer : " it is light
bottled up in the earth for tens of thousands of years,—
light, absorbed by plants and vegetables, being necessary
for the condensation of carbon during the process of
their growth, if it be not carbon in another form,—and
now, after being buried in the earth for long ages in
fields of coal, that latent light is again brought forth
and liberated, made to work as in that locomotive, for
great human purposes."

During the same visit, Mr. Stephenson one evening
repeated his experiment with blood drawn from the
finger, submitting it to the microscope in order to show
the curious circulation of the globules. He set the ex-

ample by pricking his own thumb; and the other
guests, by turns, in like manner gave up a small
portion of their blood for the purpose of ascertaining
the comparative liveliness of their circulation.   When
Sir Robert Peel's turn came, Mr. Stephenson said he was
curious to know "how the blood globules of a great
politician would conduct themselves."   Sir Robert held
forth his finger for the purpose of being pricked; but
once, and again, he sensitively shrunk back, and at
length the experiment, so far as he was concerned, was
abandoned.   Sir Robert Peel's sensitiveness to pain was
extreme, and yet he was destined, a few years after, to
die a death of the most distressing agony.

In 1847, the year before his death, Mr. Stephenson
was again invited to join a distinguished party at Drayton Manor, and to assist in the ceremony of formally
opening the Trent Valley Railway, which had been
originally designed and laid out by himself many years
before.   The first sod of the railway had been cut
by the Prime Minister, in November, 1845, during the
time when Mr. Stephenson was abroad on the business
of the Spanish railway.   The formal opening took place
on the 26th of June, 1847, the line having thus been
constructed in less than two years.

What a change had come over the spirit of the landed
gentry since the time when George Stephenson had
first projected a railway through that district!   Then
they were up in arms against him, characterising him as
the devastator and spoiler of their estates; now he was
hailed as one of the greatest benefactors of the age.
Sir Robert Peel, the chief political personage in England, welcomed him as a guest and friend, and spoke
of him as the chief among practical philosophers.   A
dozen members of Parliament, seven baronets, with all
the landed magnates of the district, assembled to cele-
brate the opening of the railway.   The clergy were there

to bless the enterprise, and to bid all hail to railway
progress, as "enabling them to carry on with greater
facility those operations in connexion with religion which
were calculated to be so beneficial to the country." The
army, speaking through the mouth of General A'Court,
acknowledged the vast importance of railways, as tending to improve the military defences of the country.
And representatives from eight corporations were there
to acknowledge the great benefits which railways had
conferred upon the merchants, tradesmen, and working
classes of their respective towns and cities.

Shortly after this celebration at Tamworth, Mr. Stephenson was invited to be present at an assemblage of
railway men in Manchester, at which a testimonial was
presented to Mr. J. P. Westhead, the former chairman
of the Manchester and Birmingham Railway. The
original Liverpool and Manchester line had now swelled
into gigantic proportions. It formed the nucleus of the
vast system now known as the London and North-
western Railway. First one line, and then another, of
which Mr. Stephenson was the engineer, had been amal-
gamated with it, until the main line extended from
London to Lancaster, stretching out its great arms to
Leeds in one direction and Holyhead in the other, and
exercising an influence over other northern lines, as far
as Glasgow, Edinburgh, and Aberdeen. On the occasion
to which we refer, Mr. Stephenson, the "father of railways," was not forgotten. It was mainly his ingenuity,
energy, and perseverance that had called forth the
commercial enterprise which issued in this magnificent
system of internal communication ; and the railway men
who assembled to do honour to Mr. Westhead did not
fail to recognise the great practical genius through
whose labours it had been established. He was "the
rock from which they had been hewn," observed Mr.
Westhead,—the father of railway enterprise,—and the

forerunner of all that had been done to extend the
locomotive system throughout England and throughout
the world.

In the spring of 1848 Mr. Stephenson was invited to
Whittington House, near Chesterfield, the residence of
his friend and former pupil, Mr. Swanwick, to meet the
distinguished American, Emerson.   It was interesting
to see those two remarkable men, so different in most
respects, and whose lines of thought and action lay in
such widely different directions, yet so quick to recognise
each other's merits.   Mr. Stephenson was not, as yet,
acquainted with Mr. Emerson as an author; and the
contemplative American might not be supposed to be
particularly interested beforehand in the English engineer, whom he knew by reputation only as a giant in
the material world.   But there was in both an equal
aspiration after excellence, each in his own sphere,—the
æsthetic and abstract tendencies of the one complement-
ing the keen and accurate perceptions of the material
of the other.   Upon being introduced, they did not
immediately engage in conversation ; but presently
Stephenson jumped up, took Emerson by the collar,
and, giving him one of his friendly shakes, asked how it
was that in England we could always tell an American ?
This led to an interesting conversation, in the course of
which Emerson said how much he had everywhere been
struck by the haleness and comeliness of the English
men and women ; and then they diverged into a further
discussion of the influences which air, climate, moisture,
soil, and other conditions exercised upon the physical
and moral development of a people.   The conversation was next directed to the subject of electricity,
upon which Stephenson launched out enthusiastically,
explaining his views by several simple and striking
illustrations.   From thence it gradually turned to the
events of his own life, which he related in so graphic a
manner as completely to rivet the attention of the

American. Afterwards Emerson said, "that it was worth crossing the Atlantic to have seen Stephenson alone; he had such native force of character and vigour of intellect." Although Emerson does not particularly refer to this interview in the interesting essay afterwards published by him, entitled 'English Traits,' embodying the results of the observations made by him in his journeys through England, one cannot help feeling that his interview with such a man as Stephenson must have tended to fix in his mind those sterling qualities of pluck, bottom, perseverance, energy, shrewdness, bravery, and freedom, which he so vividly depicts in his book as the prominent characteristics of the modern Englishman.

The rest of Mr. Stephenson's days were spent quietly at Tapton, amongst his dogs, his rabbits, and his birds. When not engaged about the works connected with his collieries, he was occupied in horticulture and farming. He continued proud of his flowers, his fruits, and his crops; and the old spirit of competition was still strong within him. Although he had for some time been in delicate health, and his hand shook from nervous affection, he appeared to possess a sound constitution. Emerson had observed of him that he had the lives of many men in him. But perhaps the American spoke figuratively, in reference to his vast stores of experience. It appeared that he had never completely recovered from the attack of pleurisy which seized him during his return from Spain. As late, however, as the 26th of July, 1848, he felt himself sufficiently well to be able to attend a meeting of the Institute of Mechanical Engineers at Birmingham, and to read to the members his paper " On the Fallacies of the Rotatory. Engine." It was his last appearance before them. Shortly after his return to Tapton, he had an attack of intermittent fever, from which he seemed to be recovering, when a sudden effusion of blood from the lungs carried him off, on the

12th of August, 1848, in the sixty-seventh year of his age. When all was over, Robert wrote to Edmund Pease, " With deep pain I inform you, as one of his oldest friends, of the death of my dear father this morning at 12 o'clock, after about ten days' illness from severe fever." Mr. Starbuck, who was also present, wrote : " The favourable symptoms of yesterday morning were towards evening followed by a serious change for the worse. This continued during the night, and early this morning it became evident that he was sinking. At a few minutes before 12 to-day he breathed his last. All that the most devoted and unremitting care of Mrs. Stephenson, and the skill of medicine could accomplish, has been done, but in vain."

George Stephenson's remains were followed to the grave by a large body of his workpeople, by whom he was greatly admired and beloved. They remembered him as a kind master, who was ever ready actively to promote all measures for their moral, physical, and mental improvement. The inhabitants of Chesterfield evinced their respect for the deceased by suspending

TRINITY CHURCH, CHESTERFIELD.

business, closing their shops, and joining in the funeral procession, which was headed by the corporation of the town. Many of the surrounding gentry also attended. The body was interred in Trinity Church, Chesterfield, where a simple tablet marks the great engineer's last resting-place.

The statue of George Stephenson, which the Liverpool and Manchester and Grand Junction Companies had commissioned, was on its way to England when his death occurred; and it served for a monument, though his best monument will always be his works. The Liverpool Board placed a minute on their books, embodying the graceful tribute of their secretary, Mr. Henry Booth, in which they recorded their admiration of the life, and their esteem for the character of the deceased. "The directors," they say, "on the present occasion look back with peculiar interest to their first connexion with Mr. Stephenson, in the construction of the Liverpool and Manchester Railway; to a period now twenty years past, when he floated their new line over Chat Moss, and cut his way through the rock-cutting at Olive Mount. Tracing the progress of railways from the first beginning to the present time, they find Mr. Stephenson foremost in urging forward the great railway movement; earning and maintaining his title to be considered, before any other man, the author of that universal system of locomotion which has effected such mighty results—commercial, social, and political—throughout the civilized world. Two years ago, the directors entrusted to Mr. Gibson, of Rome, the duty and the privilege of producing a statue that might do honour to their friend, then living amongst them. They did not anticipate that on the completion of this work of art the great original would be no more,—that they should be constrained to accept the marble effigy of the engineer in lieu of the living presence of the man."

The statue here referred to was placed in St. George's Hall, Liverpool. A full-length statue of the deceased, by Bailey, was also erected a few years later, in the noble vestibule of the London and North Western Station, in Euston Square. A subscription for the purpose was set on foot by the Society of Mechanical Engineers, of which he had been the founder and president. A few advertisements were inserted in the newspapers, inviting subscriptions; and it is a notable fact that the voluntary offerings included an average of two shillings each from 3150 working men, who embraced this opportunity of doing honour to their distinguished fellow workman.

But unquestionably the finest and most appropriate statue to the memory of George Stephenson is that erected in the course of the present year at Newcastle-upon-Tyne. It is in the immediate neighbourhood of the Literary and Philosophical Institute, to which both George and his son Robert were so much indebted in their early years; close to the great Stephenson locomotive foundry established by the shrewdness of the father; and in the vicinity of the High Level Bridge, one of the grandest products of the genius of the son. The statue is by John Lough, a sculptor whose genius is equalled by his modesty. The head of Stephenson, as expressed in this noble work, is massive, characteristic, and faithful; and the attitude of the figure is simple yet manly and energetic. It stands on a pedestal, at the respective corners of which are sculptured the recumbent figures of a pitman, a mechanic, an engine-driver, and a plate-layer. These figures are admirably executed, and their design in connection with the central figure seems to us quite original. The statue appropriately stands in a very thoroughfare of working men, thousands of whom see it daily as they pass to and from their work; and we can imagine them, as they look up to Stephenson's manly figure, applying to it the words addressed by Robert Nicoll to Robert Burns, with perhaps still greater appropriateness :—

> " Before the proudest of the earth
> We stand, with an uplifted brow ;
> Like us, thou wast a toiling man,—
> And we are noble, now ! "

The portrait prefixed to this volume gives a good indication of George Stephenson's shrewd, kind, honest, manly face. His fair, clear countenance was ruddy, and seemingly glowed with health. The forehead was large and high, projecting over the eyes ; and there was that massive breadth across the lower part which is usually observed in men of eminent constructive skill. The mouth was firmly marked, and shrewdness and humour lurked there as well as in the keen grey eye. His frame was compact, well-knit, and rather spare. His hair became grey at an early age, and towards the close of his life it was of a pure silky whiteness. He dressed neatly in black, wearing a white neckcloth ; and his face, his person, and his deportment at once arrested attention, and marked the Gentleman.

George Stephenson bequeathed to his son his valuable collieries, his share in the engine manufactory at Newcastle, and his large accumulation of savings, which, together with the fortune he had himself amassed by railway work, gave Robert the position of an engineer millionaire—the first of his race. He continued, however, to live in a quiet style ; and although he bought occasional pictures and statues, and indulged in the luxury of a yacht, he did not live up to his income, which went on rapidly accumulating until his death.

There was no longer the necessity for applying himself to the harassing business of a parliamentary engineer, in which he had now been occupied for some fifteen years. Shortly after his father's death, Edward Pease strongly recommended him to give up the more harassing work of his profession; and his reply (15th June, 1850) was as follows :—" The suggestion which your kind note contains is quite in accordance with my own feelings and intentions respecting retirement; but I find it a very difficult matter to bring to a close so complicated a connexion in business as that which has been established by twenty-five years of active and arduous professional duty.   Comparative retirement is, however, my intention; and I trust that your prayer for the Divine blessing to grant me happiness and quiet comfort will be fulfilled.   I cannot but feel deeply grateful to the Great Disposer of events for the success which has hitherto attended my exertions in life; and I trust that the future will also be marked by a continuance of His mercies."

Robert Stephenson lived long enough, however, to repeat his Tubular Bridge in the magnificent structure across the St. Lawrence at Montreal, and, in a modified form, in the two bridges across the Nile, near Damietta in Lower Egypt.   The Victoria Bridge was erected after Mr. Stephenson's designs under the immediate direction of Mr. Malcolm Ross, who acted as resident and joint engineer.   With its approaches it is only sixty yards short of two miles in length.   In gigantic strength and majestic proportions there is no structure to compare with it in ancient or modern times.   It consists of not less than twenty-five immense tubular bridges joined into one; the great central span being 330 feet, the others 242 feet in length.   In constructing these tubes, the cellular principle has been entirely dispensed with.   The weight of wrought-iron in the bridge is about 10,000 tons; the piers being of massive stone, each containing some 8000 tons of solid masonry.   This vast structure was begun in 1854, and finished in 1860; but the engineer did not live to see its completion.

The principal feature of Mr. Stephenson's Egyptian bridges was in the road being carried *upon* the tubes in-

stead of within them.   The larger of the two is over the
Damietta branch of the Nile, near Benha.   It contains
eight spans or openings of 80 feet each, and two centre
spans, formed by one of the largest swing-bridges ever
constructed—the total length of the swing-beam being
157 feet; leaving a clear water-way of 60 feet on either
side of the central pier.   The greatest difficulty encountered in the erection of the bridge was in getting in the
foundations, which were sunk 33 feet through soil of a
peculiarly shifting character.

During the later years of his life Mr. Stephenson was
frequently called upon to act as arbitrator between contractors and railway companies, or between one company and another—great value being attached to his
opinion on account of his weighty judgment, his great
experience, and his upright character, and we believe
his decisions were invariably stamped by the qualities of
impartiality and justice.   He was always ready to lend
a helping hand to a friend, and no petty jealousy stood
between him and rivals in the engineering world.   The
author remembers being with Mr. Stephenson one evening at his house in Gloucester Square, when a note was
put into his hands from his friend Brunel, then engaged
in his first fruitless efforts to launch the *Great Eastern*.
It was to ask Stephenson to come down to Blackwall
early next morning, and give him the benefit of his
judgment.   Shortly after six next morning Stephenson
was in Scott Russell's building-yard, and he remained
there until dusk.   About midday, while superintending
the launching operations, the balk of timber on which
he stood canted up, and he fell up to his middle in the
Thames mud.   He was dressed as usual, without great-
coat (though the day was bitter cold), and with only
thin boots upon his feet.   He was urged to leave the
yard, and change his dress, or at least dry himself; but
with his usual disregard of health, he replied, " Oh,
never mind me—I'm quite used to this sort of thing;"

and he went paddling about in the mud, smoking his cigar, until almost dark, when the day's work was brought to an end. The result of this exposure was an attack of inflammation of the lungs, which kept him to his bed for a fortnight.

Mr. Stephenson also took considerable interest in public affairs and in scientific investigations. In 1847 he entered the House of Commons as member for Whitby; but he does not seem to have been very devoted in his attendance, and only appeared on divisions when there was a "whip" of the party to which he belonged. He was a member of the Sanitary and Sewage Commissions, and of the Commission which sat on Westminster Bridge. The last occasions on which he addressed the House were on the Suez Canal and the cleansing of the Serpentine. He pronounced the Suez Canal to be an impracticable scheme. "I have surveyed the line," said he, "I have travelled the whole distance on foot, and I declare there is no fall between the two seas. Honourable members talk about a canal. A canal is impossible—the thing would only be a ditch."

Besides constructing the railway between Alexandria and Cairo, he was consulted respecting many important lines abroad. He was early consulted, like his father, by the King of Belgium, as to the railways of that country; and he was made Knight of the Order of Leopold because of the improvements which he had made in locomotive engines, so much to the advantage of the Belgian system of inland transit. He was consulted by the King of Sweden as to the railway between Christiana and Lake Miösen, and in consideration of his services was decorated with the Grand Cross of the Order of St. Olaf. He also visited Switzerland, Piedmont, and Denmark, to advise as to the system of railway communication best suited for those countries. At the Paris Exhibition of 1855 the Emperor of France decorated him with the Legion of Honour in consideration of his public services; and at home the University of Oxford made him a Doctor of Civil Laws. In 1855, he was elected President of the Institute of Civil Engineers, which office he held with honour and filled with distinguished ability for two years, giving place to his friend Mr. Locke at the end of 1857.

It was when on a visit to Norway in the autumn of 1859 that Robert Stephenson was seized by the illness which terminated his illustrious career. He had been for some time ailing, and was in indifferent health when he sailed. But a deep-seated disease lurked within him —an old liver-complaint which first developed itself in jaundice and then in dropsy, of which he died on the 12th of October, in the fifty-sixth year of his age.[1] He was buried by the side of Telford in Westminster Abbey, amidst the departed great men of his country, and was attended to his resting-place by many of the intimate friends of his boyhood and his manhood. Among those who assembled round his grave were some of the greatest men of thought and action in England, who embraced the sad occasion to pay the last mark of their respect to this illustrious son of one of England's greatest working men.

---

[1] In 1829 Robert Stephenson married Frances, daughter of John Sanderson, merchant, London; but she died in 1842, without issue, and Mr. Stephenson did not marry again. Writing to his friend Edward Pease, of Darlington, shortly after his wife's death, in 1842, he said :—" You have my sincere thanks for your kind expressions relative to the heavy affliction with which the Almighty in his wisdom has been pleased to visit me. It has, indeed, been severe, but I feel that the weight of the blow was much mitigated by my being mercifully permitted to witness the last moments of my beloved companion in life, which were those of a fervent and faithful Christian; and my prayer is that my last end may be like hers." Until the close of his life, Robert Stephenson was accustomed twice in every year to visit his wife's grave in Hampstead churchyard.

CHAPTER XXI.

Characteristics.

It would be out of keeping with the subject thus drawn to a conclusion, to pronounce a panegyric on the character and achievements of George Stephenson and his son. Both were emphatically true men, presenting in their lives and works a combination of those sterling qualities which we are proud to regard as essentially English.

In the old Teutonic tongue, *Steeveson*, of which Stevenson and Stephenson are but modifications, is said to mean the " Son of the Strong ;" nor did either of our engineers belie the appellation. Doubtless they owed much to their birth, belonging as they did to the hardy race of the north—a race less supple, soft, and polished than the people of more southern districts ; but, like their Danish progenitors, full of courage, vigour, ingenuity, and persevering industry. Their strong, guttural speech, which sounds so harsh and unmusical in southern ears, is indeed but a type of their nature. When George Stephenson was struggling to give utterance to his views upon the locomotive before the Committee of the House of Commons, those who did not know him supposed he was " a foreigner." Before long the world saw in him an Englishman, stout-hearted and true—one of those master minds who, by energetic action in new fields of industry, impress their character from time to time upon the age and nation to which they belong.

The poverty of his parents being such that they could not give him any, even the very simplest, education, beyond the good example of integrity and industry, he was early left to shift for himself, and compelled to be self-reliant. Having the will to learn, he soon found a way for himself. No beginning could have been more humble than his; but he persevered : he had determined to learn, and he did learn. To such a resolution as his, nothing really beneficial in life is denied. He might have said, like Sebastian Bach, "I was industrious ; and whoever is equally sedulous will be equally successful."

The whole secret of Mr. Stephenson's success in life was his careful improvement of time, which is the rock out of which fortunes are carved and great characters formed. He believed in genius to the extent that Buffon did when he said that "patience is genius;" or as some other thinker put it, when he defined genius to be the power of making efforts. But he never would have it that he was a genius, or that he had done anything which other men, equally laborious and persevering as himself, could not have accomplished. He repeatedly said to the young men about him : "Do as I have done —persevere ! "

Every step of advance which he made was conquered by patient labour. When an engineman, he systematically took his engine to pieces on Saturday afternoons while the works were at a stand, for the purpose of cleaning it thoroughly, and "gaining insight." He thus gradually mastered the mechanism of the steam-engine, so that, when opportunity offered, he was enabled to improve it, and to make it work even when its own maker was baffled. He practically studied hydraulics in the same plodding way, when acting as plugman; and when all the local pump-doctors at Killingworth were in despair, he stepped in, and successfully applied the knowledge which he had so laboriously gained. A man of such a temper and purpose could not but succeed in life.

Whether working as a brakesman or an engineer, his mind was always full of the work in hand. He gave

himself thoroughly up to it. Like the painter, he might
say that he had become great "by neglecting nothing."
Whatever he was engaged upon, he was as careful of
the details as if each were itself the whole. He did all
thoroughly and honestly. There was no "scamping"
with him. When a workman he put his brains and
labour into his work; and when a master he put his
conscience and character into it. He would have no
slop-work executed merely for the sake of profit. The
materials must be as genuine as the workmanship was
skilful. The structures which he designed and executed
were distinguished for their thoroughness and solidity;
his locomotives were famous for their durability and
excellent working qualities. The engines which he sent
to the United States in 1832 are still in good condition;
and even the engines built by him for the Killingworth
colliery, upwards of thirty years ago, are working
steadily there to this day. All his work was honest,
representing the actual character of the man.

The battle which Mr. Stephenson fought for the
locomotive—and he himself always spoke of it as a
"battle"—would have discouraged most other men;
but it only served to bring into prominence that energy
and determination which formed the back-bone of his
character. "I have fought," said he, "for the locomotive single-handed for nearly twenty years, having
no engineer to help me until I had reared engineers
under my own care." The leading engineers of the
day were against him, without exception; yet he did
not despair. He had laid hold of a great idea, and he
stuck by it; his mind was locked and bolted to the
results. "I put up," he says, "with every rebuff,
determined not to be put down." When the use of his
locomotive on the Liverpool and Manchester line was
reported against, and the employment of fixed engines
recommended instead, Mr. Stephenson implored the
directors, who were no engineers, only to afford him a

fair opportunity for a trial of the locomotive. Their common sense came to his rescue. They had immense confidence in the Newcastle engine-wright. He had already made stedfast friends of several of the most influential men amongst them, who valued his manly uprightness and integrity, and were strongly disposed to believe in him, though all the engineering world stood on the one side, and he alone on the other. His patient purpose, not less than his intense earnestness, persuaded them. They adopted his recommendation, and offered a prize of 500*l.* for the best locomotive. Though many proclaimed the Liverpool men to be as great maniacs as Stephenson, yet the result proved the practical sagacity of the directors and the skill of their engineer; but it was the determined purpose of the latter which secured the triumph of the locomotive. His resolution, founded on sound convictions, was the precursor of what he eventually achieved; and his intense anticipation was but the true presentiment of what he was afterwards found capable of accomplishing.

He was ready to turn his hand to anything—shoes and clocks, railways and locomotives. He contrived his safety-lamp with the object of saving pitmen's lives, and perilled his own life in testing it. Whatever work was nearest him, he turned to and did it. With him to resolve was to do. Many men knew far more than he; but none was more ready forthwith to apply what he did know to practical purposes. It was while working at Willington as a brakesman, that he first learnt how best to handle a spade in throwing ballast out of the ships' holds. This casual employment seems to have left upon his mind the strongest impression of what "hard work" was; and he often used to revert to it, and say to the young men about him, "Ah, ye lads! there's none o' ye know what *wark* is." Mr. Gooch says he was proud of the dexterity in handling a spade which he had thus acquired, and that he has frequently

seen him take the shovel from a labourer in some railway cutting, and show him how to use it more deftly
in filling waggons of earth, gravel, or sand.　Sir Joshua
Walmsley has also informed us, that, when examining
the works of the Orleans and Tours Railway, Mr.
Stephenson, seeing a large number of excavators filling
and wheeling sand in a cutting, at a great waste of time
and labour after the manner of foreign navvies, went
up to the men and said he would show them how to fill
their barrow in half the time.　He showed them the
proper position in which to stand so as to exercise the
greatest amount of power with the least expenditure of
strength; and he filled the barrow with comparative
ease again and again in their presence, to the great
delight of the workmen.　When passing through his
own workshops, he would point out to his men how to
save labour, and to get through their work skilfully
and with ease.　His energy imparted itself to others,
quickening and influencing them as strong characters
always do—flowing down into theirs, and bringing out
their best powers.

His deportment towards the workmen employed
under him was familiar, yet firm and consistent.　As
he respected their manhood, so did they respect his
masterhood.　Although he comported himself towards
his men as if they occupied very much the same level
as himself, he yet possessed that peculiar capacity for
governing which enabled him always to preserve
amongst them the strictest discipline, and to secure
their cheerful and hearty services.　Mr. Ingram, M.P.
for South Shields, one day went over the workshops at
Newcastle with Mr. Stephenson, and was particularly
struck with this quality of the master in his bearing
towards his men.　"There was nothing," said he, "of
undue familiarity in their intercourse, but they spoke to
each other as man to man; and nothing seemed to please
the master more than to point out illustrations of the

470 DEPORTMENT TO HIS WORKMEN. Chap. XXI.

ingenuity of his artisans. He took up a rivet, and
expatiated on the skill with which it had been fashioned
by the workman's hand—its perfectness and truth. He
was always proud of his workmen and his pupils; and,
while indifferent and careless as to what might be said
of himself, he fired up in a moment if disparagement
were thrown upon any one whom he had taught or
trained."

In manner, George Stephenson was simple, modest,
and unassuming, but always manly. He was frank and
social in spirit. When a humble workman, he had
carefully preserved his sense of self-respect. His companions looked up to him, and his example was worth
even more to many of them than books or schools. His
devoted love of knowledge made his poverty respectable,
and adorned his humble calling. When he rose to a
more elevated station, and associated with men of the
highest position and influence in Britain, he took his
place amongst them with perfect self-possession. They
wondered at the quiet ease and simple dignity of his
deportment; and men in the best ranks of life have said
of him that "He was one of Nature's gentlemen."

Probably no military chiefs were ever more beloved
by their soldiers than were both father and son by the
army of men who, under their guidance, worked at
labours of profit, made labours of love by their earnest
will and purpose. True leaders of men and lords of
industry, they were always ready to recognise and
encourage talent in those who worked for and with
them. Thus it was pleasant, at the openings of the
Stephenson lines, to hear the chief engineers attributing
the successful completion of the works to their able
assistants; whilst the assistants, on the other hand,
ascribed the entire glory to their chiefs.

A fine trait in George Stephenson's character was his
generosity, which would not permit an attack to be
made upon the absent or the weak. He would never

sanction any injustice of act or opinion towards those associated with himself. On one occasion, during the progress of the Liverpool and Manchester works, while he had a strong party to contend with at the Board, the conduct of one of his assistants was called in question, as he thought unjustly, and a censure was threatened. Rather than submit to this injustice to his assistant, Mr. Stephenson tendered his resignation; but it was not accepted, and the censure was not voted. The same chivalrous protection was on many occasions extended by him to the weaker against the stronger. Even if he were himself displeased with any one engaged about him, any attack from another quarter would rouse him in defence, not in the spirit of opposition, but from a kind and generous impulse to succour those in difficulty.

Mr. Stephenson, though a thrifty and frugal man, was essentially unsordid. His rugged path in early life made him careful of his resources. He never saved to hoard, but saved for a purpose, such as the maintenance of his parents or the education of his son. In later years, he became a prosperous and even a wealthy man; but riches never closed his heart, nor stole away the elasticity of his soul. He enjoyed life cheerfully, because hopefully. When he entered upon a commercial enterprise, whether for others or for himself, he looked carefully at the ways and means. Unless they would "pay," he held back. "He would have nothing to do," he declared, "with stock-jobbing speculations." His refusal to sell his name to the schemes of the railway mania—his survey of the Spanish lines without remuneration—his offer to postpone his claim for payment from a poor company until their affairs became more prosperous—are instances of the unsordid spirit in which he acted.

Another marked feature in Mr. Stephenson's character was his patience. Notwithstanding the strength of his convictions as to the great uses to which the locomotive might be applied, he waited long and patiently for the opportunity of bringing it into notice; and for years after he had completed an efficient engine he went on quietly devoting himself to the ordinary work of the colliery. He made no noise nor stir about his locomotive, but allowed another to take credit for the experiments on velocity and friction made with it by himself upon the Killingworth railroad.

By patient industry and laborious contrivance, he was enabled, with the powerful help of his son, to do for the locomotive what James Watt had done for the condensing engine. He found it clumsy and inefficient; and he made it powerful, efficient, and useful. Both have been described as the improvers of their respective engines; but, as to all that is admirable in their structure or vast in their utility, they are rather entitled to be described as their Inventors. While the invention of Watt increased the power, and at the same time so regulated the action, of the steam-engine, as to make it capable of being applied alike to the hardest work and to the finest manufactures, the invention of Stephenson gave an effective power to the locomotive, which enabled it to perform the work of teams of the most powerful horses, and to outstrip the speed of the fleetest. Watt's invention exercised a wonderfully quickening influence on every branch of industry, and multiplied a thousand-fold the amount of manufactured productions; and Stephenson's enabled these to be distributed with an economy and despatch such as had never before been thought possible. They have both tended to increase indefinitely the mass of human comforts and enjoyments, and to render them cheap and accessible to all. But Stephenson's invention, by the influence which it is daily exercising upon the civilisation of the world, is even more remarkable than that of Watt, and is calculated to have still more important consequences. In this respect, it is to be regarded as the grandest application of steam power that has yet been discovered.

The Locomotive, like the condensing engine, exhibits the realisation of various capital, but wholly distinct, ideas, promulgated by many ingenious inventors. Stephenson, like Watt, exhibited a power of selection, combination, and invention of his own, by which—while availing himself of all that had been done before him, and superadding the many skilful contrivances devised by himself—he was at length enabled to bring his engine into a condition of marvellous power and efficiency. He gathered together the scattered threads of ingenuity which already existed, and combined them into one firm and complete fabric of his own. He realised the plans which others had imperfectly formed ; and was the first to construct, what so many others had unsuccessfully attempted, the practical working locomotive.

If he was occasionally impatient of the opposition of professional brethren, it is scarcely to be wondered at when we look at the simple earnestness of his character, and consider that his sole aim was the establishment of his own well-founded convictions. No wonder that he should have been intolerant of that professional gladiatorship against which his life had been one prolonged struggle. Nor could he forget that the engineering class had been arrayed against him during his arduous battle for the locomotive, and that, but for his own pluck and persistency, they would have strangled it in its cradle. A man of his stern resolution might well be a little positive sometimes. Who that has made his way through so many difficulties would not be so ? Especially was he annoyed at the " quirks and quiddities " of the barristers, who subjected him to annoying cross-examinations before the Parliamentary Committees. On coming down from the witness-box on one occasion, he went up to the counsel who had been severely cross-

examining him, and said—" Oh T——, I'm ashamed of
you! You know my line's the best, and that I'm in
the right and you're in the wrong, and yet you've been
worrying me as if you did'nt know that I *was* right."

Mr. Stephenson's close and accurate observation provided him with a fulness of information on many
subjects, which often appeared surprising to those who
had devoted to them a special study   On one occasion
the accuracy of his knowledge of birds came out in a
curious way at a convivial meeting of railway men in
London. The engineers and railway directors present
knew each other as railway men and nothing more.
The talk had been all of railways and railway politics.
Mr. Stephenson was a great talker on those subjects,
and was generally allowed, from the interest of his
conversation and the extent of his experience, to take
the lead. At length one of the party broke in with—
" Come now, Stephenson, we have had nothing but railways; cannot we have a change, and try if we can talk
a little about something else?" " Well," said Mr.
Stephenson, " I'll give you a wide range of subjects;
what shall it be about?" " Say *birds' nests!*" rejoined
the other, who prided himself on his special knowledge
of this subject. " Then bird's nests be it." A long
and animated conversation ensued : the bird-nesting of
his boyhood, the blackbird's nest which his father had
held him up in his arms to look at when a child at
Wylam, the hedges in which he had found the thrush's
and the linnet's nests, the mossy bank where the robin
built, the cleft in the branch of the young tree where
the chaffinch had reared its dwelling—all rose up clear
in his mind's eye, and led him back to the scenes of his
boyhood at Callerton and Dewley Burn. The colour
and number of the bird's eggs, the period of their
incubation, the materials employed by them for the
walls and lining of their nests—were described by him
so vividly, and illustrated by such graphic anecdotes,

that one of the party remarked that, if George Stephenson had not been the greatest engineer of his day, he might have been one of the greatest naturalists.

His powers of conversation were very great. He was so thoughtful, so original, and so suggestive. There was scarcely a department of science on which he had not formed some novel and sometimes daring theory. Thus Mr. Gooch, his pupil, who lived with him when at Liverpool, informs us that when sitting over the fire, he would frequently broach his favourite theory of the sun's light and heat being the original source of the light and heat given forth by the burning coal. " It fed the plants of which that coal is made," he would say, " and has been bottled up in the earth ever since, to be given out again now for the use of man." [1] His son Robert once said of him, " My father flashed his bull's-eye full upon a subject, and brought it out in its most vivid light in an instant: his strong common sense, and his varied experience operating upon a thoughtful mind, were his most powerful illuminators."

Mr. Stephenson had once a conversation with a watchmaker, whom he astonished by the extent and minuteness of his knowledge as to the parts of a watch. The watchmaker knew him to be an eminent engineer, and asked how he had acquired so extensive a knowledge of a branch of business so much out of his sphere. " It is very easy to be explained," said Mr. Stephenson ; " I worked long at watch-cleaning myself, and when I was at a loss, I was never ashamed to ask for information."

It is Göthe, we believe, who has said that no man ever receives a new idea, at variance with his preconceived notions, after forty. But this observation, though

---

[1] Mr. W. B. Adams, in his ' Roads and Rails,' London, 1862, cites a passage from a volume of rhymes published by Effingham Wilson in 1831, from which it would appear that the same idea had occurred to other thinkers besides George Stephenson.

it may be generally, is not invariably true. There are many great minds which never close. Mr. Stephenson, to the last, was open to the reception of new ideas, new facts, new theories. He was a late learner; but he went on learning to the end. He shut his mind, however, against what he considered humbugs—especially mechanical humbugs. Thus, he said at Tamworth, that he had not been to see the atmospheric railway, because it was a great humbug. He had gone to see Pinkus's model of it, and that had determined him on the subject. He then declared the atmospheric system to be "a rope of sand;" that it could never hold together, and he would not countenance it.

When he heard of Perkins's celebrated machine, which was said to work at a tremendous pressure, without steam, but with water in the boiler almost at red heat, he went with his son to see it. The engine exhibited was of six-horse power, and the pressure was said to be not less than 1500 lbs. to the square inch. Mr. Stephenson said he thought it humbug; but he would test its power. Taking up a little oakum, and wrapping some round each hand, he firmly seized hold of the piston-rod and held it down with all his strength. The machine was at once brought to a stand, very much to Mr. Perkins's annoyance. But the humbug had been exploded to Mr. Stephenson's satisfaction.

Towards the close of his life he frequently went down to Newcastle, and visited the scenes of his boyhood. "I have been to Callerton," said he one day to a friend, "and seen the fields in which I used to pull turnips at twopence a day; and many a cold finger, I can tell you, I had."

On one occasion, he accidentally met a gentleman and his wife at an inn in Derbyshire, whom he entertained for some time with his shrewd observations and playful sallies. At length the lady requested to know the name of the remarkable stranger. "Why, madam,"

said he, "they used once to call me Geordie Stephenson; I'm now called George Stephenson, *Esquire*,
of Tapton House, near Chesterfield. And further let
me say, that I've dined with princes, and peers, and
commoners—with persons of all classes, from the highest
to the humblest; I've made my dinner off a red-herring
in a hedge bottom, and gone through the meanest
drudgery; I've seen mankind in all its phases, and the
conclusion I have arrived at is—that if we're all
stripped, there's not much difference."

His hand was open to his former fellow-workmen
whom old age had left in poverty. To poor Robert
Gray, of Newburn, who acted as his bridesman on his
marriage to Fanny Henderson, he left a pension for life.
He would slip a five-pound note into the hand of a poor
man or a widow in such a way as not to offend their
delicacy, but to make them feel as if the obligation were
all on his side. When Farmer Paterson, who married
a sister of George's first wife, Fanny Henderson, died
and left a large young family fatherless, poverty stared
them in the face. "But ye ken," said our informant,
"*George struck in fayther for them.*" And perhaps the
providential character of the act could not have been
more graphically expressed than in these simple words.

On his visits to Newcastle, he would frequently meet
the friends of his early days, occupying very nearly the
same station, while he had meanwhile risen to almost
world-wide fame. But he was no less hearty in his
greeting of them than if their relative position had
continued the same. Thus, one day, after shaking hands
with Mr. Brandling on alighting from his carriage, he
proceeded to shake hands with his coachman, Anthony
Wigham, a still older friend, though he only sat on the
box.

Robert Stephenson inherited his father's kindly spirit
and benevolent disposition. He almost worshipped his
father's memory, and was ever ready to attribute to him

the chief merit of his own achievements as an engineer. "It was his thorough training," we once heard him say, "his example, and his character, which made me the man I am." On a more public occasion he said, "It is my great pride to remember, that whatever may have been done, and however extensive may have been my own connection with railway development, all I know and all I have done is primarily due to the parent whose memory I cherish and revere."[1] To Mr. Lough, the sculptor, he said he had never had but two loves— one for his father, the other for his wife.

Like his father, he was eminently practical, and yet always open to the influence and guidance of correct theory.[2] His main consideration in laying out his lines of railway was what would best answer the intended purpose, or, to use his own words, to secure the maximum of result with the minimum of means. He was pre-eminently a safe man, because cautious, tentative, and experimental; following closely the lines of conduct trodden by his father, and often quoting his maxims.

In society Robert Stephenson was simple, unobtrusive, and modest; but charming and even fascinating in an eminent degree. Sir John Lawrence has said of him that he was, of all others, the man he most delighted to meet in England—he was so manly, yet gentle, and withal so great. While admired and beloved by men of such calibre, he was equally a favourite with women and children. He put himself upon the level of all, and charmed them no less by his inexpressible kindliness of manner than by his simple yet impressive conversation.

---

[1] Address as President of the Institution of Civil Engineers, January, 1856.

[2] Writing from Mariquita, South America, in 1826, when only twenty-three years of age, he said:—"Practical men are certainly to be esteemed as such, but I am far from attaching the importance to them which our masters appear inclined to do. In-deed, in the working of gold and silver mines in veins in this country, it is absolutely essential that theory and practice should be united and go hand in hand; not that the former should be appreciated beyond its value, and the other depreciated below it, but that both should be entitled to equal consideration and weight."

His great wealth enabled him to perform many generous acts in a right noble and yet modest manner, not letting his right hand know what his left hand did. Of the numerous kindly acts of his which have been made public, we may mention the graceful manner in which he repaid the obligations which both himself and his father owed to the Newcastle Literary and Philosophical Institute, when working together as humble experimenters in their cottage at Killingworth. The Institute was struggling under a debt of 6200*l.*, which seriously impaired its usefulness as an educational agency. Robert Stephenson offered to pay one-half of the sum, provided the local supporters of the Institute would raise the remainder; and conditional also on the annual subscription being reduced from two guineas to one, in order that the usefulness of the institution might be extended. The generous offer was accepted, and the debt extinguished.

Both father and son were offered knighthood, and both declined it. George Stephenson, however, did desire to be admitted to the membership of the Institute of Civil Engineers, the chair of which his son afterwards so ably filled. But there were two obstacles to George's admission to the Institute : the first was, that he had served no regular apprenticeship as an engineer ; and the second was, that he should go through the form required of the youngest member of the profession, and fill in a paper detailing his experience, to which he must afterwards obtain the signatures of several members of the Institute, recommending him personally and professionally for election. He could not comply with the first condition, and his son strongly recommended him not to comply with the second. The council of the Institute were willing to waive the former, but not the latter point. Probably he thought it was too much to ask of him, that he should undergo the probationary test required from comparatively unknown juniors, and

state his experience as an engineer to a society many of
whose members had been his own pupils or assistants.
And his son held the opinion that a society which had
elected many scientific gentlemen of their body as
honorary members, would not have done itself discredit
by admitting the Father of Railway Engineering on the
same terms.   As it was, he turned his back, though
reluctantly, on the Institute of Civil Engineers, and
accepted the office of President of the Institution of
Mechanical Engineers at Birmingham, which he held
until his death.

During the summer of 1847, George Stephenson was
invited to offer himself as a candidate for the representation of South Shields in Parliament.   But his
politics were at best of a very undefined sort; indeed
his life had been so much occupied with subjects of a
practical character, that he had scarcely troubled himself
to form any decided opinion on the party political topics
of the day; and to stand the cross fire of the electors
on the hustings might have been found an even more
distressing ordeal than the cross-questioning of the
barristers in the Committees of the House of Commons.
" Politics," he used to say, " are all matters of theory—
there is no stability.in them; they shift about like the
sands of the sea; and I should feel quite out of my
element amongst them."   He had accordingly the good
sense respectfully to decline the honour of contesting
the representation of South Shields.

We have, however, been informed by Sir Joseph
Paxton, that although George Stephenson held no
strong opinions on political questions generally, there
was one question on which he entertained a decided
conviction, and that was the question of Free-trade.
The words used by him on one occasion to Sir Joseph
were very strong.   "England," said he, " is, and must
be a shopkeeper; and our docks and harbours are only
so many wholesale shops, the doors of which should

always be kept wide open." It is curious that his son Robert should have taken precisely the opposite view of this question, and acted throughout with the most rigid party amongst the protectionists, supporting the Navigation Laws and opposing Free Trade, even to the extent of going into the lobby on the 26th November, 1852, with the famous " cannon-balls." [1]

But Robert Stephenson will be judged in after times by his achievements as an engineer, rather than by his acts as a politician; and happily these last were far outweighed in value by the immense practical services which he rendered to trade, commerce, and civilisation, through the facilities which his railways afforded for free intercommunication between men in all parts of the world. Speaking in the midst of his friends at Newcastle, in 1850, he observed :—

" It seems to me but as yesterday that I was engaged as an assistant in laying out the Stockton and Darlington Railway. Since then, the Liverpool and Manchester and a hundred other great works have sprung into existence. As I look back upon these stupendous undertakings, accomplished in so short a time, it seems as though we had realised in our generation the fabled powers of the magician's wand. Hills have been cut down and valleys filled up; and when these simple expedients have not sufficed, high and magnificent viaducts have been raised, and, if mountains stood in the way, tunnels of unexampled magnitude have pierced them through, bearing their triumphant attestation to the indomitable energy of the nation, and the unrivalled skill of our artisans."

---

[1] For the origin of this term see the 'Times' leader of November 29th, 1852. The division took place on Lord Palmerston's motion as to the results of the Free-trade policy adopted by Sir Robert Peel in 1846, on which there appeared 486 for, and 53 against it. The Noes included Robert Stephenson, Colonel Sibthorp, Mr. Spooner, &c. Mr. Stephenson felt very strongly the "betrayal of the protectionist party" by their Parliamentary leader; and he even went so far as to say that he " could never forgive Peel."

As respects the immense advantages of railways to mankind, there cannot be two opinions. They exhibit, probably, the grandest organisation of capital and labour that the world has yet seen. Although they have unhappily occasioned great loss to many, the loss has been that of individuals; whilst, as a national system, the gain has already been enormous. As tending to multiply and spread abroad the conveniences of life, opening up new fields of industry, bringing nations nearer to each other, and thus promoting the great ends of civilisation, the founding of the railway system by George Stephenson and his son must be regarded as one of the most important events, if not the very greatest, in the first half of this nineteenth century.

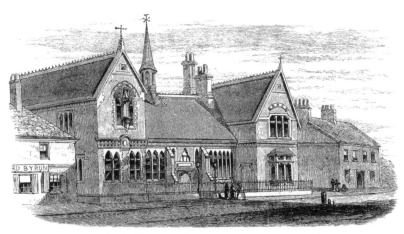

THE STEPHENSON MEMORIAL SCHOOLS, WILLINGTON QUAY.

[By R. P. Leitch.]

NARRATIVE

OF

GEORGE STEPHENSON'S INVENTIONS AND IMPROVEMENTS

IN CONNECTION WITH THE

LOCOMOTIVE ENGINE AND RAILWAYS.

BY HIS SON, ROBERT STEPHENSON.

ROBERT STEPHENSON'S NARRATIVE OF
HIS FATHER'S INVENTIONS, &c.

"When my father commenced his improvements upon the loco-motive engine, two comparatively successful attempts had already been made—one by Mr. Blenkinsop, of Leeds, and the other by Mr. Blackett, of Wylam.

"Mr. Blenkinsop's engine consisted of two cylinders working upon cranks at right angles to each other, and communicating their joint action to a cog-wheel which worked into a cog-rail. The wheels which supported the engine were entirely inde-pendent of the working parts of the engine, and therefore merely supported its weight upon the rails, the progress being made by means of the cog-wheel working into the cog-rail. Mr. Blenkinsop was induced to resort to this contrivance from the conviction (then prevalent in the minds of all engineers) that the adhesion between a smooth wheel and a smooth rail was not sufficient to resist the action of the engines—that is, the wheel would slip round upon the rail, and consequently no progress would be made. These engines of Mr. Blenkinsop's worked for some time with apparent success.

"The other attempt by Mr. Blackett also consisted of two engines combined; but their action was communicated to the wheels by which the entire engine was supported, and therefore depended entirely upon the adhesion between the wheels and the rails for making progress. This experiment of Mr. Blackett's was made upon what is called a tramroad, the flange being upon the rail, instead of (as it is at present in the ordinary rails) upon the wheel.

"When my father began his first engine he was convinced that the adhesion between a smooth wheel and an edge-rail would be as efficient as Mr. Blackett had found it to be between the wheel and the tramroad. Although every one at that time argued that the adhesion upon a tram-rail was by no means a

criterion of what the adhesion would be on an edge-rail, my
father felt sure that there was no essential difference between
the one and the other.

"The construction of my father's first engine was very much
after the same plan as that made by Mr. Blenkinsop; but the
combined power of the two cylinders was communicated to the
wheels which supported the engine on the rail instead of to
the cog-wheel, which, in Mr. Blenkinsop's engine, acted on
a cogged-rail independently of the four supporting wheels.
This engine was completed and tried upon the Killingworth
railway on the 25th July, 1814. It performed its duties with
comparative success; but, having to compete with horses, was
considered barely economical. At the end of the year, however, the steam-power and horse-power were found to be very
nearly on a par with each other in regard to cost. A few
months of experience and careful observation upon the operation of this engine convinced my father that the complication
arising out of the action of the two cylinders being combined
by spur-wheels would prevent their coming into practical application. He then directed his attention to an entire change in
the construction and mechanical arrangements, and in the following year took out a patent, dated February 28th, 1815, for
an engine which combined in a remarkable degree the essential
requisites of an economical locomotive—that is to say, few
parts, simplicity in their action, and great simplicity in the
mode by which the power was communicated to the wheels supporting the engine.

"This second engine consisted as before of two vertical cylinders, which communicated directly with each pair of the four
wheels that supported the engine by a cross-head and a pair of
connecting rods; but in attempting to establish a direct communication between the cylinders and the wheels that rolled
upon the rails, considerable difficulties presented themselves.
The ordinary joints could not be employed to unite the engine,
which was a rigid mass, with the wheels rolling upon the irregular surface of the rails; for it was evident that the two rails
of the line of railway could not always be maintained at the
same level with respect to each other—that one wheel at the
end of the axle might be depressed into a part of the line which
had subsided, whilst the other would be elevated. In such a
position of the axle and wheels it was clear that a rigid communication between the cross-head and the wheels was im-

practicable. Hence it became necessary to form a joint at the top of the piston-rod where it united with the cross-head, so as to permit the cross-head always to preserve complete parallelism with the axle of the wheels with which it was in communication.

" In order to obtain the flexibility combined with direct action which was essential for ensuring power and avoiding needless friction and jars from irregularities in the rail, my father employed the "ball and socket" joint for effecting a union between the ends of the cross-heads where they united with the connecting-rods, and between the end of the connecting-rods where they were united with the crank-pins attached to each driving-wheel. By this arrangement the parallelism between the cross-head and the axle was at all times maintained, it being permitted to take place without producing jar or friction upon any part of the machine.

" The next important point was to combine each pair of wheels by some simple mechanism, instead of the cog-wheels which had formerly been used. My father began by inserting each axle into two cranks at right angles to each other, with rods communicating horizontally between them. An engine was made on this plan, and answered extremely well. But at that period (1815) the mechanical skill of the country was not equal to the task of forging cranked axles of the soundness and strength necessary to stand the jars incident to locomotive work; so my father was compelled to fall back upon a substitute which, though less simple and less efficient, was within the mechanical capabilities of the workmen of that day, either for construction or repair. He adopted a chain which rolled over indented wheels placed on the centre of each axle, and so arranged that the two pairs of wheels were effectually coupled and made to keep pace with each other. But these chains after a few years' use, became stretched, and then the engines were liable to irregularity in their working, especially in changing from working back to forward again. Nevertheless, these engines continued in profitable use upon the Killingworth Colliery Railway for some years. Eventually the chain was laid aside, and the front and hind-wheels were united by rods on the *outside*, instead of by rods and crank-ankles *inside*, as specified in the original patent; and this expedient completely answered the purpose required, without involving any expensive or difficult workmanship.

"Another important improvement was introduced in this engine. The eduction steam had hitherto been allowed to escape direct into the open atmosphere; but my father, having observed the great velocity with which the waste-steam escaped, compared with the velocity with which the smoke issued from the chimney of the same engine, thought that by conveying the eduction steam into the chimney and there allowing it to escape in a vertical direction, its velocity would be imparted to the smoke from the engine, or to the ascending current of air in the chimney. The experiment was no sooner made than the power of the engine became more than doubled; combustion was stimulated, as it were, by a blast; consequently the power of the boiler for generating steam was increased, and, in the same proportion, the useful duty of the engine was augmented.

"Thus in 1815 my father had succeeded in manufacturing an engine which included the following important improvements on all previous attempts in the same direction: simple and direct communication between the cylinder and the wheels rolling upon the rails; joint adhesion of all the wheels, attained by the use of horizontal connecting-rods; and, finally, a beautiful method of exciting the combustion of fuel by employing the waste-steam which had formerly been allowed uselessly to escape. It is, perhaps, not too much to say that this engine, as a mechanical contrivance, contained the germ of all that has since been effected. It may be regarded, in fact, as a type of the present locomotive engine.

"In describing my father's application of the waste-steam for the purpose of increasing the intensity of combustion in the boiler, and thus increasing the power of the engine without adding to its weight, and while claiming for this engine the merit of being a type of all those which have been successfully devised since the commencement of the Liverpool and Manchester Railway, it is necessary to observe that the next great improvement in the same direction, the "multitubular boiler," which took place some years later, could never have been used without the help of that simple expedient *the steam-blast,* by which power only the burning of coke was rendered possible.

"I cannot pass over this last-named invention of my father's without remarking how slightly, as an original idea, it has been appreciated; and yet how small would be the comparative value of the locomotive engine of the present day without the application of that important invention!

" Engines constructed by my father in the year 1818 upon the principles just described are in use on the Killingworth Colliery Railway to this very day (1857), conveying, at the speed of perhaps five or six miles an hour, heavy coal-trains, probably as economically as any of the more perfect engines now in use.

" There was another remarkable piece of ingenuity in this machine, which was completed so many years before the possibility of steam-locomotion became an object of general commercial interest and parliamentary inquiry. I have before observed that up to and after the year 1818 there was no such class of skilled mechanics, nor were there such machinery and tools for working in metals, as are now at the disposal of inventors and manufacturers. Among other difficulties of a similar character, it was not possible at that time to construct springs of sufficient strength to support the improved engines. The rails then used being extremely light, the roads became worn down by the traffic, and occasionally the whole weight of the engine, instead of being uniformly distributed over four wheels, was thrown almost diagonally upon two. In order to avoid the danger arising from such irregularities in the road, my father arranged the boiler so that it was supported upon the frame of the engine by four cylinders which opened into the interior of the boiler. These cylinders were occupied by pistons with rods which passed downwards and pressed upon the upper side of the axles. The cylinders opening into the interior of the boiler allowed the pressure of steam to be applied to the upper side of the piston, and that pressure being nearly equal to the support of one-fourth of the weight of the engine, each axle, whatever might be its position, had the same amount of weight to bear, and consequently the entire weight was at all times nearly equally distributed amongst the wheels. This expedient was more necessary in this case, as the weight of the new locomotive engines far exceeded that of the carriages which had hitherto been used upon colliery railways, and therefore subjected the rails to much greater risk from breakage. And this mode of supporting the engine remained in use until the progress of spring-making had considerably advanced, when steel springs of sufficient strength superseded this highly ingenious mode of distributing the weight of the engine uniformly among the wheels.

" Having advanced the locomotive engine to this stage of improvement, my father next turned his attention to the state of the road; as he perceived, and said, that the extended use of the locomotive must depend upon the perfection of the road upon which it was to move. Even at this early date he was in the habit of considering the road and the locomotive as one machine. All railways at that time were laid in a careless and loose manner, and great inequalities of level were permitted to take place without much attention to repairs, the result being that great loss of power and great wear-and-tear of machinery were incurred.

" My father therefore now began to direct his close attention to the improvement of the road, and to making it more substantial and solid. With that object he applied his mind particularly to removing the inequalities produced by the imperfect junction between rail and rail. The rails were then made of cast-iron, each being three feet long. Care was not taken to maintain the points of junction on the same level with each other; and the chair or cast iron pedestal into which the rails were inserted being flat on the bottom, it happened that whenever any disturbance took place in the stone blocks or sleepers upon which they were supported, the flat base upon which the rails rested being tilted by unequal subsidence, the end of one rail became depressed, while that of the other was elevated. This was most seriously felt, since, in the condition in which railways were then kept, very little attention was paid to maintaining a uniform surface or permanent way.

" My father's first improvement in the construction of the rail consisted in this:—instead of adopting the *butt joint* which had hitherto been used in all cast-iron rails, he adopted the *half-lap joint*, by which means the rails extended a certain distance over each other at the ends, somewhat like a scarf joint; and these ends, instead of resting upon the flat chair as had hitherto been the practice, were made to rest upon the apex of a curve forming the bottom of the chair. The supports were extended from 3 feet to 3 feet 9 inches or 4 feet apart. These rails were substituted for the old ones on the Killingworth Colliery Railway, and were found to be a great improvement, adding both to the efficiency of the horse-power and to the smooth action of the locomotive, but more particularly adding to the efficiency of the latter.

" My father's endeavours having been marked by so much success in the adaptation of locomotive engines to railways, his attention was, about this period, called by many of his friends to the subject of the application of steam to common roads; but

the accuracy with which he estimated the resistances to which
loads were exposed on railways arising from friction and gravity,
led him at a very early stage to reject the idea of successfully
applying steam power to common roads. In October, 1818, in
conjunction with Mr. Nicholas Wood, he made a series of
experiments on the resistances to which carriages are exposed
on railways, with a dynamometer of his own contrivance. This
dynamometer was chiefly remarkable for its simplicity,—but it
will not bear comparison with others that have been contrived
and made use of subsequently; it is, however, interesting as the
first systematic attempt to determine the precise amount of
resistance of carriages moving on railways. It was by this
machine for the first time ascertained, that the friction was a
constant quantity at all velocities. Although this fact had
been long before developed by Coulomb and was well known to
scientific men as an established fact, yet at the time when my
father made these experiments, the deductions of philosophers
were neither believed in nor acted upon by practical engineers.
Indeed, although the experiments of my father went directly to
corroborate the deductions of philosophers, it required a considerable space of time to overcome the prejudices which then
existed among practical men.

"It was maintained by many, that the results of the experiments led to the greatest possible mechanical absurdities. For
instance, it was maintained, that if friction were constant at all
velocities upon a level railway, when once a power was applied
to a carriage which exceeded the friction of that carriage by the
smallest possible amount, that same small excess of power would
be able to convey the carriage along a level railway at all
conceivable velocities. When this position was put by those
who opposed the conclusions at which my father had arrived,
he felt great hesitation in maintaining his own views; for it
appeared to him at first sight really to be—as it was put by his
opponents—an absurdity. Frequent repetition, however, of the
experiments to which I have alluded, left no doubt upon his
mind, that his conclusion that friction was uniform at all
velocities was a fact which must be received as positively
established; and he soon afterwards boldly maintained that
that which was an apparent absurdity was, instead, a necessary
consequence. I well remember the ridicule which was thrown
upon this view by many of those persons with whom he was
associated at the time. Nevertheless it is undoubted, that could

you practically be always applying a power in excess of the resistance, a constant increase of velocity would of necessity follow without any limit. This is so obvious to most professional men of the present day and is now so axiomatic, that I only allude to the discussion which took place when these experiments of my father were announced, for the purpose of showing how small was the amount of science at that time blended with engineering practice. A few years afterwards, an excellent pamphlet was published by Mr. Silvester on this question; he took up the whole subject and demonstrated in a very simple and beautiful manner the correctness of all the views at which my father had arrived by his course of experiments.

"The other resistances to which carriages were exposed were also investigated experimentally by my father. He perceived that these resistances were mainly three: the first being upon the axles of the carriage; the second, which may be called the rolling resistance, being between the circumference of the wheel and the surface of the rail; and the third being the resistance of gravity.

"The amount of friction and gravity he accurately ascertained; but the rolling resistance was a matter of greater difficulty, for it was subject to great variation. He, however, satisfied himself that it was so great, when the surface presented to the wheel was of a rough character, that the idea of working steam-carriages economically on common roads was out of the question. Even so early as the period alluded to he brought his theoretical calculations to a practical test; he scattered sand upon the rails when an engine was running, and found that a small quantity was quite sufficient to retard and even stop the most powerful locomotive engine that he had at that time made. And he never failed to urge this conclusive experiment upon the attention of those who were wasting their money and time upon the vain attempt to apply steam to common roads.

"The following were the principal arguments which influenced his mind to work out the use of the locomotive in a directly opposite course to that pursued by a number of ingenious inventors, who between 1820 and 1836 were engaged in attempting to apply steam-power to turnpike roads. Having ascertained that resistance might be taken as represented by 10 lbs. to a ton weight on a level railway, it became obvious to him that so small a rise as 1 in 100 would diminish the useful effort of a locomotive by upwards of fifty per cent. This fact called my father's attention to the question of gradients in future locomotive lines. He then became convinced of the vital importance, in an economical point of view, of reducing the country through which a railway was intended to pass, to as near a level as possible. This originated in his mind the distinctive character of railway works as contradistinguished from all other roads,—for in railroads he early contended that large sums would be wisely expended in perforating barriers of hills with long tunnels, and in raising low ground with the excess cut down from the adjacent high ground. In proportion as these views fixed themselves upon his mind, and were corroborated by his daily experience, he became more and more convinced of the hopelessness of applying steam locomotion to common roads, —for every argument in favour of a level railway was an argument against the rough and hilly course of a common road. He never ceased to urge upon the patrons of road steam-carriages that if by any amount of ingenuity an engine could be made which could by possibility traverse a turnpike road at a speed at least equal to that obtainable by horse-power and at a less cost, such an engine if applied to the more perfect surface of a railway would have its efficiency enormously enhanced. For instance, he calculated that if an engine had been constructed, and had been found to travel uniformly between London and Birmingham at an average speed of 10 miles an hour,—conveying say 20 or 30 passengers at a cost of 1s. per mile, it was clear that the same engine if applied to a railway, instead of conveying 20 or 30 people, would have conveyed 200 or 300 people, and instead of a speed of 10 or 12 miles an hour, a speed of at least 30 to 40 miles an hour would have been obtained. It is difficult now to understand how it was that this obvious inference never occurred to the minds of those who so long persisted in vain attempts to apply locomotive power to turnpike roads.

"Identified as my father at this period had become with every step made towards increased utility in the locomotive engine, he did not allow his enthusiasm to carry him away into costly mistakes. He most carefully drew a broad line between those cases in which the locomotive could be advantageously employed, and those in which stationary engines were more economical. This led him, when called upon to execute railways over rough countries where gradients within the compass of the locomotive engine could not be obtained, to apply stationary engines most extensively. Many instances of the successful application of this mixed power endure to this day in the north of England. The railway from the Hetton Colliery to Sunderland was perhaps the earliest and most remarkable work in which these two powers were most successfully combined. The Stockton and Darlington Railway is another instance where they were combined with most success and efficiency, and although subsequently the application of stationary power on that line was partially superseded by tunnelling and locomotive power, the change has only been justified by the traffic having become so enormous, that a uniform and uninterrupted system was alone applicable—the stationary engine system being one that is limited by the necessity of reciprocating the trains over a short piece of railway at limited intervals.

" In 1820 my father established in conjunction with two friends of capital a manufactory of locomotive engines at Newcastle-upon-Tyne. Before the opening of that establishment all the locomotive engines which he had constructed had been made by ordinary mechanics, working amongst the collieries in the north of England. But my father felt that the accuracy and style of their workmanship admitted of great improvement, and that upon improvement of workmanship the perfect action of the engine was greatly dependent. One great object that he had in view in establishing this factory, was to concentrate a number of good workmen for the purpose of realising and carrying out the improvements in detail which he was constantly making. This was the only manufactory at which locomotive engines were made until after the opening of the Liverpool and Manchester Railway in 1831. After that great event other mechanics began to devote their attention as a matter of regular business to the construction of locomotive engines for railway purposes. At the Newcastle factory all the engines that were employed upon the Stockton and Darlington Railway were made; and for some time after the Liverpool and Manchester Railway was opened this establishment alone supplied the engines for working the traffic between those two important commercial towns.

" The writer of an article on Railways which appeared in the ' Edinburgh Review ' in 1832, founded a charge of monopoly in favour of the Newcastle factory, against the Liverpool and Manchester Railway Directors, upon the fact that all the engines on the Liverpool and Manchester Railway were made after my father's plans, and in his factory; the simple truth being that

that was the only source at that period from which efficient engines could be obtained. The Directors were fully alive to the importance of inducing competition in this new kind of manufacture. They offered every inducement with a view to extending the field from which they could draw their supplies of engines, and as soon as they could rely upon the quality of the article supplied they distributed their orders indiscriminately and impartially. Since the opening of the Liverpool and Manchester Railway, works for the manufacture of engines have gradually extended themselves into every part of Great Britain, the continent of Europe, and the United States of America. But the main object of my father in establishing the Newcastle Works was to educate a class of workmen in skilled labour, who should be able to execute the many ideas which presented themselves to his inventive and practical mind.

" After the opening of the Stockton and Darlington, and before that of the Liverpool and Manchester, my father directed his attention to various methods of increasing the evaporative power of the boiler of the locomotive engine. Amongst other attempts he introduced tubes (as had been done before in other engines) —small tubes, containing water, by which the heating surface was materially increased. Two engines with such tubes were constructed for the St. Etienne Railway in France, which was in process of construction in the year 1828; but the expedient was not successful—the tubes became furred with deposit and burned out.

" Other engines with boilers of a variety of construction were made, all having in view the increase of the heating surface, as it then became obvious to my father that the speed of the engine could not be increased without increasing the evaporative power of the boiler. Increase of surface was in some cases obtained by inserting two tubes, each containing a separate fire, into the boiler; in other cases, the same result was obtained by returning the same tube through the boiler, but it was not until my father was engaged in making some experiments during the progress of the Liverpool and Manchester Railway in conjunction with Mr. Henry Booth, the well known secretary of that line, that any great movement in this direction was effected, and that the present multitubular boiler assumed a practicable shape. It was in conjunction with Mr. Booth that my father constructed the " Rocket " engine which obtained the prize at the celebrated competition which took place a little prior to the opening of the Liverpool and Manchester Railway.

" At this stage of the locomotive engine we have in the multi-tubular boiler the only important principle of construction introduced, in addition to those which my father had brought to bear at a very early stage (between 1815 and 1821) on the Killingworth Colliery Railway. In the " Rocket" engine the power of generating steam was prodigiously increased by the multitubular system. Its efficiency was further augmented by narrowing the orifice by which the waste steam escaped into the chimney; for by this means the velocity of the air in the chimney, in other words the draught of the fire, was increased to an extent that surpassed the expectations even of those who had been the authors of the combination.

"From the date of running the " Rocket" on the Liverpool and Manchester Railway, the locomotive engine has received many minor improvements in detail and especially in accuracy of workmanship, but in no essential particular does the existing engine differ from that which obtained the prize at the celebrated competition at Rainhill.

" In this instance as in every other important step in science or art, various claimants have arisen for the merit of having suggested the multitubular boiler as a means of obtaining the necessary heating surface. Whatever may be the value of their respective claims, the public, useful, and extensive application of it must certainly bear date from the experiments made at Rainhill. M. Seguin, for whom engines had been made by my father some few years previously, states that he patented a similar multitubular boiler in France some years before. A still prior claim is made by Mr. Stevens of New York, who was all but a rival to Fulton in the introduction of steamboats on the American rivers. It is stated that so early as 1807 he used the multitubular boiler. These claimants may all be entitled to great and independent merit, but certain it is, that the perfect establishment of the success of the multitubular boiler is more immediately owing to the suggestion of Mr. Henry Booth, the Secretary to the Liverpool and Manchester Railway, and to my father's practical knowledge in carrying it out."

INVENTION OF THE STEAM BLAST.

—◆◇◆—

SINCE the publication of 'The Life of George Stephenson,' in 1857, several claims have been set up to the merit of having invented the steam-blast. Trevithick's friends have claimed it for him. Mr. O. D. Hedley, in his book entitled 'Who Invented the Locomotive?' claims it for Mr. William Hedley, viewer, Wylam, as well as the invention of the locomotive itself.[1] Then Mr. John Hackworth, in a series of letters published in the 'Engineer,' claimed it for his relation, Mr. Timothy Hackworth; and following all these came, lastly, Mr. Goldsworthy Gurney, who, in his pamphlet entitled 'An Account of the Invention of the Steam Jet or Blast,' claimed the invention for himself, ignoring the claims of all the others.

In the pamphlet last mentioned Mr. Gurney says, "Mr. Stephenson himself never claimed this invention;" and he further alleges, that in the 'Life of George Stephenson' it was "advanced for the first time." Mr. William Fairbairn, however, in his 'Useful Information for Engineers,' second series (p. 241), says: "I have every reason to believe that it belongs to Stephenson, as I have heard him claim its introduction, and have no reason to doubt his veracity, or that he was quite equal to the task." And so far from the claim having been made "for the first time" in the 'Life of George Stephenson,' it will be found distinctly made in the seventh edition of the 'Encyclopedia Britannica,' published in 1836, while all the parties interested were alive; and, so far as the present writer is aware, it was never contradicted. The author of the article 'Railways' in that publication was Mr. Lecount, one of the engineers employed on the London and Birmingham Railway; and his words were these: "Some writers have assigned to

---

[1] In his preface Mr. Hedley says: "The author disputes the truth of Mr. Smiles's statements, which assign to Mr. Stephenson the invention of the locomotive engine [no such statement was ever made]; and he is in a position to prove that the credit of this great achievement is due, and due only, to the late Mr. William Hedley"!

Trevithick the merit of inventing the steam-blast up the
chimney, which may be termed the life-blood of the locomotive-
engine. Trevithick has laurels enough, and has no need to
borrow a single leaf from the crown of another. The steam-
blast was invented by George Stephenson, and used by him
certainly prior to 1815; while in June, 1815, Trevithick—so
far from using the waste steam to increase the draught—took
out a patent in which, among other improvements, he included
a method of urging his fire *by fanners,* similar to a winnowing
machine." The writer of the article on the same subject of
' Railways ' in the eighth edition of the ' Encyclopedia Britannica '
is Mr. D. K. Clark—a gentleman who has made the history of
the locomotive the subject of his special study, and probably
knows more of it than any man living,—and he is equally
explicit on the point. He says: " The blast-pipe thus designed
and applied was undoubtedly the invention of George Stephenson ; in conjunction with the multitubular flue it altered and
vastly improved the range and capacity of the locomotive ; and,
in further conjunction with the direct connection of the steam-
cylinder to one axle and pair of wheels, it was tantamount to
a new and original machine."

Robert Stephenson, writing to the author on the controversy
which arose on this subject in 1857, said: "Nothing can be
so clear as that George Stephenson was the real inventor of
the steam-blast, and nothing confounds me more than to see
that a question is raised upon it." As claims have, however,
been seriously set up on behalf of William Hedley, Jonathan
Hackworth, and Goldsworthy Gurney, as the authors of this
invention, a brief examination of the grounds on which their
respective claims are founded is here rendered necessary.

We will take the Wylam claim first. From what has been
said in the text, it will readily be understood that the claim
attempted to be set up in behalf of William Hedley as the inventor of the locomotive engine is mere moonshine. Trevithick
and Blenkinsop's engines preceded those at Wylam colliery by
many years ; and even the Wylam engines were not made
after the designs of Hedley, but after those of Trevithick, and
afterwards of Jonathan Foster, the engineer of the colliery.
But it is further alleged that William Hedley invented the
blast-pipe. This is effectually contradicted by the fact that
the Wylam engine had *no* blast-pipe. " I remember the Wylam
engine," Robert Stephenson wrote to us in 1857, " and I am

positive there was no blast-pipe." But Mr. O. D. Hedley says there was one; and he gives a representation of the engine from the first edition of Mr. Nicholas Wood's 'Practical Treatise on Railroads,' published in 1825, in proof of what the engine actually was. The illustration, however, entirely confutes the assertion that the Wylam engine contained any blast at all. In fact, it embodied a contrivance for the express purpose of *preventing* a blast. Mr. Wood explains clearly enough how this object was secured, contrasting it with Stephenson's Killingworth engine to the disadvantage of the latter, which *had* a blast, whilst the other had none. For it is a curious fact that Mr. Wood at that time did not approve of the steam-blast, and he referred to the Wylam engine in illustration of how it might be avoided.

The evidence contained in Mr. Wood's book, published as it was in 1825, is especially valuable as showing the express purpose for which George Stephenson invented and adopted the steam-blast in the Killingworth engines. Describing their action, Mr. Wood says: " The steam is admitted to the top and bottom of the piston by means of a sliding valve, which, being moved up and down alternately, opens a communication between the top and bottom of the cylinder and the pipe that is *open into the chimney and turns up within it*. The steam, after performing its office within the cylinder, is thus thrown into the chimney, and the power with which it issues will be proportionate to the degree of elasticity ; and *the exit being directed upwards, accelerates the velocity of the current of heated air accordingly* " (p. 147). And again, at another part of the book, he says: "There is another great objection urged against locomotives, which is, the noise that the steam makes in escaping into the chimney; this objection is very singular, as it is not the result of any inherent form in the organisation of such engines, but an accidental circumstance. When the engines *were first made*, the steam escaped into the atmosphere, and made comparatively little noise; *it was found difficult then to produce steam in sufficient quantity to keep the engine constantly working, or rather to obtain an adequate rapidity of current in the chimney to give sufficient intensity to the fire. To effect a greater rapidity, or to increase the draught of the chimney, Mr. Stephenson thought that by causing the steam to escape into the chimney through a pipe with its end turned upwards, the velocity of the current would be accelerated, and such was the effect ;* but, in remedying one evil another has been produced, which, though objectionable in some

2 K 2

places, was not considered as objectionable on a private rail-road. The tube through the boiler having been increased, there is now no longer any occasion for the action of the steam to assist the motion of the heated air in the chimney. The steam thrown in this manner into the chimney acts as a trumpet, and certainly makes a very disagreeable noise. Nothing, however, is more easy to remedy, and the very act of remedying this defect will also be the means of economising the fuel" (pp. 292-3).[1]

Mr. Wood then proceeds to show how the noise caused by the blast, how in fact the blast itself, might be effectually prevented by adopting the expedient employed in the Wylam engine; which was, to send the exhaust steam, not into the chimney (where alone the blast could act with effect by stimulating the draught), but into a steam-reservoir expressly provided for the purpose. His words are these: "Nothing more is wanted to destroy the noise than *to cause the steam to expand itself into a reservoir, and then allow it to escape gradually to the atmosphere through the chimney*. Upon the Wylam railroad the noise was made the subject of complaint by a neighbouring gentleman, and they adopted this mode, which had the effect above mentioned" (p. 294).

We think this ought to be perfectly conclusive as to the Wylam engine—that it had no blast, and that it contained an arrangement for the express purpose of preventing any blast. And thus we dismiss Mr. Hedley's claim.

It is curious to find that Mr. Nicholas Wood continued to object to the use of the steam-blast down even to the time when the Liverpool and Manchester Railway Bill was before Parliament in 1825. Hence Mr. Wood, in his evidence before the Committee on that Bill in 1825, said: " Those engines [at Killingworth] *puff very much*, and *the object is to get an increased draught in the chimney*. Now (by enlarging the flue-tube and giving it a double turn through the boiler) we have got a suf-ciency of steam without it, and I have no doubt by allowing

---

[1] These passages will be found in the first edition of Mr. Wood's work, published in 1825. The subsequent editions do not contain them. A few years' experience wrought great changes of opinion on many points connected with the practical working of railways, and Mr. Wood altered his text accordingly. But it is most im-portant for our present purpose that, in the year 1825, long before the Liverpool and Manchester line was opened, Mr. Wood should have so clearly described the steam-blast which had been in regular use for more than ten years in all Stephenson's locomo-tives employed in the working of the Killingworth railway.

the steam to exhaust itself in a reservoir it would pass quietly into the chimney without that noise." In fact, Mr. Wood was still in favour of the arrangement adopted in the Wylam engine, by which the steam-blast had been got rid of altogether.

The claim made on behalf of Timothy Hackworth is, that he invented the steam-blast for the "Sanspareil" locomotive in the year 1829—that is, fourteen years after George Stephenson had been making regular use of the invention in every engine constructed by him. Timothy Hackworth had been employed in Stephenson's locomotive workshops at Newcastle, and was appointed by George Stephenson the foreman of the locomotive department of the Stockton and Darlington Railway. He was, therefore, quite familiar with all George Stephenson's arrangements, including his blast-pipe. That he sharpened it there is no doubt, and we believe that this is claimed as the gist of his "invention;" but even of this he is deprived by Mr. Goldsworthy Gurney, who affirms that it was he who "furnished Mr. Hackworth with the steam-jet, to fix on the eduction-pipe of his engine, the 'Sanspareil.'"[1] Mr. Gurney claims to have made the invention in the year 1820, about six years after the date at which George Stephenson regularly employed it for the express purpose of producing a draught in his Killingworth engines. Mr. Gurney says he first used it to obtain a more intense heat in the decomposing furnaces, when engaged as lecturer at the Surrey Institution; that he next applied it to steam-boats in 1824; and subsequently to steam-carriages run upon common roads. We are ready to believe all this, and yet it does not in the slightest degree invalidate George Stephenson's claim to priority in the invention as above explained.

The following narrative relative to the blast-pipe of the "Rocket" and the sharpening of the blast in that engine and the "Sanspareil" (about which a controversy was raised in 'The Engineer' journal) was written by Robert Stephenson, and communicated to the author in January, 1858:—

ROBERT STEPHENSON'S NARRATIVE.

"CERTAINLY not many weeks had elapsed after the first travelling engine was placed on the Killingworth waggon-way in

---

[1] P. 8 of Mr. Goldsworthy Gurney's ' Account of the Invention of the Steam-jet, or Blast.' London, 1859.

1814, before the steam-blast was introduced by my father into
the chimney, and it was uniformly employed in every subsequent engine that was built; but the orifice of the blast-pipe
was, I believe, in no instance contracted so as to give a less
area than that of the steam-ports.

" The Stockton and Darlington Railway was opened in 1825,
but as I was absent from England at the time, I cannot state
whether the engines constructed in Forth-street or at the
factory, for that line, had contracted blast-pipes or not. Shortly
after my return from America, I was frequently in the habit, as
a matter of business, of visiting the line alluded to, the superintendence of the locomotive engines being then under Timothy
Hackworth, with whom I was constantly in the habit of discussing
the remarkable effects produced by the blast in the chimney. It
was about that time, I believe, Mr. Hackworth had found that an
increased effect was obtained by contracting the orifice of the
blast-pipe. Considerable doubt was, however, then entertained
whether such contraction would be attended with any actual
economy in the working of the engine, for, although the combustion was a little more excited, and a more copious amount of
steam was generated, it was believed that the negative pressure
produced on the piston counterbalanced in a great measure the
advantages mentioned.

" During the construction of the 'Rocket' a series of experiments was made with blast-pipes of different diameters, and
their efficiency was tested by the amount of vacuum that was
formed in the smoke-box. The degree of rarefaction was
determined by a glass tube fixed to the bottom of the smoke-box
and descending into a bucket of water,—the tube being open at
both ends. As the rarefaction took place the water would of
course rise in the tube, and the height to which it rose above
the surface of the water in the bucket was made the measure of
the amount of rarefaction.

" These experiments certainly showed that a considerable
increase of draught was obtained by contracting the orifice, and
accordingly the two blast-pipes in the 'Rocket' were contracted
slightly below the area of. the steam ports,—and before she left
the factory the water rose in the glass tube three inches above
the water in the bucket.

" I was quite aware at the time that Mr. Hackworth's
'Sanspareil' was being constructed in the same manner, with

the exception that the two eduction-pipes were brought into one blast-pipe in the centre of the chimney. The two engines might therefore be considered as precisely alike in principle.

" With respect to the objection which has been made in ' The Engineer,' to two separate orifices, I must affirm that no remark could possibly be more unfounded. The writer states that when two separate orifices are employed, the blast produced by one is neutralized by the other. The ' Rocket' worked perfectly well with the double blast-pipe, and, to the best of my recollection, the prize was won without any alteration having been made in that part of the engine.

" The experiments already mentioned proved that the double blast-pipe in the 'Rocket' was capable of producing a considerable rarefaction in the chimney, and the alteration from two blast-pipes to one was made by myself rather with a view of lessening the space occupied by them in the chimney.

"The writer in ' The Engineer' completely ignores the fact of the single steam-blast having been in existence eleven years prior to the opening out of the Stockton and Darlington Railway, and seems to argue that the ' Sanspareil' was the first engine to which the steam-blast was ever applied with effect; whereas it had actually been in regular use since the year 1814, and the only alteration which it underwent, was the contraction of the orifice made on the Stockton and Darlington Railway some time between the years 1825 and 1827.

" Whatever merit or value may attach to this alteration I believe to be due to Timothy Hackworth, but nothing beyond it, I am quite certain; and even this was decidedly much overrated by him: in fact, he carried the contraction to such an extent that nearly half of the fuel was thrown out of the chimney unconsumed, as many can testify who witnessed the experiments at Rain Hill.

" But surely such an alteration is not to deprive George Stephenson of the merit of the invention of the steam-blast. Moreover the contraction in many of our best locomotive engines is totally unnecessary and rather disadvantageous than otherwise; for, since the speed of the engines has been increased, the velocity of the eduction steam is quite sufficient to produce the needful rarefaction in the chimney without any contraction whatever. In the early engines, when the speed of the piston was slow, the contraction was undoubtedly advantageous, but now that the boilers have been increased in size—the heating surfaces thereby being extended—a less intense blast is required. The orifices of the blast-pipes of many engines running at the present day are as large as the steam ports. Consequently they cannot be said to be contracted at all. In fact the greater apparent efficiency of the steam-blast, as at present used, is entirely due to the greater velocity of the piston."

In a subsequent letter to us, Mr. Stephenson added the following remarks:—

" In conclusion, let us suppose that Hackworth really did first contract the blast-pipe, does that at all affect the claim of George Stephenson to have been the first discoverer of the fact that throwing the eduction steam in the form of a vertical jet into the chimney, greatly increased the power of the locomotive engine? As well might it be contended that James Watt had no merit for his invention of the steam-engine, because its effectual performance has been so greatly improved since his death.

" The value of an invention does not consist merely in the results which are immediately produced by it, but in those which quickly follow. But in the case of that modification of the invention in question, it has been found that other circumstances have attended the progress of the locomotive engine, which have rendered the contraction of the blast-pipe comparatively unnecessary, and in some cases positively objectionable."

We trust that the explanations thus given will have made it sufficiently clear to the reader that the claims respectively made on behalf of Trevithick, Hedley, Hackworth, and Gurney, of having invented the steam-blast, are without foundation; and that George Stephenson, and no other person, was its sole inventor.

INDEX.

THE END.

LONDON: PRINTED BY WILLIAM CLOWES AND SONS, STAMFORD STREET, AND CHARING CROSS.

Milton Keynes UK
Ingram Content Group UK Ltd.
UKHW032320161024
449665UK00001B/23